THE MOLECULAR THEORY
OF FLUIDS

THE MOLECULAR THEORY OF FLUIDS

BY

HERBERT S. GREEN,
A.R.C.S., Ph.D., D.Sc.

Professor of Mathematical Physics at the University of Adelaide;
formerly Visiting Professor at the Dublin Institute for Advanced Studies;
formerly Member of the Princeton Institute for Advanced Study

DOVER PUBLICATIONS, INC.
NEW YORK

Published in Canada by General Publishing Company, Ltd., 30 Lesmill Road, Don Mills, Toronto, Ontario.

Published in the United Kingdom by Constable and Company, Ltd., 10 Orange Street, London WC 2.

This Dover edition, first published in 1969, is an unabridged and slightly corrected republication of the work originally published in 1952. It is reprinted by special arrangement with the North-Holland Publishing Company, Amsterdam, publisher of the original edition.

Standard Book Number: 486-6233-5
Library of Congress Catalog Card Number: 69-20422

Manufactured in the United States of America
Dover Publications, Inc.
180 Varick Street
New York, N. Y. 10014

PREFACE

This volume is intended to provide, for rheologists and the many other scientific workers with interests in the field, an account of the general properties of fluids, in terms of molecular structure.

Though a rigorous quantitative treatment of the subject must inevitably involve a certain amount of mathematical formalism, an attempt has been made by the author to render this intelligible to the reader with a modest mathematical equipment, and also to provide, side by side with the mathematical development, a qualitative or semi-quantitative description in purely physical terms. It is hoped that the result will prove to be of use to a wide class of readers.

The scope of the book is consciously different from that of any existing volume on the molecular theory of fluids. Attention has been focussed on the liquid state, though its relations to the gaseous and crystalline states have always been kept in mind. The molecular model adopted is the same throughout; in the author's judgement it is the simplest which adequately describes the physical situation and at the same time admits the understanding of most of the range of fluid phenomena. There are simpler models which are useful for the quantitative treatment of particular properties; but, since these have for the most part been sufficiently described elsewhere, they have been referred to only where their relation to the more systematic treatment has been properly established. What is thus lost by way of numerical results — whose main purpose is after all the vindication of a theory the premises of which are still in doubt — is gained in rigour and coherent understanding.

Though the treatment of the subject as a whole is systematic, an effort has been made to render the volume suitable for reference purposes. It will be found possible to read sections on most particular topics without reference to the bulk of what has gone before. Where necessary, cross-references between different sections have been provided.

References to other literature are collected at the end of the book; only key references are given, and they are indicated in the text only in connection with more specialized topics which the reader might wish to study in greater detail.

Foremost among those to whom this book owes its existence, the author wishes to mention Prof. MAX BORN, whose early inspiration has been a constant source of guidance. The author is indebted also to members of the Princeton Institute for Advanced Study and the Dublin Institute for Advanced Studies for stimulating discussions, and to the Directors of these Institutes for hospitality while the book was being written. Many thanks are due to Prof. J. M. BURGERS and Prof. J. J. HERMANS for reading the manuscript and making a considerable number of suggestions which the author was very glad to adopt, and to Mr. G. Szekeres for reading the proofs.

Thanks are tendered also to the publishers for their courteous cooperation.

H. S. GREEN

University of Adelaide, S. Australia

CONTENTS

CHAPTER I: INTRODUCTION
Page
1. The Physical Properties of Fluids 1
2. The Microscopic Picture of a Fluid 5
   2.1 The Picture of a Gas, p. 8; 2.2 Condensation and the Condensed State, p. 10; 2.3 Fluids in Motion, p. 12; 2.4 Fluid Mixtures, p. 14
3. The Quantum Theory 14
4. Mathematical Technique and Notation 16
5. List of Symbols 23

CHAPTER II: STATISTICS OF MOLECULAR MOTION
1. Probability and Averages 25
2. Fluid Density 27
3. Distribution of Molecular Velocities 30
4. Molecular Mechanics of Fluids 34
5. Statistical Thermodynamics 40
   5.1 Proof of Boltzmann's Law 43; 5.2 Some Thermodynamical Relations 47
6. Statistical Mechanics of Fluids 49
7. The Theory of Partitions 53

CHAPTER III: THE STRUCTURE OF FLUIDS AT REST
1. The Scattering of X-Rays by Fluids 57
2. Optical Scattering and Density Fluctuations 62
3. The Structure of Fluids 64
4. Distribution Functions in General 67
5. Some Properties of the Entropy 70
6. Radial Distribution 74
   6.1 Radial Distribution in Liquids 75; 6.2 More Exact Methods . 80
7. Intermolecular Forces 82

CHAPTER IV: CONDENSATION AND THE LIQUID STATE
1. The Nature of Condensation 87
2. The Cluster Integrals 91
3. The Virial Series 94

	Page
4. The Theory of Condensation	99
5. Approximate Methods	104
6. The Solid State	108
7. The Theory of Freezing	112
8. The Cell Model for Liquids	116

CHAPTER V: THE STRUCTURE OF FLUIDS IN MOTION

1. General Considerations	122
2. The Evolution of Molecular Distributions	126
3. The Theory of Flow	130
4. Viscosity and Thermal Conduction	135
4.1 Viscosity, p. 136; 4.2 Thermal Conduction, p. 141	
5. Theory of Non-Uniform Fluids	145
6. The Friction Constant	151
7. Applications of the Theory	156
8. Motion of Molecular Clusters	160

CHAPTER VI: COMPLEX FLUIDS AND FLUID MIXTURES

1. Complex Structures	163
2. Fluid Mixtures	168
3. Theory of Diffusion	174

CHAPTER VII: FURTHER EQUILIBRIUM PROPERTIES

1. Elasticity	181
2. Surface Tension	186
3. The Brownian Motion	195
4. The Second Virial Coefficient	200
5. Gravity and Boundary Forces	203
6. The Dielectric Constant	207

CHAPTER VIII: THE KINETIC THEORY OF FLUIDS

1. Introductory	214
2. Boltzmann's Equation	216
2.1 Liouville's Theorem for Gases, p. 217; 2.2 Derivation of Boltzmann's Equation, p. 218	
3. Corrections to Boltzmann's Equation	221
4. Solution of Boltzmann's Equation	225
4.1 Viscosity and Thermal Conduction	225

CHAPTER IX: THE QUANTUM THEORY OF FLUIDS

1. General Principles	231
1.1 The Classical Theory as a Limit, p. 235; 1.2 Quantum Mechanical States and Operators, p. 237; 1.3 Transition Probabilities, p. 240	

		Page
2.	Equilibrium in the Quantum Theory	242
	2.1 Very Low Temperatures	247
3.	Quantum Statistics	252
4.	The Quantum Theory of Flow	258
REFERENCES		262

CHAPTER I

INTRODUCTION

1. The Physical Properties of Fluids

It is proposed in the following pages to make a study of a general type of substances, called *fluids*, which are capable of *flow*, from the point of view of their internal structure. The most obvious property of a fluid is its state, which may be solid, liquid or gas. It must of course be admitted that a large number of solids are not fluids at all since they cannot flow without first changing their state; this is true particularly of all crystalline solids, which are therefore excluded from consideration in this book. Nevertheless, many solids which do not appear to flow at all under ordinary conditions can be made to flow by subjecting them to heavy pressure over a long period of time, and are therefore classed as fluids; this is true for example of most kinds of glassy and plastic substances. The distinguishing feature of such solids is that they do not exhibit any visible kind of structure even under the closest scrutiny, and that they are isotropic in their natural state, that is, physically indistinguishable when examined from different directions in space. There is generally no sharp distinction in the glassy substances between solid and liquid; on heating they do not melt suddenly, like ice, at a definite temperature, but gradually become softer and more pliable; they are therefore to be considered as *liquids* with a very high viscosity.

All liquids are fluids, and have in common the property that they show no visible structure; they are generally distinguished from gases by the fact that they normally exhibit a free surface, which is formed in the process of condensation. Above what is known as the critical temperature, however, there is no condensation, and a gas in compression passes continuously through a succession of states which may end with a density characteristic of the liquid state. It is, in fact, possible to bring *any* substance from the gaseous to the liquid state without condensation by heating it to a temperature above the critical point. The distinction between liquid and gas is not, therefore, as clearly defined as one might suppose, and it is quite fitting that gases, as well as non-crystalline solids, should be regarded as fluids.

It is the ultimate object of the molecular theory of fluids to give a complete account of their physical properties. This should be of a quantitative as well as of a qualitative nature, to enable one to calculate and predict the result of any physical measurement. Now the measurable characteristics of a fluid are rather numerous, as a brief consideration will show. A great deal of information is summarized in what is known as the *equation of state*, which describes the behaviour of the substance under variations in temperature and pressure. It is a matter of experience that the state of a fluid at rest may be completely determined by just two measurements; if the temperature and pressure are measured, the equation of state determines the density; if the density and temperature are fixed, so is the pressure; and if the density and pressure are known, the temperature can be read from the appropriate tables. From a knowledge of the relation between temperature, pressure, and density, many other data can be derived. For example, the compressibility is obtained directly as the rate of change of density with variation of pressure, and the coefficient of expansion as the rate of change of density with variation of temperature. If one adds to this information a knowledge of the specific heat at constant density *or* constant pressure, the thermodynamical behaviour of the fluid is completely determined.

Another question of considerable interest concerns the elastic properties of fluids. It is clear that quasi-solid substances like the glasses have a shear modulus of elasticity as well as the compressibility which can be derived from the equation of state; but it is not obvious that mobile liquids and gases are elastic in this sense at all. If one considers what happens to the glasses when their temperature is raised beyond the zone of fusion, however, one sees that at no stage do these substances actually lose their shear elasticity, but that it is masked by the increasing fluidity. Recent experiments conducted with rapidly oscillating stresses have shown how to measure the shear modulus of mobile liquids which do not normally betray this kind of elasticity at all. It is one of the tasks of the molecular theory of flow to show how to compute the elastic constants, and also to predict the deviations from the 'Newtonian' behaviour described by Hooke's Law.

A task of considerably greater difficulty is encountered when one proceeds to consider fluids in motion. Although from a macroscopic point of view there is no great difference in principle between the theory of elasticity and viscosity, the calculations required to derive elastic constants from molecular theory are simple compared with those encountered in the attempt to obtain coefficients of viscosity. One source of embarrassment has been the fact that the viscosities of liquids and gases appear to obey quite different laws: while the coefficient of viscosity of a gas increases as the temperature is raised, and is not very sensitive to changes in density, the opposite is found to hold for liquids. It has been recognized for some time that the molecular mechanism of viscosity must be considerably different for the two states; but, although the nature of viscous processes in gases has been known for many years, it is only recently that a complete understanding has been reached of the corresponding processes in liquids. Another process which from an empirical point of view seems quite unrelated to viscosity, but presents exactly the same difficulty, is the conduction of heat by fluids. The thermal conductivity of gases and liquids obey what appear to be very different laws, and there has been considerable difficulty in extending the molecular theory of thermal conduction in gases to meet the requirements of a kinetic theory of liquids. From the theoretical point of view, there is indeed little difference in the problems presented by viscosity and thermal conduction, and the same methods can be applied to both.

The properties of fluids so far considered are most conveniently studied in connection with *simple* fluids, consisting of molecules which are all of the same kind, but there is no essentially new principle involved in the theory of fluid mixtures as long as the concentrations of the various components of the mixture are supposed to be the same throughout the fluid. If the concentrations are not uniform, however, the phenomenon of *diffusion* appears; and, even if the concentrations are uniform initially, they will not remain so when there is a gradient of temperature in the fluid. Owing to the interesting process of *thermal* diffusion, there is a tendency for different kinds of molecules to separate along the direction of the temperature gradient; this has an important application to the practical problem of the separation of different isotopes of the same element. Diffusion and thermal diffusion are not easy to understand in terms of a purely macroscopic theory, and the bearing of the molecular theory of flow on these phenomena is therefore of special importance. As for viscosity and thermal conduction, the study of diffusion and thermal diffusion in gases does not present very great difficulties, but it is only very recently that much progress has been made towards the understanding of the corresponding processes in liquids.

An account of the properties of fluids proceeding from the molecular hypothesis would be very incomplete without some consideration of optical and electromagnetic phenomena. For not only are these of considerable intrinsic interest, but they are the source of the most direct experimental evidence concerning the ultra-microscopic structure of fluids. The most powerful method of investigating the molecular structure of matter arises from the scattering of electromagnetic radiation by the constituent electrons. When the wave-length of the incident radiation is of the same order as the distance between neighbouring molecules, the scattered radiation forms a pattern which can be photographed and from which the configuration of the molecules is easily inferred. This property of X-radiation has been used to determine molecular distributions in a wide variety of fluids, and the results have been of the utmost value in securing an understanding of many aspects of molecular flow. For almost every measurable property of fluids depends in some way on their internal structure, which is not, as one might suppose, invariable for the same substance, but changes with temperature, density, and the state of motion of the substance. It is now known that the wide diversity of the molecular structure of fluids at different temperatures and pressures is responsible for the complex behaviour of macroscopic quantities like viscosity and thermal conductivity.

Some interesting questions arise in connection with the boundary layers of fluids, whether they are supposed to be in contact with other substances, or, for liquids, in contact with the saturated vapour. In this region the density of the fluids and other properties such as the energy density, undergo rapid changes in an almost immeasurably short distance; this manifests itself physically in the phenomenon of surface tension. Among the facts to be explained is that, although drops of liquid form only with great difficulty in a condensing gas which is perfectly free of the impurities which normally act as nuclei of condensation, so that supersaturation normally occurs, once a drop has formed exceeding a certain critical size it will continue to grow indefinitely. This is accounted for by the macroscopic theory of surface tension in a qualitative way, but an exact treatment based on the molecular theory introduces features of considerable complexity. The explanation given by molecular theory of the phenomenon of condensation may indeed be regarded as one of its major achievements.

While it cannot be claimed that the theory of fluids based on the known molecular structure of matter is in any sense complete, it is now possible to give a qualitative, and often a quantitative account of all the principal properties which have been considered. A rigorous derivation can be given of the laws of viscosity and thermal conduction, for example, and the limits of their validity assigned. It is rather disconcerting to find, therefore, that there is a fluid which does not obey these laws at all: liquid helium II, the modification of liquid helium discovered at temperatures less than 2 degrees from absolute zero. This indicates that one of the premises on which the molecular theory of the transport processes is based cannot be correct in all circumstances. Fortunately it is not difficult to locate the error, which consists in the assumption that classical mechanics can give an adequate account of the macroscopic properties of matter. It is generally agreed that the explanation of the anomalous properties of liquid helium is to be found in the quantum theory of matter, though much still has to be done in the application of this theory to the phenomena of very low temperatures. The most reassuring result so far obtained is that at ordinary temperatures classical and quantum mechanics give practically identical results, thus establishing a firm theoretical foundation in modern physics for the numerous successful applications of the molecular theory of flow.

2. The Microscopic Picture of a Fluid

The facts concerning the molecular constitution of matter are today so widely known that it is no longer considered necessary to regard the concept of molecules as an hypothesis which must be tested by reference to the experimental evidence. Nevertheless it is still useful to consider what is the justification for supposing that a fluid consists of billions of tiny molecules, all in rapid motion relative to one another. This conception did not originally arise from physics, and found its first support in chemistry where it was useful in explaining the laws of combination of different substances. It was soon observed, however, that the perfect gas laws could be simply explained on the same hypothesis, and it is reasonable to suppose that the kinetic theory of gases and its extension to liquids would have developed in the same way even if no independent evidence for the molecular structure of matter had been forthcoming. The very fact that a large number of properties of fluids could be correlated by a single assumption would have been regarded as very strong evidence that the basic assumption was correct. The additional fact that many other branches of physics have need of the same assumption increases the confidence with which it can be used, but does not essentially alter the position. For all the experimental evidence for phenomena which cannot be apprehended by the senses is in some way indirect. Some of the evidence, like that afforded by the X-ray scattering photographs, is very suggestive indeed, but always a process of inference is required to proceed from the experimental results to the theoretical explanation.

That this is so, however, need not lead anyone to suppose that the interpretation of experimental data in the light of molecular theory is in any way less certain than the interpretation of macroscopic events in the light of past experience of macroscopic phenomena; the two processes are indeed of the same kind. The processes of inference from crude sense-data to the imaginative reconstruction of everyday objects cannot be distinguished from those involved in passing from experimentally derived sense-data to the imaginative constructions of physics.

In fact, considerably more than the simple molecular hypothesis is involved in the modern ultra-microscopic picture of a fluid. The fact that a molecule is not a point, nor a rigid structure, but is made up of individual electrons and atomic nuclei, must be taken into account if anything more than the crudest results are required. Yet to attempt to employ the detailed knowledge accumulated by atomic physics concerning the internal structure of molecules would lead to considerations so complicated that little progress could be achieved. In practice only the configuration and properties of the outmost shells of electrons surrounding the molecules have any observable effect on the macroscopic properties of the fluid made up of such molecules. Even the problem of the interaction of these outer electronic shells is not at all simple to treat exactly except for very simple molecules, and some degree of idealization is necessary. Experience has shown that the interaction between the molecules can be represented very well by supposing that they are surrounded by a conservative field of force, not necessarily symmetrical in character, and that the resultant force acting on any given molecule is the vector sum of the forces exerted separately by the other molecules. It is not necessary to suppose that every molecule has the same internal configuration provided that it is recognized that the mutual potential energy of two molecules will depend on the configuration of each. For specially simple molecules, like those of the inert gases, however, where the outer shell of electrons is spherical in shape, the internal configuration does not matter, and the force field of a molecule is spherically symmetrical. Even molecules which are not perfectly symmetrical may frequently be treated as such, as for many purposes the effect of a slight asymmetry disappears on taking a statistical average over all orientations of the molecules.

Although the interaction between different molecules is complicated in detail, two main features of the force field may be distinguished. At all distances greater than what may be loosely described as a 'molecular diameter', the force between two neighbouring molecules is *attractive*. This may be understood qualitatively as due to the mutual polarization of the charges of which the different molecules are compounded. It is clear, also, that such an attractive field might be expected to lead to the process of condensation which is a feature of the behaviour of all substances in passing from the gaseous to the liquid phase. At short distances, however, less than one molecular diameter, between the molecules, the interaction becomes repulsive. This is due to the intense electrostatic repulsion between the electrons in the outer shells of the two molecules when they are brought close together. The repulsion is not exactly of the type associated with the contact of two ideal rigid bodies, but it is nevertheless very 'hard', and is in fact responsible for the reactions between macroscopic solids. The combination of a rather soft medium-range attractive field and a hard short-range repulsive field is characteristic of the interaction between nearly all molecules, and is only modified by the existence of forces of a *chemical* nature between certain kinds of molecules; where the latter exist, temporary or permanent associations of two or more molecules are liable to arise of quite a different nature from the rather loose association of extremely large numbers of molecules which occurs in condensation.

The nature of the intermolecular forces is extremely important in determining the observable properties of fluids; an equally important feature of the microscopic picture of a fluid, however, is the thermal motion of the molecules. It has been recognized for some time that the temperature of any substance is a measure of the mean kinetic energy of the molecules, so that only at absolute zero they might be imagined to be at rest, and as the temperature is raised the velocities of the molecules increase. The fact that the molecules of a substance apparently at rest are in rapid motion is convincingly indicated by an effect discovered by Brown: if very tiny particles, suspended in a liquid, are observed through the microscope, they are seen to be in constant random motion which can only be ascribed to a succession of encounters with the individual molecules of the liquid. It cannot be supposed, however, that the velocities of the molecules are all the same; in magnitude as well as in orientation they range over all possible values. The distribution of molecular velocities is, in fact, very much like the distribution of shots about the centre of the target, approximating to what is known as the 'normal error' law. Very high velocities are rather improbable, and, just as shots tend to cluster about the centre of the target, the molecular velocities tend to cluster about the velocity of motion of the fluid as a whole.

It is not easy, on the basis of normal macroscopic experience, to gain a reliable impression of the magnitudes involved in the molecular domain. The effective diameter of a molecule may range from 10^{-8} to 10^{-7} centimetres; there are about 2.7×10^{19} molecules per cubic centimetre in a gas at normal temperature and pressure, and thousands of times more in a liquid. The average speed of a molecule at a given temperature is inversely proportional to the square root of the molecular weight; for hydrogen at $0°$ Centigrade it is 169,000 centimetres per second, and increases with temperature.

2. 1 The Picture of a Gas

In a gas, the molecules have an existence which is made up of brief periods of uninterrupted rectilinear motion, punctuated even more briefly by collisions with one another. The mean distance travelled by a hydrogen molecule between two successive collisions is 1.1×10^{-5} centimetres at normal temperature and pressure—almost 1000 times the molecular diameter—and the mean interval of time between successive collisions is 6.6×10^{-11} seconds.

The classical picture of the individual collisions in a gas is based on Newtonian mechanics. According to this picture, when two molecules approach closely to one another, they are each influenced by the other's field of force — which, it may be recalled, is attractive at moderately short, and repulsive at very short distances. Their normal rectilinear motion is progressively modified till, at close approach, their trajectories are orbits somewhat similar to the hyperbolic orbits described by some celestial bodies. At the end of the encounter, the two molecules pass from one another's sphere of attraction and resume a state of rectilinear motion. In the course of such a collision, the total energy, and the resultant momentum and angular momentum of the pair of molecules remains unchanged. The direction of motion of both molecules is, however, usually changed by the encounter.

Similar collisions may be imagined to take place between the molecules of the gas and those of the wall of the containing vessel. A gas molecule approaching the wall comes within the sphere of influence of the molecules of the wall, with the result that its rectilinear motion is changed, and, after describing a curved trajectory, it passes back into the gas in another direction. In the course of this process, there is an exchange of momentum between the incident molecule and the molecules of the wall, contributing to the pressure which the gas is observed to exert on the wall. There may also be an interchange of energy between the gas molecule and the wall, but, unless the wall has a different temperature from the gas itself, as much energy will be lost as is gained by this process, on the average. Of course, if the temperature of the wall is higher than that of the gas in the same neighbourhood, energy will be lost from the wall to the gas molecules in the encounters which take place, and, conversely, if the wall is cooler than the gas, energy will be gained by the wall. In this way, thermal energy is communicated between the gas and its surroundings.

If a molecule originating in the gas loses energy to the molecules of the wall, it may not be able to escape back into the gas, and necessarily remains within the attractive field of the molecules of the wall. This corresponds to the macroscopically observable phenomenon of the adsorption of the gas by the wall. Normally a state of equilibrium is reached in which as many molecules escape back from the adsorbed layer into the gas as are captured by the layer from the gas. If energy is lost continuously by the gas molecules to the wall, however, a state is ultimately reached in which the number of gas molecules collecting at the wall increases indefinitely, and the gas is observed to condense, forming drops of liquid on the wall.

A similar process may occur if solid particles are present within the gas. The gas is merely adsorbed by the solid, so long as there is no general loss of energy by the molecules of the gas; but if energy is lost by radiation, for example, the gas molecules may accumulate around the solid particle, which then becomes a nucleus of condensation.

The internal energy of a gas is accounted for mainly by the kinetic energy of its molecules. If the gas is in motion, its internal energy is approximately the difference between the total kinetic energy of the molecules and the kinetic energy of the visible motion of the fluid as a whole. On account of the very high speeds of the gas molecules, however, the latter is generally a very small correction to the total kinetic energy of the molecules of the gas. Another correction to the internal energy estimated in this way arises from the fact that, in the course of a collision between two molecules, their total kinetic energy is increased (or decreased) at the expense of the mutual potential energy of the molecules at short distances. Since, in a gas, the duration of an encounter is short compared with the interval between two successive encounters, the total mutual potential energy of the molecules is always small compared with the total kinetic energy; nevertheless, it should properly be included in the internal energy of the gas.

2. 2 Condensation and the Condensed State

If a gas is compressed, additional momentum is transferred to the molecules colliding with the wall, their average speed is increased, and the temperature tends to rise. The temperature can, however, be kept steady by cooling the gas, so that kinetic energy is lost by the molecules to compensate for that gained from collisions with the wall. Under such circumstances, the mean speed of the molecules can be kept constant while they are made to occupy an increasingly smaller amount of space. In consequence, one of two things may happen. If the temperature is above the critical temperature, the density of the gas can be increased continuously in such a way that the mean time between successive collisions of a molecule becomes very small indeed, and finally a molecule is always within the sphere of interaction of one or more other molecules. Then the gas has passed without discontinuous change from the gaseous to the liquid state. This, however, can only happen if the temperature, and consequently the mean speed of the molecules, is sufficiently high. If the temperature is low enough, a density is reached where molecules impinge on the wall in such quick succession that they are unable to escape at the same rate, and accumulate there in the form of liquid droplets. During this process, the density of the gas itself remains unchanged. Alternatively, a large number of molecules may fortuitously become involved in a complex collision, with the result that they are unable to escape from the bond of their mutual attraction, and a liquid droplet is formed within the gas itself. Any further compression of the gas will then cause further molecules to coalesce with those already in the droplet, which continues to grow, leaving the density of the surrounding vapour unchanged. In either event, the gas passes to the liquid state through a series of states in which the fluid is inhomogeneous, consisting in fact of a variable mixture of vapour and liquid droplets.

Inside the droplets, each molecule is continuously within the sphere of attraction of several other molecules; this is the distinguishing feature of the liquid state. There the motion of the molecules is no longer rectilinear; each molecule pursues an intricate, irregular path among the other molecules, attracted by the polarization which it induces among them at fairly short distances, but repelled more violently by the interaction of the electron shells at shorter distances still.

The mean speed of the molecules in the liquid is the same as that of the molecules in the vapour, at the same temperature; the liquid and vapour are distinguished mainly by the difference in density between them. The mean energy of a molecule is, however, much less in the liquid than in the vapour, owing to the negative potential energy of its interaction with the other molecules. In order that a molecule should be able to escape from a droplet, it is necessary that it should possess sufficient kinetic energy to overcome the attraction of the neighbouring molecules. The energy required to liberate a molecule from the liquid is the same as that released, as latent heat of condensation, when the liquid drop is formed.

A liquid drop of macroscopic size consists of billions of molecules. Near the periphery of the drop, where it is in contact with the vapour, there is a surface zone of microscopic depth where the molecular density changes imperceptibly from its value in the liquid to the value in the gas. Where a liquid droplet is in contact with a containing wall, the mean density of the fluid molecules also falls off quickly from the comparatively large value in the 'surface' of the droplet to zero within the wall.

The pressure exerted by a liquid drop on the wall of the containing vessel is, of course, the same as that exerted by the surrounding vapour. It would, however, be incorrect to suppose that this pressure can be accounted for in terms of the momentum transferred in collisions of the molecules of the liquid with those of the wall. In the liquid, and its interface with the wall, it is indeed impossible to speak of a collision in any absolute sense; the fluid molecules are in continuous interaction with the molecules of the wall, and the pressure is simply the mean force per unit area exerted on the wall as a result of this interaction. This explains why the perfect gas law fails so badly if applied to condensed fluids.

2. 3 Fluids in Motion

Little requires to be added to this picture to obtain a description of fluids in motion. The characteristic feature of a moving fluid is that the molecular velocities in any part of the fluid are not grouped symmetrically about the state of rest, but cluster instead about a velocity which is identifiable with the macroscopic velocity of the fluid at that point. A situation of interest arises when this macroscopic velocity varies from point to point in the fluid, so that the 'central', or mean velocity varies in a similar way. Then it can be seen that molecular processes will tend ultimately to remove such variations. In a gas, the molecules are always travelling from point to point in the fluid, in between collisions with one another. A molecule travelling from a certain point in the fluid is more likely to have a velocity equal to [1] the macroscopic velocity of the fluid at that point than any other; so the interchange of molecules between two points of the fluid has the ultimate effect of eliminating any difference between the macroscopic velocities at those points. This is, essentially, the explanation of the phenomenon of viscosity in gases. The effectiveness of the process obviously depends on the ability of the molecules to travel freely from point to point in the fluid. In liquids, however, owing to the high density, the molecules are not able to move freely, and some other process must be sought to account for the transfer of momentum. The explanation of viscosity in liquids rests, in fact, on

---

[1] The velocity of a molecule or a fluid is a vector quantity, and the equality of two vectors implies the equality of all three components. To say that two velocities are equal is a much stronger statement than to assert that the corresponding speeds are the same.

a study of the effect of the intermolecular forces. Between neighbouring parts of the fluid with different macroscopic velocities, the molecules exert an attraction on one another, and this attraction has a tendency to equalize the mean velocities of groups of molecules. In this way a lack of uniformity in the visible motion of the liquid tends to be eliminated through the interaction of the molecules. Such an effect would naturally be very small in gases, because the molecules are generally at comparatively large distances from one another, where the interaction is negligible.

The transfer of energy in liquids and gases can be explained by very similar considerations. If one part of a gas is warmer than another, the average speed of the molecules in the first locality is greater than in the second. Consequently the interchange of molecules between the two parts of the gas will tend to distribute the kinetic energy of the molecules more evenly, and to equalize the temperature of the fluid between them. In liquids, the motion of the molecules is very restricted, and such a process could not be very effective. However, energy can be communicated from one molecule to another through the action of the intermolecular forces, and this is undoubtedly the principal mode of thermal conduction in liquids.

Viscosity and thermal conduction are both irreversible processes, and this fact can easily be understood by consideration of the molecular processes involved. It is not *impossible* for the normal redistribution of momentum and energy among the molecules to be reversed, and, on the microscopic scale, may happen quite frequently by chance. The likelihood of such an event on an observable scale is, however, so small that in practice it never happens. The irreversibility of natural processes is the outcome of enormous statistical preponderances rather than any inviolable law.

Not all the molecular processes in moving fluids are irreversible. For example, a pure dilatation of a gas is a reversible process. When the vessel containing a gas is expanded, the molecules receive less momentum in their collisions with the wall, the mean kinetic energy of the molecules is reduced, and the temperature falls. To compensate for the energy thus lost by the gas, mechanical work has been done in moving the wall. A similar amount of work, used to compress the gas, will restore the temperature to its former value. In the course of this process, the gas behaves as a perfectly elastic medium. Liquids probably do not possess this property of perfect elasticity under dilatation to the same degree. If a liquid suffers compression or decompression, the mutual potential energy of the molecules is changed, and if the change is accomplished at all quickly, part of the energy of compression disappears, being irreversibly converted into internal energy. The fluid will then exhibit volume viscosity. On the other hand, the operation of ordinary shear viscosity is often accompanied by a reversible elastic process. Even very mobile liquids and gases may exhibit shear elasticity, in which the molecules are all systematically displaced under the influence of an internal stress system. This displacement is unfortunately very difficult to observe, except in very viscous fluids, because it is obscured by the viscous motion of the fluid which follows the elastic displacement.

2.4 Fluid Mixtures

What has been said so far is independent of whether the molecules of the fluid are all of the same kind, or whether the fluid is a mixture of two or more different kinds of molecules. In fluid mixtures, certain additional features present themselves in the microscopic domain, associated with the macroscopic process of diffusion. If, in certain regions of the fluid, molecules of one kind are more numerous than elsewhere, the motion of the molecules from point to point will generally lead to a gradual disappearance of the irregularities in the distribution. This is obviously an irreversible process, of the same nature as that which is responsible for viscosity and thermal conduction in gases. There is, however, no counterpart in the theory of diffusion to the transfer of momentum and energy by molecular interaction, which explains the high viscosity and thermal conductivity of liquids. Diffusion in liquids is therefore a rather slow process: the molecules impede one another in their motion from place to place in the liquid, and the rate of mixing by diffusion of two different species of molecules is severely limited.

3. The Quantum Theory

The picture of the molecular structure of a fluid developed in the preceding section is readily comprehensible to the physical intuition, and enables one to draw both qualitative and quantitative conclusions concerning the macroscopic behaviour of a fluid which are, on the whole, well verified by experience. Up till 1925, it is unlikely that any well informed person could be found to doubt the essential correctness of this description. Since that time, it has, however, become evident that molecules may not be regarded in the same way, and are not subject to the same mechanical laws, as ordinary macroscopic bodies. It is incorrect to picture a molecule as occupying a definite region of space and moving with a definite velocity in accordance with determinate physical laws. It is, in principle, possible to assign a definite position to a molecule, and it is also possible, in principle, to measure its velocity exactly; but it is *impossible* to determine both its position and its velocity with any precision. This difficulty is more than an experimental one; it is, in fact, inconceivable, on the basis of modern quantum theory, that a molecule should have both its position and velocity precisely determined. It follows that Newtonian mechanics, which is founded on such a materialistic concept of matter, must be inapplicable in the molecular domain. Instead, the principles of what is known as quantum mechanics have to be introduced.

There is a further subtle difficulty which arises in the quantum theory, connected with the identity of two similar molecules. Classically, one may imagine that, although two molecules are similar in all respects, it has a meaning to say, for example, that *one of them* is at rest, and *the other* in motion. According to quantum statistics, this, however, would involve a completely invalid separation of the identity of the two molecules. It is possible for two or more similar molecules to merge their identity in a way which is quite outside our normal macroscopic experience.

The question naturally arises whether all the conclusions drawn on the basis of what is now known to be an incorrect model of a fluid have to be abandoned, and if so, why it is that the classical theory was so successful in predicting macroscopic properties of fluids. The answer is that, although the classical model may not be accepted as providing a faithful representation of molecular phenomena, all the arguments based on it are essentially correct. Some classical statements relating to the probability of molecular events, require a somewhat different interpretation, but are by no means invalid in the quantum theory. In its quantitative aspects also the classical theory requires little modification, except at very low temperatures. So long as only temperatures far removed from absolute zero are considered, the classical model and Newtonian mechanics may be applied with complete confidence. Liquid hydrogen and liquid helium are, in fact, virtually the only fluids for which the quantum theory has any practical importance; all other substances solidify before quantum effects can become significant.

Because of the exceptional nature of liquid hydrogen and liquid helium, and also in the interests of obtaining an exact picture of any fluid on the molecular scale, it is, notwithstanding, desirable not to leave quantum mechanics out of account altogether. In a gas, the concept of a molecular encounter has to be modified. Two molecules may be supposed to have definite velocities, but then their precise position in the fluid cannot be known. One can only specify a certain probability that they will collide, and that their final velocities will have assigned values if they do collide; these probabilities can be calculated on the basis of the quantum theory, if desired. Further, if one is prepared to admit some uncertainty in the values of the velocities of the molecules, one may imagine them to be vaguely localized at certain points in the gas. The same is true of the molecules in a liquid; so long as one is concerned only with probabilities, one may adhere quite closely to the classical description of molecular phenomena.

The modifications of the classical model required by quantum mechanics and quantum statistics do have some observable consequences, even in gases, at low temperatures. The second virial coefficient derived from the experimentally determined equation of state requires quantum mechanics for its accurate prediction. However, there are no important *qualitative* deviations from the predictions of classical theory, except in liquid helium at less than 2 degrees from absolute zero. The anomalous fluid behaviour which has been observed experimentally presents a very interesting challenge to the quantum theory of the molecular constitution of fluids at very low temperatures.

4. Mathematical Technique and Notation

The quantitative analysis of molecular phenomena in three dimensions demands the use of the concise notation of vector analysis. The particular form of vector analysis employed in this book is an adaptation of that used by CHAPMAN & COWLING [1939] in their standard treatise on the molecular theory of non-uniform gases. When this notation has been understood, even the more complicated formulae appearing in the following pages will be found to have an easily comprehensible physical significance. For convenience the method of interpretation of vector and tensor equations is briefly explained in the present section.

A vector is defined as a quantity possessing both magnitude and direction; such a quantity is represented in Clarendon type. As examples, one may cite the vector displacement of a molecule from a fixed point, represented by **x**; the velocity of a molecule, represented by **ξ**; the acceleration, represented by **η**; and the external force acting on a molecule, represented by **F**. A vector may be specified by its three components, which are the projections of the vector on three mutually perpendicular lines, called the coordinate axes. Thus, the components x_1, x_2 and x_3 of the vector **x** are proportional to the cosines of the angles made by the vector with the coordinate axes. The magnitude $|\mathbf{x}|$ of the vector **x** is represented by the corresponding letter x, in italics; by Pythagoras' theorem, one has

(4.1) $$x^2 = x_1^2 + x_2^2 + x_3^2$$

Unlike the components x_1, x_2 and x_3, the magnitude x of a vector is an essentially positive quantity.

The assertion that two vectors are equal implies the equality of all three components. Thus the vector equations

$$\mathbf{x}^{(2)} = \mathbf{x}^{(1)} + \mathbf{r} \quad , \quad \mathbf{x} = l\,\mathbf{y}$$

may be regarded as a concise way of writing

$$x_1^{(2)} = x_1^{(1)} + r_1 \quad , \quad x_1 = l\,y_1$$
$$x_2^{(2)} = x_2^{(1)} + r_2 \quad , \quad x_2 = l\,y_2$$
$$x_3^{(2)} = x_3^{(1)} + r_3 \quad , \quad x_3 = l\,y_3,$$

or

$$x_k^{(2)} = x_k^{(1)} + r_k \quad , \quad x_k = l\,y_k \quad (k = 1, 2, 3).$$

On the other hand, it is very desirable to keep in mind the concept of a vector as a single physical entity, in the light of which the above equations express the composition of two vectors $\mathbf{x}^{(1)}$ and $\mathbf{r}$ to form a third vector $\mathbf{x}^{(2)}$ and the multiplication of the vector **y** by a number l to form the vector **x**, respectively.

The scalar product of any two vectors **x** and **y** is represented by $\mathbf{x}\cdot\mathbf{y}$; it is not itself a vector, but represents the number

(4.2) $$\mathbf{x}\cdot\mathbf{y} = x_1 y_1 + x_2 y_2 + x_3 y_3 = \sum_{k=1}^{3} x_k y_k.$$

It is equal to the product of the magnitudes x and y of the two vectors, and the cosine of the angle between them. From (4. 1) it follows that

(4. 3) $$\mathbf{x}^2 = \mathbf{x} \cdot \mathbf{x} = x^2.$$

An expression of the type $n(\mathbf{x})$ represents a variable n whose value depends on the position of a point whose vector displacement from a fixed point is $\mathbf{x}$. Then $n(\mathbf{x})$ is a function of the three components x_1, x_2 and x_3 of the vector $\mathbf{x}$; and it may be said that n depends on the position of the point $\mathbf{x}$.

The vector function denoted by $\frac{\partial n}{\partial \mathbf{x}}$ is defined as having the components $\frac{\partial n}{\partial x_1}$, $\frac{\partial n}{\partial x_2}$ and $\frac{\partial n}{\partial x_3}$; it has the interpretation of the gradient of the function n at the point $\mathbf{x}$. Similarly $f(\boldsymbol{\xi}, \mathbf{x})$ is a function of the components of both of the vectors $\boldsymbol{\xi}$ and $\mathbf{x}$, and $\frac{\partial f}{\partial \boldsymbol{\xi}}$ represents the vector whose components are $\frac{\partial f}{\partial \xi_1}$, $\frac{\partial f}{\partial \xi_2}$ and $\frac{\partial f}{\partial \xi_3}$. It will not always be necessary to indicate explicitly that f depends on $\mathbf{x}$; under such circumstances one may use the abbreviation $f(\boldsymbol{\xi})$ for $f(\boldsymbol{\xi}, \mathbf{x})$.

The integral of a function $f(\boldsymbol{\xi})$ of the velocity $\boldsymbol{\xi}$, over all values of ξ_1, ξ_2 and ξ_3, will be represented by

(4. 4) $$\int f(\boldsymbol{\xi}) \, d\boldsymbol{\xi} = \int_{-\infty}^{\infty} \int_{-\infty}^{\infty} \int_{-\infty}^{\infty} f(\xi) \, d\xi_1 \, d\xi_2 \, d\xi_3.$$

This indicates that the elementary range of values of $\boldsymbol{\xi}$ with components between ξ_1 and $\xi_1 + d\xi_1$, ξ_2 and $\xi_2 + d\xi_2$, ξ_3 and $\xi_3 + d\xi_3$ respectively, is represented by

(4. 5) $$d\boldsymbol{\xi} = d\xi_1 \, d\xi_2 \, d\xi_3,$$

and that the triple integral is replaced by a single one, with the omission of the infinite limits of integration. If θ and φ are the polar angles, and ξ is as usual the magnitude of the vector $\boldsymbol{\xi}$, (4. 4) may be written in the form

(4. 6) $$\int f(\boldsymbol{\xi}) \, d\boldsymbol{\xi} = \int_0^{\infty} \int_0^{\pi} \int_0^{2\pi} f(\boldsymbol{\xi}) \, d\varphi \sin \theta \, d\theta \, \xi^2 \, d\xi.$$

In considering a fluid which is contained in a vessel of finite volume V, any special integral of the form

$$\int n(\mathbf{x}) \, d\mathbf{x}$$

is understood to mean

$$\int \int \int n(\mathbf{x}) \, dx_1 \, dx_2 \, dx_3$$

with limits of integration *determined by the wall of the containing vessel.* Thus the integration is carried out only over the interior of the vessel.

If an expression of the form of a gradient is integrated over a finite region of space, it is possible to transform the integral to an integral on the surface of the region. The surface of the region, whatever its shape, can be divided into elements which are very nearly plane. A typical surface element can be represented by a vector **dS**, with magnitude dS equal to the area of the element, and direction along the outward normal to the surface at that point. Then one has the vector equation

$$(4.7) \qquad \int \frac{\partial p}{\partial \mathbf{x}} \, d\mathbf{x} = \int p(\mathbf{x}) \, \mathbf{dS}$$

where $p(\mathbf{x})$ is the value of the variable p at the point $\mathbf{x}$ on the surface element **dS**, and it is understood that the surface integral extends over the whole surface of the region. The formula (4.7), which will be referred to as Gauss' theorem, is easily proved by considering each component in turn. The component dS_1 of **dS** is simply the projection $dx_2 \, dx_3$ of the surface element on a plane normal to the x_1-axis, counted positive if the angle between the outward normal to the surface and the x_1-axis is acute, and negative otherwise. Thus

$$\int p(\mathbf{x}) \, dS_1 = \iint_1 p(\mathbf{x}) \, dx_2 \, dx_3 - \iint_2 p(\mathbf{x}) \, dx_2 \, dx_3$$

where the first integral on the right-hand side covers those parts of the surface whose outward normal makes an acute angle with the x_1-axis, and the second integral covers the remainder. However, one has by a straightforward integration

$$\iiint \frac{\partial p}{\partial x_1} \, dx_1 \, dx_2 \, dx_3 = \iint_1 p(\mathbf{x}) \, dx_2 \, dx_3 - \iint_2 p(\mathbf{x}) \, dx_2 \, dx_3$$

so (4.7) is verified.

The numerical quantity

$$(4.8) \qquad \frac{\partial}{\partial \mathbf{x}} \cdot \mathbf{q} = \frac{\partial q_1}{\partial x_1} + \frac{\partial q_2}{\partial x_2} + \frac{\partial q_3}{\partial x_3} = \sum_{k=1}^{3} \frac{\partial q_k}{\partial x_k}$$

is often called the divergence of the vector **q**. The original form of Gauss' theorem,

$$(4.9) \qquad \int \frac{\partial}{\partial \mathbf{x}} \cdot \mathbf{q} \, d\mathbf{x} = \int \mathbf{q}(\mathbf{x}) \cdot \mathbf{dS}$$

is an obvious consequence of the formulae obtained by substituting the components of **q** for p in (4.7).

Integrals in velocity 'space' can also be transformed by Gauss' theorem, if the infinite domain of integration is replaced by a very large but finite region. For example, it may be proved by this means that

$$\int \frac{\partial f}{\partial \boldsymbol{\xi}} d\boldsymbol{\xi} = 0$$

if $f(\boldsymbol{\xi})$ tends to zero more rapidly than ξ^{-2} for large values of $\xi = |\boldsymbol{\xi}|$. By restricting the region of integration to a large 'sphere' of radius R in velocity space, one obtains

$$\int \frac{\partial f}{\partial \xi_3} d\boldsymbol{\xi} = \int_0^\pi \int_0^{2\pi} f(\boldsymbol{\xi}) \, R^2 \cos\theta \, d\varphi \sin\theta \, d\theta$$

where θ is the angle between the vector $\boldsymbol{\xi}$ and the x_3-axis, and φ is the other polar angle. If $R^2 f(\boldsymbol{\xi})$ tends to zero as R becomes very large, this integral also vanishes in the limit.

The outer product of two vectors is an example of what is known as a tensor. For example, **x y** is a tensor whose components are

$$\begin{array}{ccc} x_1 y_1 & x_1 y_2 & x_1 y_3 \\ x_2 y_1 & x_2 y_2 & x_2 y_3 \\ x_3 y_1 & x_3 y_2 & x_3 y_3 \end{array}$$

and can be represented concisely by $x_k y_l$ ($k, l = 1, 2, 3$). Some tensors cannot be expressed as the outer product of two vectors, and these are represented in *sans serif* type [1]).

For example, **a** represents a tensor whose components are a_{kl} ($k, l = 1, 2, 3$). If $a_{kl} = a_{lk}$, the tensor is said to be symmetrical. Some examples of expressions involving tensors, together with their significance, are listed below:

$$(4.10) \begin{cases} (\boldsymbol{\xi} - \mathbf{u})(\boldsymbol{\xi} - \mathbf{u}) & : \; (\xi_k - u_k)(\xi_l - u_l) \quad (k, l = 1, 2, 3) \\ \mathbf{u} \cdot \left(\dfrac{\partial}{\partial \mathbf{x}} \mathbf{u}\right) & : \; \sum_{l=1}^{3} u_l \dfrac{\partial u_k}{\partial x_l} \quad (k = 1, 2, 3) \\ \mathbf{p} : \left(\dfrac{\partial}{\partial \mathbf{x}} \mathbf{u}\right) & : \; \sum_{k=1}^{3} \sum_{l=1}^{3} p_{kl} \dfrac{\partial u_k}{\partial x_l} \\ \int \dfrac{\partial}{\partial \mathbf{x}} \cdot \mathbf{p} \, d\mathbf{x} - \int \mathbf{p} \cdot d\mathbf{S} & : \; \int \sum_{k=1}^{3} \dfrac{\partial p_{kl}}{\partial x_k} d\mathbf{x} - \int \sum_{k=1}^{3} p_{lk} \, dS_k \quad (l = 1, 2, 3). \end{cases}$$

---

[1]) An exception is made for the unit tensor **δ**, which for typographical convenience is represented in Clarendon.

The last expression vanishes, by Gauss' theorem, provided that $\boldsymbol{p}$ is a symmetrical tensor.

The unit tensor $\boldsymbol{\delta}$ is defined as having diagonal components equal to unity, and non-diagonal components equal to zero, thus:

(4. 11) $$\begin{cases} \delta_{11} = \delta_{22} = \delta_{33} = 1, \\ \delta_{23} = \delta_{32} = 0, \quad \delta_{31} = \delta_{13} = 0, \quad \delta_{12} = \delta_{21} = 0. \end{cases}$$

It has the property

(4. 12) $$\mathbf{y} \cdot \boldsymbol{\delta} = \mathbf{y} = \boldsymbol{\delta} \cdot \mathbf{y}$$

for any vector $\mathbf{y}$. Given any tensor $\boldsymbol{a}$, a numerical quantity a can be obtained by taking one third of the sum of its diagonal components:

(4. 13) $$a = \tfrac{1}{3}(a_{11} + a_{22} + a_{33})$$

The tensor

(4. 14) $$\boldsymbol{a}' = \boldsymbol{a} - a\boldsymbol{\delta}$$

is then said to be non-divergent, as the sum of its diagonal components vanishes identically. If $\boldsymbol{a}'$ does not depend on the vector $\boldsymbol{\xi}$, and ω is any function of the magnitude $\xi = |\boldsymbol{\xi}|$ of $\boldsymbol{\xi}$, it is not hard to establish the identity

(4. 15) $$\int \boldsymbol{\xi} \cdot \boldsymbol{a}' \cdot \boldsymbol{\xi}\, \omega(\xi)\, \xi^{-2}\, \boldsymbol{\xi}\boldsymbol{\xi}\, d\boldsymbol{\xi} = \tfrac{2}{15} \boldsymbol{a}' \int \omega(\xi)\, \xi^2\, d\xi.$$

For let θ be the angle made by the vector $\boldsymbol{\xi}$ with the x_3-axis, and φ be the other polar angle. Then, omitting terms which will obviously vanish, one has

$$\int \boldsymbol{\xi} \cdot \boldsymbol{a}' \cdot \boldsymbol{\xi}\, \omega(\xi) \xi^{-2} \xi_3 \xi_3\, d\boldsymbol{\xi} = \int_0^\infty \int_0^\pi \int_0^{2\pi} \{a'_{33} \cos^2\theta + $$
$$+ (a'_{11} \cos^2\varphi + a'_{22} \sin^2\varphi) \sin^2\theta\} \times \xi^2 \cos^2\theta\, \omega(\xi)\, d\varphi \sin\theta\, d\theta\, \xi^2\, d\xi$$
$$= \int_0^\infty \int_0^\pi \int_0^{2\pi} a'_{33} (\cos^2\theta - \tfrac{1}{2}\sin^2\theta)\, \xi^2 \cos^2\theta\, \omega(\xi)\, d\varphi \sin\theta\, d\theta\, \xi^2\, d\xi,$$

using the fact that

$$a'_{11} + a'_{22} = -a'_{33}. \ .$$

This integral reduces to

$$\tfrac{2}{15} a'_{33} \int_0^\infty \int_0^\pi \int_0^{2\pi} \omega(\xi)\, \xi^2\, d\varphi \sin\theta\, d\theta\, \xi^2\, d\xi,$$

as required. Other components of the tensor equation (4. 15) can be verified in a similar way.

It is sometimes necessary to use a form of Taylor's theorem adapted to three-dimensional analysis. If $f(\boldsymbol{\xi})$ is any function of the vector $\boldsymbol{\xi}$, one may use the expansion

$$(4.\,16) \quad f(\boldsymbol{\xi} + \boldsymbol{\rho}) = f(\boldsymbol{\xi}) + \boldsymbol{\rho} \cdot \frac{\partial f}{\partial \boldsymbol{\xi}} + \frac{1}{2!}(\boldsymbol{\rho}\boldsymbol{\rho}) : \frac{\partial^2 f}{\partial \boldsymbol{\xi} \, \partial \boldsymbol{\xi}} + \frac{1}{3!}(\boldsymbol{\rho}\boldsymbol{\rho}\boldsymbol{\rho}) \vdots \frac{\partial^3 f}{\partial \boldsymbol{\xi} \, \partial \boldsymbol{\xi} \, \partial \boldsymbol{\xi}}$$

for values of $\boldsymbol{\rho}$ within a certain 'ellipsoid of convergence' of the series. This result is easily obtained from the one-dimensional form of Taylor's theorem by making successive expansions of $f(\boldsymbol{\xi} + \boldsymbol{\rho})$ with respect to ϱ_1, ϱ_2 and ϱ_3. The latter may, of course, be regarded as independent variables.

A list of the symbols used most frequently in the sequel, together with their physical significance, is appended.

5. List of Symbols

The following is the common significance of some of the symbols used most frequently in the succeeding chapters.

D coefficient of diffusion
E energy, $\quad E_1$ mean energy per molecule
f velocity distribution function
F free energy, $\quad F_1$ free energy per molecule
F external force
g acceleration distribution function
h Planck's constant $= 2\pi \hbar$
k Boltzmann's constant
m molecular mass
n number density, $\quad n_q$ molecular distribution function
N number of molecules in fluid
p pressure tensor, $\quad p^0$ hydrostatic pressure
q number of molecules in cluster
q thermal flux vector
r relative position vector
S entropy, $\quad S_1$ entropy per molecule
T absolute temperature
u macroscopic velocity of flow
U internal energy, $\quad U_1$ internal energy per molecule
v velocity of molecule relative to flow
V volume of fluid
x position vector
$\beta = 1/(kT)$, $\quad \beta_s$ irreducible cluster integral
δ unit tensor
η viscosity, $\quad$ **η** molecular acceleration
ζ friction constant
$\lambda = (2\pi/\beta m)$

ξ molecular velocity
ϱ density matrix, ρ relative molecular velocity
τ time interval
ϕ intermolecular potential energy
Φ total potential energy.

CHAPTER II

STATISTICS OF MOLECULAR MOTION

1. Probability and Averages

The idea of probability is one which is important not only in microphysics, but also in many other aspects of human experience. It is essentially of a very simple nature, arising wherever a complete knowledge of some physical situation either cannot be obtained, or can be obtained only with prohibitive difficulty. Under all such circumstances, it is natural to try to assess the unknown features of the situation with the help of the limited information available. To obtain a complete knowledge of the state of a fluid, for example, it would be necessary to measure simultaneously and exactly the position and velocity of every molecule in the fluid—an undertaking not only obviously impossible in practice, but also impossible in principle according to the principle of uncertainty in quantum mechanics. In practice the most which can be assumed to be measurable is the distribution of such macroscopic quantities as the density, temperature, and velocity of flow in the fluid; and molecular distributions have therefore to be described by means of probabilities.

To avoid considerations of a metaphysical nature, it is necessary to emphasize that a probability is a *calculated* quantity which depends on the amount of information available concerning the physical situation as well as the physical situation itself. On the other hand, it represents a definite prediction concerning the properties of the physical system which can be tested, to any desired degree of accuracy, by experiment: not generally by a single experiment, but by a large number of similar experiments conducted under invariable conditions. In this context the conditions will be judged to be invariable if the measurable characteristics on which the calculation of the probability is based are the same in each experiment.

To be precise, suppose that there is a number of different possible results of the experiment, which will be called $R_1, R_2, R_3, \ldots R_k$; then to say that the probability of the result R_i is p_i is to make the following prediction. If the experiment is repeated a large number of times (N, say) under the same conditions, and the result R_i is obtained N_i times, then the ratio N_i/N will be almost equal to p_i, and the difference between these two quantities can be made arbitrarily small by sufficiently increasing the number of experiments N. The probability p_i is calculated from a knowledge of the invariable conditions under which the experiments were performed; to assert that the probability of a result R_i is p_i represents a definite prediction concerning the results of the experiments, based on the relevant physical theories.

Several general statements can be made at once concerning probabilities. Probabilites are *additive* if the events to which they refer are *mutually exclusive*. Thus, assuming that the results R_1 and R_2 of an experiment are mutually exclusive, i.e., that they cannot both occur, the probability that the result is either R_1 or R_2 is the sum of the probabilities of the results R_1 and R_2 separately. Further, the sum of the probabilities of a set of mutually exclusive events which is also *exhaustive* is unity. Thus, assuming that the results $R_1, R_2, \ldots R_k$ are both mutually exclusive and exhaustive, i.e., that one and only one of these results must be obtained, then the sum of the probabilities $p_1, p_2, \ldots p_k$ is 1.

It may be supposed that, the initial experiment having been performed and a definite result obtained, another experiment or observation is made, the possible results of which will be called $S_1, S_2, \ldots S_k$. It is not necessary that the second experiment should be of the same kind as the first, or even that the results should be of the same nature. For example, one might suppose the first experiment to consist in the measurement of the position of a molecule, and the second in the measurement of the velocity of another molecule. The probability that the result of the first experiment is R_i has been called p_i; the combined probability that the result of the first experiment is R_i *and* the result of the second experiment is S_j will be called p_{ij}. Now suppose the first experiment has been performed, and that the result has been found to be R_i; then the probability that the result of the second experiment will be S_j is the ratio p_{ij}/p_i. This is true whether the outcome of the first experiment affects the second or not. The probability p'_j that the result of the second experiment will be S_j, determined *without knowing the result of the first*, is not, however, generally p_{ij}/p_i: only when the result of the first experiment does not affect the second, so that the results are *independent* of one another, are p_{ij}/p_i and p'_j the same. So one arrives at the conclusion that probabilities are multiplicative only for events which do not affect one another. Assuming, however, that the results R_i and S_j of two experiments are independent of one another, the probability p_{ij} of obtaining *both R_i and S_j* is the product of the probabilities p_i and p'_j of the results R_i and S_j separately.

From the knowledge of the probabilities of results of a numerical nature, averages are easily predicted. Supposing that the results $R_1, R_2, \ldots R_k$ are numbers derived from the measurement of some physical quantity R, then the predicted average value of R is derived by multiplying the results by their respective probabilities, and adding the numbers so obtained. Like the probabilities, such an average is a calculated quantity, representing a prediction concerning the results of a large number of measurements. The average value derived from the measurements themselves is $(N_1 R_1 + N_2 R_2 + \ldots N_k R_k)/N$, assuming that the result R_1 is obtained N_1 times, the result R_2 is obtained N_2 times, and so on, in N experiments. However, it is obvious that this will agree with the predicted value $p_1 R_1 + p_2 R_2 + \ldots p_k R_k$ provided $p_1 = N_1/N$, $p_2 = N_2/N$, and so on, sufficiently nearly, as has been assumed.

2. Fluid Density

An elementary application of the principles of probability arises in the definition of the density of a fluid from a microscopic point of view. Assuming that the fluid consists of a large number of identical molecules each with a given mass (m), the mass density of the fluid is obtained by multiplying this molecular mass with the number (n) of molecules per unit volume. Unless the fluid is homogeneous, however, the latter quantity varies from place to place, and owing to the discontinuous nature of the medium it is obviously not correct to suppose that the number of molecules per unit volume within a small region approaches a well defined limit as the volume of the region becomes indefinitely small. It is not obvious, therefore, what precise meaning can be given to the number density n. In overcoming this difficulty the concept of probability is particularly useful.

A molecule will be considered to occupy an element of volume provided its mass centre lies within the element. If the volume element is extremely small, the possibility that it may be occupied by more than one molecule can be discounted, since the presence of the electronic shells surrounding the molecular nuclei will prevent the mass centres of two molecules from approaching to within less than a certain distance from one another. On the other hand, the probability that at any given time (t) a volume element is occupied by just one molecule is clearly proportional to the volume ($d\mathbf{x}$) and will be called $n\,d\mathbf{x}$. In practice, the coefficient n may be calculated with the help of the equation of state from a knowledge of the temperature and pressure at the point considered, or by other means; in principle one could determine whether or not a very small volume element contains a molecule, and check the calculated value by a series of experiments of the kind considered in the previous section. It is easy to see, however, that the factor n is what one generally understands to be the number density of the molecules at that point in the fluid. For suppose N molecules are contained in a large volume V surrounding the volume element. *If the density is uniform*, every molecule is equally likely to be found anywhere inside this region, and the probability of finding a *given* molecule in the volume element is therefore $d\mathbf{x}/V$. It follows that the probability of finding *any one* of the N molecules in the volume element is $Nd\mathbf{x}/V$. So, when the density is uniform, the factor n may be identified with the number N/V of molecules per unit volume; but it may further be regarded as the definition of number density even when this varies from point to point in the fluid.

In further illustration of the principles of probability, one may proceed to consider *two* exceedingly small elements of volume located at different points in the fluid. The probability that at any time (t) *both* of these volume elements are occupied by molecules is proportional to the volume of each ($d\mathbf{x}^{(1)}$ and $d\mathbf{x}^{(2)}$) and will be called $n_2 d\mathbf{x}^{(1)} d\mathbf{x}^{(2)}$. The factor n_2 clearly depends on the positions of both volume elements in the fluid. Now, the probability that the first volume element contains a molecule, independently of whether the second contains one or not, is $n^{(1)} d\mathbf{x}^{(1)}$, where $n^{(1)}$ represents the value of the number density n at the position of the first volume element. Hence, if an experiment had been performed which ascertained that the first volume element $d\mathbf{x}^{(1)}$ was actually occupied at the time in question, the probability that the second volume element $d\mathbf{x}^{(2)}$ was also occupied at that time would be the ratio of $n_2 d\mathbf{x}^{(1)} d\mathbf{x}^{(2)}$ to $n^{(1)} d\mathbf{x}^{(1)}$. The following assertion may therefore be made: the probability that the second volume element is occupied, on the assumption that the first is definitely occupied, is $n_2 d\mathbf{x}^{(2)}/n^{(1)}$. In accordance with the principle already explained, this is in general quite different from the probability $n^{(2)}d\mathbf{x}^{(2)}$ of the second volume element being occupied when nothing is known about the first. For the fact that the first is occupied will influence the chances of the second being occupied, especially if they are very near together in the fluid. Only when the two volume elements are sufficiently far apart to make it impossible for conditions in the neighbourhood of one of them to influence the region in which the other is situated, will the occupational probabilities satisfy the condition of independence, and will n_2 reduce to the product $n^{(1)}n^{(2)}$. In a uniform fluid at rest, n_2 can depend, apart from the state of the fluid (i.e., its density and temperature), only on the distance (r) between the two volume elements; it has been determined experimentally for a number of fluids as a function of r, and is found to rise

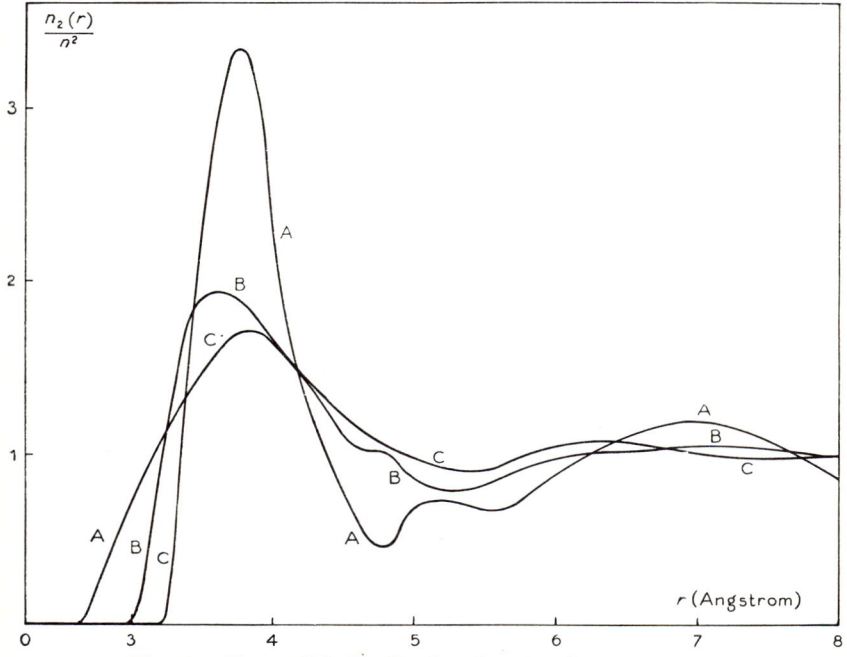

Fig. 1. The radial distribution function in liquid argon.
Curve A: 84.4° K., 0.8 atm.
B: 126.7° K., 18.3 atm.
C: 149.3° K., 43.8 atm.
— all at the saturated vapour pressure, and normalized to unity at great distances.

from the value zero when this distance is very small to one or more maxima at distances greater than a 'molecular diameter', finally approaching the constant value n^2 when the distance is sufficiently great. Some examples of the radial distribution function, as n_2 is called when expressed as a function of the distance r, are shown in fig. 1. It is of great importance in determining the structure and consequently very many of the macroscopic properties of liquids at rest. One of the most important and at the same time one of the most difficult tasks of the statistical mechanical theory of fluids is the calculation of this function from a knowledge of the intermolecular force.

3. Distribution of Molecular Velocities

The molecules of a fluid are in continual motion, even when the fluid as a whole is at rest. This fact is most convincingly demonstrated in the phenomenon called the Brownian motion, which is readily observed under a microscope: tiny particles of foreign matter suspended in a liquid are continually moving in exactly the way one would expect from the hypothesis that they are subject to continual collisions from the molecules of the liquid. The velocities of the molecules are not all the same, and are in fact distributed over a wide range of magnitude; on the average, they tend to increase with increasing temperature. A complete knowledge of the state of a fluid, according to the conception of classical mechanics, would require the simultaneous measurement of the velocities as well as of the positions of all the molecules; as this would be impossible, the introduction of the idea of probability is again essential.

A molecule will be considered to have a velocity in the range $\boldsymbol{\xi}$, $d\boldsymbol{\xi}$ if the components of its velocity in three mutually perpendicular directions have values between ξ_1 and $\xi_1 + d\xi_1$, ξ_2 and $\xi_2 + d\xi_2$, and ξ_3 and $\xi_3 + d\xi_3$ respectively. When the velocity range is sufficiently small, the probability that a particular molecule has a velocity in this range is clearly proportional to $d\boldsymbol{\xi} = d\xi_1 d\xi_2 d\xi_3$; supposing that this probability is $\varphi \, d\boldsymbol{\xi}$, the coefficient φ, which naturally depends on the velocity $\boldsymbol{\xi}$, is called the velocity distribution function. In a uniform fluid at rest, it is clear from considerations of symmetry that φ can depend, apart from the state of the fluid, only on the magnitude $\xi = \sqrt{(\xi_1^2 + \xi_2^2 + \xi_3^2)}$ of the velocity $\boldsymbol{\xi}$; it has a maximum at very small velocities, and decreases uniformly with increasing velocity, as shown in fig. 2.

Like the radial distribution function, the velocity distribution function is of considerable importance in determining the macroscopic properties of fluids, and especially of gases. It turns out, in fact, that for rare gases the velocity distribution function is of greater importance, and the radial distribution function relatively unimportant,

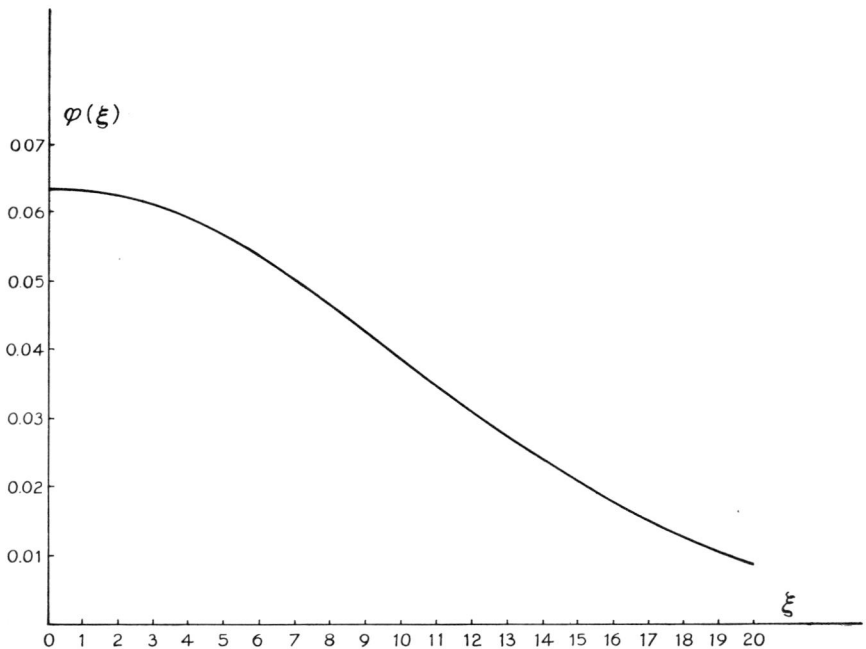

Fig. 2. The velocity distribution function for a fluid in thermal and mechanical equilibrium at the temperature $(m/k)°$ K. (where k is Boltzmann's constant and m the molecular mass).

whilst for liquids below the critical point, the radial distribution is of greater importance, in deciding the macroscopic behaviour of the substance.

The probabilities of finding a particular molecule in different velocity ranges are, according to the principles of § 1, additive, since the events to which they refer are mutually exclusive. Also the sum $[\int\varphi(\xi)d\xi]$ of the probabilities of finding the molecule in each of the infinite number of possible velocity ranges $(\xi, d\xi)$ must be unity, since the molecule must lie in one of these ranges. In symbols, therefore,

$$(3.1) \qquad \int \varphi(\pmb{\xi}) \, d\pmb{\xi} = 1,$$

where $\int \ldots d\pmb{\xi}$ represents the triple integral $\int_{-\infty}^{\infty} \int_{-\infty}^{\infty} \int_{-\infty}^{\infty} \ldots d\xi_1 d\xi_2 d\xi_3$ over all velocities. Further, the average velocity $\mathbf{u}$ is obtained by multiplying the same probabilities $\varphi(\pmb{\xi}) d\pmb{\xi}$ by the corresponding velocities ($\pmb{\xi}$), and again summing over all velocity ranges, thus:

$$(3.2) \qquad \mathbf{u} = \int \pmb{\xi} \varphi(\pmb{\xi}) \, d\pmb{\xi}.$$

For a fluid in non-uniform motion, the probability of finding a particular molecule in any velocity range, and therefore the velocity distribution function φ, varies according to the position of the molecule in the fluid, and also probably with the time. The average molecular velocity $\mathbf{u}$, which depends on φ, therefore also varies in a corresponding way. These facts are expressed by writing

$$(3.3) \qquad \mathbf{u}(\mathbf{x}, t) = \int \pmb{\xi} \varphi(\pmb{\xi}, \mathbf{x}, t) \, d\pmb{\xi},$$

where $\varphi(\pmb{\xi}, \mathbf{x}, t) d\pmb{\xi}$ is the probability that a molecule at the point $\mathbf{x}$ at time t has a velocity in the range $\pmb{\xi}$, $d\pmb{\xi}$.

> It may be noticed in passing that, from the point of view of quantum mechanics, "the velocity of a molecule at a point $\mathbf{x}$" is an illegitimate concept, as it is impossible in principle to measure the velocity of a molecule whose position is known. It should not be supposed that in consequence the function $\varphi(\pmb{\xi}, \mathbf{x}, t)$ does not exist; it can, however, be calculated easily only on the basis of classical mechanics, and predictions based on such calculations will be correct only where classical mechanics is applicable. Fortunately, as will appear in the chapter dealing with the quantum mechanics of fluids, deviations from classical behaviour are negligible except at very low temperatures—effectively in liquid hydrogen and liquid helium. Elsewhere predictions based on the classical theory may be made with complete confidence.

The average molecular velocity expressed by (3.3) above is particularly important as it is identical with the macroscopic velocity of flow at the point $\mathbf{x}$ and the time t. For, the mass centre of a large group of molecules in the same region will move with this average velocity, and the macroscopic velocity is defined in terms of the mass motion of the fluid. It has thus been seen how to define two important macroscopic quantities — density and velocity of flow — in terms of the corresponding microscopic concepts. The temperature, another macroscopic quantity, will next be defined in a similar way.

The average kinetic energy of a molecule at the point $\mathbf{x}$ and the time t is obtained by multiplying its kinetic energy $\tfrac{1}{2}m\boldsymbol{\xi}^2$ when moving with the velocity $\boldsymbol{\xi}$ by the probability $\varphi(\boldsymbol{\xi})d\boldsymbol{\xi}$ that it has a velocity in the range $\boldsymbol{\xi}$, $d\boldsymbol{\xi}$ and summing over all possible ranges; it is, therefore, $\tfrac{1}{2}m\int \boldsymbol{\xi}^2 \varphi(\boldsymbol{\xi}, \mathbf{x}, t)d\boldsymbol{\xi}$. This is actually greater than the kinetic energy $\tfrac{1}{2}m\mathbf{u}^2$ of a molecule moving with the average molecular velocity $\mathbf{u}$. For, calling the difference between the two quantities $\tfrac{3}{2}kT_1$, one has, with the help of (3. 1) and (3. 2),

(3. 4) $\quad \begin{cases} \tfrac{3}{2}kT_1 = \tfrac{1}{2} m \left\{ \int \boldsymbol{\xi}^2 \varphi(\boldsymbol{\xi}, \mathbf{x}, t) \, d\boldsymbol{\xi} - \mathbf{u}^2(\mathbf{x}, t) \right\} \\ \qquad\quad = \tfrac{1}{2} m \int (\boldsymbol{\xi} - \mathbf{u})^2 \varphi(\boldsymbol{\xi}) \, d\boldsymbol{\xi}, \end{cases}$

and the final expression is obviously a positive quantity. The constant k appearing in the equation (3. 4) is called Boltzmann's constant, and has the value 1.381×10^{-16} erg/deg. C. When k is given this value, the quantity T_1 appearing in (3. 4) is identical with the absolute temperature measured in degrees Centigrade on Kelvin's scale. This will be proved in § 5 of this chapter, but it can be understood qualitatively in the following way. The right-hand side of (3. 4) represents the difference between the mean kinetic energy of the molecules in the fluid and the macroscopic kinetic energy per molecule; it is therefore a contribution to the *internal energy* of the fluid, which increases with the temperature. Indeed, in a rare gas where the potential energy between the molecules can be neglected, there is no other contribution to the internal energy, so that the quantity which has been called $\tfrac{3}{2}kT_1$ must be the internal energy per molecule; but, again, for a rare gas, the internal energy is proportional to the temperature. Hence (3. 4) may be regarded provisionally as a definition of the temperature in microscopic terms; that it is equivalent to the thermodynamic definition will be proved later.

In a fluid at rest, the fluid velocity $\mathbf{u}$ everywhere vanishes, and (3. 4) then states simply that the mean kinetic energy of the molecules is proportional to the absolute temperature. The general statement is that the mean kinetic energy of the molecules, as observed by an observer moving with the fluid, is proportional to the absolute temperature.

4. Molecular Mechanics of Fluids

It is now possible to consider the mechanics of simple fluids from the microscopic point of view, with the ultimate object of obtaining expressions for the pressure and internal energy in terms of the forces which are exerted between the molecules.

Let attention be confined to a well defined region of the fluid, whose boundary is carried along with the macroscopic motion of the fluid at each point. The velocity of any point on the imagined boundary of the region is then the same as the mean velocity **u** of a molecule in that part of the fluid. The resultant momentum of the molecules contained within the region is continually changing for two reasons. The first is that, although the *mean* velocity of molecules relative to the boundary is zero, individual molecules are continually entering and leaving the region, carrying their own momenta with them. The second reason is that the molecules outside the region may attract or repel those inside, in such a way that there is a resultant force on the region as a whole; this is additional to any external forces, such as gravitation, which may be present. To apply Newtonian mechanics to the fluid, it is necessary to know the magnitude of each of these effects.

The resultant momentum of the molecules within the region R considered, at any given time, is

$$\int_R m\,\mathbf{u}\,n\,d\mathbf{x}$$

since $m\,\mathbf{u}$ is the mean momentum of a single molecule, and n is the molecular number density. The suffix R to the integral indicates that the integration with respect to the spatial co-ordinates is to be carried out only over the region R. To obtain the rate of change of the momentum within the region, due to the passage of molecules across the boundary, consider first the molecules crossing a small area dS of the boundary, around the point **x**. The probability that a molecule at the point **x** has a velocity in the range $\boldsymbol{\xi}$, $d\boldsymbol{\xi}$ is $\varphi(\boldsymbol{\xi},\mathbf{x})\,d\boldsymbol{\xi}$, and the velocity of a molecule in this range relative to the boundary is $\boldsymbol{\xi} - \mathbf{u}$. If **dS** is the vector along the outward normal to the surface element at **x** (shown in fig. 3), with magnitude equal to the area dS, a molecule with velocity $\boldsymbol{\xi} - \mathbf{u}$ relative to the surface is entering or leaving the region R according as $(\boldsymbol{\xi} - \mathbf{u})\cdot\mathbf{dS} < 0$ or $(\boldsymbol{\xi} - \mathbf{u})\cdot\mathbf{dS} > 0$. Assuming $(\boldsymbol{\xi} - \mathbf{u})\cdot\mathbf{dS} > 0$, the probability per unit time that a molecule with velocity in the range $\boldsymbol{\xi}$, $d\boldsymbol{\xi}$ crosses the surface element in an outward direction is $n(\boldsymbol{\xi} - \mathbf{u}) \cdot \mathbf{dS}\, \varphi\,(\boldsymbol{\xi}, \mathbf{x})\, d\boldsymbol{\xi}$; and, assuming $(\boldsymbol{\xi} - \mathbf{u}) \cdot \mathbf{dS} < 0$, the probability per unit time that a molecule in the velocity range $\boldsymbol{\xi}$, $d\boldsymbol{\xi}$ crosses the surface element in an inward direction is $- n\,(\boldsymbol{\xi} - \mathbf{u}) \cdot \mathbf{dS}\, \varphi\,(\boldsymbol{\xi}, \mathbf{x})\, d\boldsymbol{\xi}$. The momentum carried by such molecules is $m\,\boldsymbol{\xi}$.

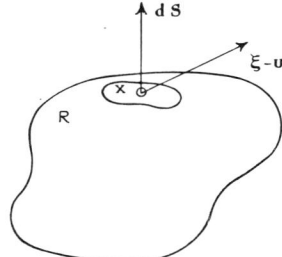

Fig. 3. The boundary of the region R moves at each point with the velocity of the fluid at that point. A molecule with velocity $\boldsymbol{\xi}$ crossing the surface element $\mathbf{dS}$ will therefore have velocity $\boldsymbol{\xi} - \mathbf{u}$ relative to the surface.

Hence, the rate of *outflow* of momentum from within the region as a result of molecules both entering and leaving through this particular surface element is

$$\int m\,\boldsymbol{\xi}\, n\,(\boldsymbol{\xi} - \mathbf{u}) \cdot \mathbf{dS}\, \varphi\,(\boldsymbol{\xi}, \mathbf{x})\, d\boldsymbol{\xi},$$

the integration being carried out over all velocities. The rate of outflow across the whole boundary is therefore

(4.1) $$m \iint_S n\,\varphi\,(\boldsymbol{\xi}, \mathbf{x})\, \boldsymbol{\xi}\,(\boldsymbol{\xi} - \mathbf{u}) \cdot \mathbf{dS}\, d\boldsymbol{\xi}$$

where the suffix S indicates integration over the closed surface.

The momentum within the region R is also changed as a result of the forces exerted by the molecules outside the boundary on those within. To form a quantitative estimate of this effect, it is convenient to determine first the mean force exerted on a single molecule at the point $\mathbf{x}^{(1)}$ due to the action of all the surrounding molecules. Assuming that the internal structure of the molecules can be ignored, the mutual potential energy of two molecules at the points $\mathbf{x}^{(1)}$ and $\mathbf{x}^{(2)}$ is a function only of the distance $r = |\mathbf{x}^{(2)} - \mathbf{x}^{(1)}|$ between them, and will be represented by $\phi(r)$. Then the force acting on the molecule at the point $\mathbf{x}^{(1)}$ due to the action of a second molecule at the point $\mathbf{x}^{(2)}$ is

$$-\frac{\partial \phi(r)}{\partial \mathbf{x}^{(1)}} = \phi'(r)\,\mathbf{r}/r,$$

where $\mathbf{r}$ is the vector $\mathbf{x}^{(2)} - \mathbf{x}^{(1)}$ directed from $\mathbf{x}^{(1)}$ to $\mathbf{x}^{(2)}$. Further, the probability of finding a second molecule in a volume element $d\mathbf{x}^{(2)}$

at $\mathbf{x}^{(2)}$ is $n_2(\mathbf{x}^{(1)}, \mathbf{x}^{(2)}) \, d\mathbf{x}^{(2)}/n(\mathbf{x}^{(1)})$, so the average force on the molecule at $\mathbf{x}^{(1)}$ due to molecules in this volume element alone is

$$n_2(\mathbf{x}^{(1)}, \mathbf{x}^{(2)}) \, \phi'(r) \, (\mathbf{r}/r) \, d\mathbf{x}^{(2)}/n(\mathbf{x}^{(1)}).$$

The average force due to the action of all the other molecules is therefore

(4. 2) $$\int n_2(\mathbf{x}^{(1)}, \mathbf{x}^{(2)}) \, \phi'(r) \, (\mathbf{r}/r) \, d\mathbf{x}^{(2)}/n(\mathbf{x}^{(1)}).$$

The molecular distribution function n_2 is normally expressed as a function $\bar{n}_2$ of $\mathbf{r}$ and the mean centre $\mathbf{x} = \frac{1}{2}(\mathbf{x}^{(1)} + \mathbf{x}^{(2)})$ of the two positions, defined by

(4. 3) $$\bar{n}_2(\mathbf{r}, \mathbf{x}) = n_2(\mathbf{x} - \tfrac{1}{2}\mathbf{r}, \mathbf{x} + \tfrac{1}{2}\mathbf{r}).$$

For the present purpose, however, it is convenient to express $\bar{n}_2(\mathbf{r}, \mathbf{x})$ as a function of $\bar{n}_2(\mathbf{r}, \mathbf{x}^{(1)})$ and its derivatives with respect to $\mathbf{x}^{(1)}$; this is accomplished by means of a Taylor's expansion about the point $\mathbf{x}^{(1)}$:

(4. 4) $$\bar{n}_2(\mathbf{r}, \mathbf{x}) = \bar{n}_2(\mathbf{r}, \mathbf{x}^{(1)}) + (\mathbf{x} - \mathbf{x}^{(1)}) \cdot \frac{\partial}{\partial \mathbf{x}^{(1)}} \bar{n}_2(\mathbf{r}, \mathbf{x}^{(1)}) + \ldots$$

As $\bar{n}_2(\mathbf{r}, \mathbf{x})$ generally varies only very slowly with $\mathbf{x}$ in regions of molecular dimensions, only the first two terms of this series need be taken into account. Substituting into (4. 2), and changing the variable of integration to $\mathbf{r}$, one obtains, for the mean force acting on the molecule at $\mathbf{x}^{(1)}$,

$$\{\int \bar{n}_2(\mathbf{r}, \mathbf{x}^{(1)}) \, \phi'(r) \, (\mathbf{r}/r) \, d\mathbf{r} + \frac{\partial}{\partial \mathbf{x}^{(1)}} \cdot \int \tfrac{1}{2} \mathbf{r} \, \bar{n}_2(\mathbf{r}, \mathbf{x}^{(1)}) \, \phi'(r) \, (\mathbf{r}/r) \, d\mathbf{r}\}/n(\mathbf{x}^{(1)}).$$

Consider the behaviour of this expression when the variable vector $\mathbf{r}$ is replaced by $\mathbf{r}' = -\mathbf{r}$. Since $n_2(\mathbf{x}^{(1)}, \mathbf{x}^{(2)}) = n_2(\mathbf{x}^{(2)}, \mathbf{x}^{(1)})$, it is clear from (4. 3) that $\bar{n}_2(\mathbf{r}, \mathbf{x}^{(1)})$ will be unchanged. Also, since $\int_{-\infty}^{\infty} dr_1$, for example, becomes $\int_{\infty}^{-\infty}(-dr_1) = \int_{-\infty}^{\infty} dr_1$, the integration $\int d\mathbf{r}$ is unchanged. Hence the whole of the above expression becomes

$$\{- \int \bar{n}_2(\mathbf{r}, \mathbf{x}^{(1)}) \, \phi'(r) \, (\mathbf{r}/r) \, d\mathbf{r} + \frac{\partial}{\partial \mathbf{x}^{(1)}} \cdot \int \tfrac{1}{2} \mathbf{r} \, \bar{n}_2(\mathbf{r}, \mathbf{x}^{(1)}) \, \phi'(r) \, (\mathbf{r}/r) \, d\mathbf{r}\}/n(\mathbf{x}^{(1)}).$$

Now, as the two expressions are equal, each must be equal to half their sum, which is

(4. 5) $$\frac{1}{n(\mathbf{x}^{(1)})} \frac{\partial}{\partial \mathbf{x}^{(1)}} \cdot \int \tfrac{1}{2} \mathbf{r} \, \bar{n}_2(\mathbf{r}, \mathbf{x}^{(1)}) \, \phi'(r) \, (\mathbf{r}/r) \, d\mathbf{r}.$$

This is the mean force acting on a molecule at the point $\mathbf{x}^{(1)}$. To obtain the mean force acting on the whole of the region R, one multiplies by the probability $n(\mathbf{x}^{(1)}) d\mathbf{x}^{(1)}$ of finding a molecule in the volume element $d\mathbf{x}^{(1)}$, and sums over all volume elements in R. The result,

$$\int_R \frac{\partial}{\partial \mathbf{x}^{(1)}} \cdot \int \tfrac{1}{2} \mathbf{r}\, \bar{n}_2(\mathbf{r}, \mathbf{x}^{(1)})\, \phi'(r)\, (\mathbf{r}/r)\, d\mathbf{r}\, d\mathbf{x}^{(1)}.$$

can be transformed by Gauss' theorem to the integral

(4. 6) $$\iint_S \bar{n}_2(\mathbf{r}, \mathbf{x})\, \phi'(r)\, (\mathbf{r}/r)\, \tfrac{1}{2} \mathbf{r} \cdot d\mathbf{S}\, d\mathbf{r}$$

over the surface S.

By combining the two results (4. 1) and (4. 6), it is seen that the rate of change of momentum within the region R due both to the flux of molecules across the boundary, and to the action of the molecules outside the boundary on those within, is

$$- m \iint_S n\varphi(\boldsymbol{\xi}, \mathbf{x})\, \boldsymbol{\xi}(\boldsymbol{\xi} - \mathbf{u}) \cdot d\mathbf{S}\, d\boldsymbol{\xi} + \tfrac{1}{2} \iint_S \bar{n}_2(\mathbf{r}, \mathbf{x})\, \phi'(r)\, (\mathbf{r}/r)\, \mathbf{r} \cdot d\mathbf{S}\, d\mathbf{r}.$$

Now this force is ascribed macroscopically to the action of the pressure across the boundary; in fact, if **p** is the pressure tensor [1]), the rate of change of momentum within a surface S due to the action of the boundary forces is

$$- \int \mathbf{p} \cdot d\mathbf{S}.$$

It follows that the pressure tensor must be expressed by

(4. 7) $$\mathbf{p} = m \int n\varphi\, \boldsymbol{\xi}(\boldsymbol{\xi} - \mathbf{u})\, d\boldsymbol{\xi} - \tfrac{1}{2} \int \bar{n}_2\, \phi'(r)\, (\mathbf{r}/r)\, \mathbf{r}\, d\mathbf{r}.$$

The hydrostatic pressure p is defined as one third of the sum of the diagonal elements of the pressure tensor, and is therefore

$$p = \tfrac{1}{3} m \int n\varphi\, \boldsymbol{\xi} \cdot (\boldsymbol{\xi} - \mathbf{u})\, d\boldsymbol{\xi} - \tfrac{1}{6} \int \bar{n}_2\, \phi'(r)\, r\, d\mathbf{r}.$$

The first term can be simplified with the help of the formulae (3. 3) and (3. 4); hence p reduces to

(4. 8) $$p = nkT_1 - \tfrac{1}{6} \int \bar{n}_2\, \phi'(r)\, r\, d\mathbf{r}.$$

This is the general formula for the pressure inside a fluid, whether moving or at rest. It is clearly separated into two parts, one due to the thermal motion of the molecules and the other due to the intermolecular forces. The first is proportional to the density and to the absolute temperature; the second—through the radial distribution function $\bar{n}_2$—varies in a rather complicated way with both density

---

[1]) See § 4 of the Introduction for the definition of tensor quantities.

and temperature, but at low densities is roughly proportional to the square of the density and decreases slowly with increasing temperature. This being so, in the gaseous state, which is distinguished by high temperatures and, more important, by low densities, the first term is much larger and more important than the second. At liquid densities, on the other hand, though the two terms are of comparable magnitude, the behaviour of the pressure is dominated by the variation of the second.

The pressure on the walls of a vessel containing a fluid in equilibrium is naturally the same as the pressure within the fluid. In spite of this numerical equality, it would be incorrect to suppose that the formula (4. 8) is still valid at the wall. The reason is that the density varies very rapidly in the immediate neighbourhood of the wall, and also the attractions and repulsions between the molecules of the wall and those of the fluid are different from the mutual interactions between the molecules of the fluid. To describe the contributions to the pressure on the wall in terms of local molecular phenomena one would have to enter into considerations rather more complicated than those advanced in this section. It is clear, however, that, apart from the numerical equality already noticed, the composition of the boundary pressure must be essentially the same as that inside the fluid: the pressure must consist of two terms, one of which is $n_s k T_1$, where n_s is the local molecular density at the wall, and the other of which represents the mean resultant force per unit area exerted between the molecules of the wall, and those of the fluid.

The internal energy of a fluid whose density and temperature are everywhere known is easier to calculate than the pressure. It is defined as the mean, or most probable, value of the total energy of the molecules, not counting macroscopic energy, such as the energy of the mass motion or the energy due to the presence of external forces. Supposing there are no external forces, therefore, to obtain the internal energy one needs only to determine the difference between the mean total energy and the energy of the mass motion. Now the total energy of the molecules is the sum of their kinetic energies, and their mutual potential energies. The average value of the kinetic energy $\frac{1}{2}m\xi^2$ of a molecule at the point **x**, according to the considerations of the previous section, is simply

$$\tfrac{1}{2} m \int \varphi(\boldsymbol{\xi}, \mathbf{x}) \, \boldsymbol{\xi}^2 \, d\boldsymbol{\xi}.$$

To determine the mean total kinetic energy, it is necessary only to multiply by the probability $n(\mathbf{x})d\mathbf{x}$ of finding a molecule in the volume element $d\mathbf{x}$, and to sum over all volume elements. The result is

$$\tfrac{1}{2} m \iint n(\mathbf{x})\, \varphi(\boldsymbol{\xi}, \mathbf{x})\, \boldsymbol{\xi}^2\, d\boldsymbol{\xi}\, d\mathbf{x}.$$

Similarly, to determine the mean total potential energy, one multiplies the potential energy $\phi(\mathbf{x}^{(2)} - \mathbf{x}^{(1)})$ between two molecules at the points $\mathbf{x}^{(1)}$ and $\mathbf{x}^{(2)}$ by the probability $n_2(\mathbf{x}^{(1)}, \mathbf{x}^{(2)})\, d\mathbf{x}^{(1)} d\mathbf{x}^{(2)}$ of finding two molecules in the volume elements $d\mathbf{x}^{(1)}$ and $d\mathbf{x}^{(2)}$, and sums over all positions of the volume elements. The result has to be divided by 2, because otherwise the contribution of each pair $d\mathbf{x}, d\mathbf{x}'$ of volume elements would be counted twice—once when $d\mathbf{x}^{(1)}$ coincides with $d\mathbf{x}$ and $d\mathbf{x}^{(2)}$ with $d\mathbf{x}'$, and once when $d\mathbf{x}^{(2)}$ coincides with $d\mathbf{x}$, and $d\mathbf{x}^{(1)}$ with $d\mathbf{x}'$. The mean potential energy of the fluid is therefore

$$\tfrac{1}{2} \iint n_2(\mathbf{x}^{(1)}, \mathbf{x}^{(2)})\, \phi(\mathbf{x}^{(2)} - \mathbf{x}^{(1)})\, d\mathbf{x}^{(1)}\, d\mathbf{x}^{(2)},$$

and the mean total energy

(4.9) $$\tfrac{1}{2} m \iint n\varphi\, \boldsymbol{\xi}^2\, d\boldsymbol{\xi}\, d\mathbf{x} + \tfrac{1}{2} \iint n_2 \phi\, d\mathbf{x}^{(1)}\, d\mathbf{x}^{(2)}.$$

To obtain the internal energy, one has still to subtract the macroscopic energy of the visible motion of the fluid. The velocity of flow at any point $\mathbf{x}$ is $\mathbf{u}(\mathbf{x})$, and since $m n(\mathbf{x})$ is the mass density, the macroscopic kinetic energy associated with the volume element $d\mathbf{x}$ is $\tfrac{1}{2} m n(\mathbf{x})\{\mathbf{u}(\mathbf{x})\}^2 d\mathbf{x}$. The total macroscopic kinetic energy is therefore

$$\tfrac{1}{2} m \int n \mathbf{u}^2 d\mathbf{x}, \quad \text{or} \quad \tfrac{1}{2} m \iint n\varphi\, \mathbf{u}^2\, d\boldsymbol{\xi}\, d\mathbf{x},$$

according to (3.1). Hence, finally, the internal energy (U) of the fluid is

(4.10) $$U = \tfrac{1}{2} m \int n\varphi(\boldsymbol{\xi}^2 - \mathbf{u}^2)\, d\boldsymbol{\xi}\, d\mathbf{x} + \tfrac{1}{2} \iint n_2 \phi\, d\mathbf{x}^{(1)}\, d\mathbf{x}^{(2)}.$$

Again it is possible to simplify the result a little with the help of (3.4), thus:

(4.11) $$U = \tfrac{3}{2} \int n k T_1\, d\mathbf{x} + \tfrac{1}{2} \iint n_2\, \phi\, d\mathbf{x}^{(1)}\, d\mathbf{x}^{(2)},$$

which reduces further to

(4.12) $$U = \tfrac{3}{2} N k T_1 + \tfrac{1}{2} V \int \bar{n}_2\, \phi\, d\mathbf{r}$$

for a *uniform* fluid containing N molecules within a volume V.

It has been shown in this section that if the radial distribution function $\bar{n}_2(r)$ is known, the pressure p and internal energy U can be obtained by simple quadratures. In principle this provides a solution to all thermodynamical problems inasmuch as, when the pressure and internal energy are known, the free energy (F) can be obtained as a function of T_1 and V by integrating the differential equation

(4. 13) $$d(F/T_1) = -(U/T_1^2)\, dT_1 - p\, dV$$

and it is well known that $F(T_1, V)$ is a generating function from which all thermodynamical quantities can be derived. The only obstacle to this programme is the fact that $\bar{n}_2(r)$ is so far undetermined. In the next section methods will be developed, which, although they apply properly only to systems in thermal and mechanical equilibrium, provide a solution in principle to the problem of the determination of the function $\bar{n}_2(r)$.

5. Statistical Thermodynamics

It has been seen already that many important macroscopic properties of fluids at rest depend directly on two molecular statistical distributions, described by the radial distribution function $n_2(r)$ and the velocity distribution function $\varphi(\xi)$ respectively. In this section a simple method of calculating these distribution functions will be explained, requiring some considerations of statistical thermodynamics. The result, of which a simple proof will soon be given, is due to Maxwell and Boltzmann.

It is supposed that one has a system which is thermally isolated, and subject to no external force, except possibly a conservative field of force such as an electrostatic field or gravity. It is supposed further that the system *has reached thermal and mechanical equilibrium*, and that its temperature is known. This could be determined, for example, by bringing it into weak thermal contact with a small test body (thermometer) of the same temperature, and verifying that no significant exchange of energy resulted. In its most general form, Boltzmann's law may be stated as follows: the probability that such a system will be found in a state of total energy E, is $a \exp(-\beta E)$, where β is a constant inversely proportional to the temperature, and a is some other constant depending on the system considered. Hence, if the number of states of the system with energy E is $g(E)$, the probability $p(E)$ that the system has total energy E is

(5. 1) $$p(E) = a\, g(E) \exp(-\beta E).$$

As the energy of the system must have some value, the sum of the probabilities $p(E)$ for all values of the energy must be unity:

(5. 2) $$a \sum_E g(E) \exp(-\beta E) = 1.$$

— a relation which can be used to determine a in terms of β.

Stated thus, Boltzmann's law applies to any thermally isolated system, such as a fluid in a container which does not transmit heat. It also applies, however, to any part of the system which can be separated from the rest, such as a molecule which forms part of the fluid in the insulating container. Before proceeding to the proof of these statements, their application to the determination of the distribution functions will be briefly considered.

The energy of a molecule with velocity $\boldsymbol{\xi}$ is $\tfrac{1}{2}m\xi^2$, so that according to the above formulation of Boltzmann's law, the probability that it will be found in a 'state', i.e., a velocity range, with velocity $\boldsymbol{\xi}$ is proportional to $\exp(-\tfrac{1}{2}\beta m\xi^2)$. The 'number of states' with velocity in the range $\boldsymbol{\xi}$, $d\boldsymbol{\xi}$ is proportional to $d\boldsymbol{\xi}$; hence (5. 1), in this application, reduces to

(5. 3) $$\varphi(\xi)\, d\boldsymbol{\xi} = a_1 \exp(-\tfrac{1}{2}\beta m\xi^2)\, d\boldsymbol{\xi},$$

and (5. 2) — of which (3. 1) is a particular case—reduces to

(5. 4) $$a_1 \int \exp(-\tfrac{1}{2}\beta m\xi^2)\, d\boldsymbol{\xi} = 1.$$

Evaluation of the integral [1]) in this equation shows that the constant a_1 is equal to $(\beta m/2\pi)^{3/2}$. The remaining constant β may also be determined by substituting the velocity distribution function $\varphi(\xi)$ as given by (5. 3) into (3. 4). Since, for a fluid at rest, the velocity of flow vanishes, the latter equation reduces in this instance to [1])

(5. 5) $$\begin{cases} \tfrac{3}{2} kT_1 = \tfrac{1}{2} m a_1 \int \xi^2 \exp(-\tfrac{1}{2}\beta m\xi^2)\, d\boldsymbol{\xi} \\ \qquad = \tfrac{1}{2} m a_1 \cdot 3(2\pi/\beta m)^{3/2}/(\beta m). \end{cases}$$

Hence, in (5. 3) one may substitute

(5. 6) $$\beta = 1/(kT_1), \qquad a_1 = (m/2\pi kT_1)^{3/2}.$$

The velocity distribution function is then obtained as a function of ξ and the quantity T_1 which will, later in this section, be proved identical with the thermodynamic temperature on the absolute scale.

---

[1]) To evaluate such integrals, $\int \ldots d\boldsymbol{\xi}$ is replaced by $\int_0^\infty \ldots 4\pi\xi^2 d\xi$ and use is made of the formula $\int_0^\infty \xi^n \exp(-\lambda\xi^2)d\xi = \tfrac{1}{2}\lambda^{-(n+1)/2}\, \Gamma(\tfrac{1}{2}n+\tfrac{1}{2})$, where Γ is the Γ-function, with the values $\Gamma(\tfrac{1}{2}) = \sqrt{\pi}$, $\Gamma(0) = 1$, $\Gamma(n+1) = n\Gamma(n)$.

As a second application of the formula (5. 2), one may consider two molecules situated at a distance r from one another in a rare gas. Suppose that their mutual potential energy is $\phi(r)$; then the probability that they will be found at the distance r from one another must be proportional to $\exp\{-\beta\phi(r)\}$, and the probability that they occupy two volume elements $d\mathbf{x}^{(1)}$ and $d\mathbf{x}^{(2)}$ at distance r from one another is

(5. 7) $\qquad n_2(r)\, d\mathbf{x}^{(1)}\, d\mathbf{x}^{(2)} = a_2 \exp\{-\beta\phi(r)\}\, d\mathbf{x}^{(1)}\, d\mathbf{x}^{(2)},$

according to (5. 1). The constant a_2 is of course distinct from a_1 appearing in (5. 3), although β is the same. In fact, a_2 can be determined easily by examining the behaviour of (5. 7) when the distance r between the two volume elements is very large. At large distances the potential energy $\phi(r)$ between the molecules will be almost zero, and $n_2(r)$ will have very nearly its limiting value n^2, where n is the number density; the constant a_2 must therefore be equal to n^2. Even so, the formula (5. 7) for the radial distribution function is correct only for a rare gas, because when the density is appreciable the potential energy of the two molecules is not simply their mutual potential energy $\phi(r)$; their interaction with all the other molecules round about has to be taken into account. It might be thought that this interaction would be independent of the distance r, but this is not so in fact; a detailed study shows that the relative position of the two molecules under consideration affects the probable positions of the others, and therefore also their contributions to the potential energy. In general the mutual potential energy $\phi(r)$ has to be replaced in (5. 7) by a perturbed potential $\phi^*(r)$, which includes the interaction with the neighbouring molecules, and is not easy to calculate. In a rare gas, it is very unlikely that a third molecule will be anywhere near, so that (5. 7) is there substantially correct.

A further important application of the formula (5. 1) can be made to the fluid under consideration as a whole. The fluid is considered to be made up of a very large number (N) of molecules, and the total potential energy Φ is then a function of their positions $\mathbf{x}^{(1)}, \mathbf{x}^{(2)}, \ldots \mathbf{x}^{(N)}$. If $\boldsymbol{\xi}^{(1)}, \boldsymbol{\xi}^{(2)}, \ldots \boldsymbol{\xi}^{(N)}$ represent the corresponding velocities, the total energy, kinetic and potential, is then

(5. 8) $\qquad E = \tfrac{1}{2} m \sum_{i=1}^{N} \xi^{(i)2} + \Phi(\mathbf{x}^{(1)}, \mathbf{x}^{(2)}, \ldots \mathbf{x}^{(N)}).$

It follows directly from (5. 1) that the probability of finding the molecules in the volume elements $d\mathbf{x}^{(1)}, d\mathbf{x}^{(2)}, \ldots d\mathbf{x}^{(N)}$ with velocities in the ranges

$$\xi^{(1)}, d\xi^{(1)}; \xi^{(2)}, d\xi^{(2)}; \ldots \xi^{(N)}, d\xi^{(N)},$$

respectively, is

(5.9) $\quad a' \exp\{-\beta(\tfrac{1}{2} m \sum \xi^{(i)2} + \Phi)\} d\mathbf{x}^{(1)} \ldots d\mathbf{x}^{(N)} d\xi^{(1)} \ldots d\xi^{(N)}.$

This result, from which both (5.3) and (5.7) can be deduced, is fundamental in the statistical thermodynamics of fluids.

It should be noticed that all the considerations of this section apply only to uniform fluids at rest. When there are irregularities of temperature or density within the fluid, or when the fluid is in motion, the formulae (5.1) and (5.9), of which a proof will now be given, are no longer correct.

5.1 PROOF OF BOLTZMANN'S LAW

Several different proofs have been given that the expression $a \exp(-\beta E)$ for the probability of finding a system in a state with energy E is correct, but essentially these reduce to only two procedures. The first is based on purely statistical principles, and consists in the enumeration of the different ways in which a very large number of systems, all identical with the one under consideration, can be distributed over the various available energy states. It can be shown that, among all the possible distributions, those which allot approximately $A \exp(-\beta E)$ systems, where A is a constant proportional to the total number of systems, to the state with energy E, are overwhelmingly more numerous than the others. This argument, due to Darwin and Fowler, can be found in most books on statistical thermodynamics (cf. FOWLER & GUGGENHEIM [1939] or SCHRÖDINGER [1946]), but though it is extremely elegant mathematically and interesting, it is not completely convincing and will not be reproduced here.

The second of the procedures referred to above does not require the artificial construct of a large number of systems identical with that under consideration. The proof divides naturally into two parts. The first part consists in the demonstration that if two states have the same energy, the probabilities that the system will be found in these states are the same, under conditions of thermal equilibrium. This shows that the probability of finding the system in any state depends only on the energy of that state. The second part of the proof then establishes that the particular functional dependence must be of the type $a \exp(-\beta E)$.

The system will first be considered during the period before thermal and mechanical equilibrium is reached. It might be supposed that the fluid was stirred or heated, and then left to subside gradually into a condition without motion or irregularities of temperature and density. During this period, it is essential that no further energy, whether thermal or mechanical, should be imparted to or withdrawn from the system; then all changes which take place are *adiabatic* in character. Suppose that at time t the probability of finding the system in the state L is $f(L, t)$. Suppose also that the probability per unit time of a transition of the system, assumed to be in the state L_1, to the state L_2 is $p(L_1, L_2)$. Then it is an elementary consequence of both classical and quantum mechanics—of which a demonstration will be given in the appropriate place (*vide* §§ 1.3 of chapter IX)—that $p(L_1, L_2) = p(L_2, L_1)$, i.e., that the probability per unit time of a transition from one state to another is the same as that for the reverse process. With this single assumption based on mechanical considerations, the desired result is easily established.

For, owing to the possibility of transitions from the state L_1 to other states L_2, the probability $f(L_1)$ of finding the system in the state L_1 at time t will be diminished at time $t + dt$ by the amount $\sum_{L_2} f(L_1) p(L_1, L_2) dt$; and, owing to the possibility of transitions from the other states to the state L_1, there will be a compensating increase $\sum_{L_2} f(L_2) p(L_2, L_1) dt$; hence

$$(5.10) \qquad \frac{df(L_1)}{dt} = \sum_{L_2} \{f(L_2) - f(L_1)\} \, p(L_1, L_2).$$

Now consider the quantity

$$(5.11) \qquad S = -k \sum_{L_1} f(L_1) \ln f(L_1)$$

which, it will presently appear, is the entropy of the system in conventional units, if the constant k is Boltzmann's constant which first appeared in (3.4). As the logarithms of the probabilities $f(L_1)$ are all negative, this is an essentially positive quantity. Moreover, it must always increase or remain stationary in time; this is proved in the following way. The derivative of S with respect to time is

$$\text{(5.12)} \quad \begin{cases} \dfrac{dS}{dt} = -k \sum_{L_1} \{1 + \ln f(L_1)\} \dfrac{df(L_1)}{dt} \\ \qquad = k \sum_{L_1 L_2} \{1 + \ln f(L_1)\} \{f(L_1) - f(L_2)\} \, p(L_1, L_2). \end{cases}$$

Replacing L_1 by L_2 and L_2 by L_1 respectively in the last expression merely interchanges the order of the summations with respect to L_1 and L_2, and does not affect its value, so

$$\frac{dS}{dt} = k \sum_{L_1 L_2} \{1 + \ln f(L_2)\} \{f(L_2) - f(L_1)\} \, p(L_1, L_2).$$

Adding this to (5.12), one obtains

$$\text{(5.13)} \quad 2 \frac{dS}{dt} = k \sum_{L_1 L_2} \ln \{f(L_1)/f(L_2)\} \{f(L_1) - f(L_2)\} \, p(L_1, L_2).$$

Now since $\ln \{f(L_1)/f(L_2)\}$ always has the same sign as $f(L_1) - f(L_2)$, the factor $\ln \{f(L_1)/f(L_2)\} \{f(L_1) - f(L_2)\}$, as well as $p(L_1, L_2)$, is positive for all pairs of states L_1 and L_2; hence dS/dt is essentially positive. This proves that S cannot decrease in time. Further, S will remain constant in time only if one of the two quantities $f(L_1) - f(L_2)$ and $p(L_1, L_2)$ vanishes for every pair of states L_1 and L_2. *It is thus a necessary condition for thermal and mechanical equilibrium that $f(L_1) = f(L_2)$ or $p(L_1, L_2) = 0$ for every pair of states.*

This condition is also *sufficient* to ensure equilibrium, for, according to (5.10), all the probabilities $f(L_1)$ will remain unchanged if it is satisfied.

Now consider the conditions under which $p(L_1, L_2) = 0$, i.e., the conditions under which it is impossible for the system in the state L_1 to transit to the state L_2. Normally it is always possible for a fluid to go from one state to another. This is not necessarily on account of the internal interactions of the system, which for a completely isolated system would be unobservable and therefore unpredictable, but because of the action of the external environment: the confining walls or vapour surrounding a fluid, external radiations, etc. These will continually induce transitions between different states of the fluid. Under the circumstances at present considered, however, there are certain forbidden transitions, arising from the fact that strictly *adiabatic* conditions have been presupposed, in order that the system should ultimately attain a state of equilibrium. As no energy can be interchanged between the system and its environment, transitions between states with different energy cannot occur, and $p(L_1, L_2) = 0$

when the states L_1 and L_2 differ with regard to their energy. According to the theorem italicized above, therefore, $f(L_1)$ and $f(L_2)$ are not necessarily the same when L_1 and L_2 are states with different energies.

The question now arises whether it is possible for $p(L_1, L_2)$ to vanish even when the states L_1 and L_2 have the same energy. The answer is that it is possible, but does not generally happen physically. Assuming, then, that transitions *can* occur between any two states L_1 and L_2 with the same energy, $p(L_1, L_2) \neq 0$ and it follows that $f(L_1)$ and $f(L_2)$ are the same. Thus, *f(L) can depend only on the energy of the state L.*

> Supposing, however, that for some reason transitions cannot occur between states with the same energy, the above conclusion does not follow, and the formulae (5. 1) and (5. 9) which were to be proved are in fact themselves incorrect. As an example of this one may consider a fluid free to rotate within a cylindrical vessel with perfectly frictionless walls. Under these conditions transitions cannot occur between states with different angular momentum about the central axis, and $f(L)$ will then depend on the angular momentum about this axis as well as on the energy in the state L. This is not an entirely academic example, as it is possible for liquid helium II to move without friction over the walls of its containing vessel. It is clear, however, that the exceptions to the theorem just established represent very special situations, and can generally be left out of consideration.

The first part of the proof of Boltzmann's law having been accomplished, it remains to show that the probability $f(E)$ of finding the system in a state L with energy E is $a \exp(-\beta E)$. For this purpose, suppose that the system is in thermal and mechanical equilibrium with a second system, with which it interacts only very weakly. It might be considered, for example, that a fluid is separated from another fluid by a thin wall which does not completely exclude the passage of thermal energy, when there is a temperature difference between the two sides. Then, if E^* is the energy of the second system, the total energy of the two systems is effectively $E + E^*$. Also, if $f^*(E^*)$ is the probability of finding the second system in a state L^* with energy E^*, the joint probability that the first system is in the state L and the second system in the state L^* must be $f(E)f^*(E^*)$, since the two events are virtually independent. But, since the two systems are in mechanical and thermal equilibrium with one another, the theorem already proved shows that the probability of finding the composite system in a state with energy $E + E^*$ can depend only on the value of $E + E^*$. Hence

(5.14) $$f(E) f^*(E^*) = F(E + E^*),$$

where F is some function of $E + E^*$.

The equation (5.14) suffices to determine the form of f, f^* and F. For, differentiating with respect to E, keeping E^* constant, one has

$$f'(E) f^*(E^*) = F'(E + E^*),$$

and, dividing now by (5.14), it follows that

$$f'(E)/f(E) = F'(E + E^*)/F(E + E^*).$$

This is true for all values of E and E^*. Giving E any constant value E_0, and writing $f'(E_0)/f(E_0) = -\beta$, where β is some constant to be determined, one obtains the differential equation

$$F'(E^* + E_0) = -\beta F(E^* + E_0)$$

to determine F as a function of $E^* + E_0$. The most general solution is

$$F(E^* + E_0) = A \exp\{-\beta(E^* + E_0)\};$$

hence

$$F(E + E^*) = A \exp\{-\beta(E + E^*)\};$$

(5.15) $$f(E) = a \exp(-\beta E); \quad f^*(E^*) = a^* \exp(-\beta E^*)$$

where $aa^* = A$. The second of these results is what was required to be proved.

5.2 Some Thermodynamical Relations

A secondary result of the foregoing argument, expressed by the equation (5.15), is that if two systems are in thermal and mechanical equilibrium with one another, the constant β which appears in their probability distribution functions is the same for each. It will now be shown that this constant β is directly related to the thermodynamical temperature.

It will be supposed that a small amount of external work is done on the system, so that, assuming the system to be in a state of energy

E originally, it is finally in a state of energy $E + dE$. It is supposed further that, although the change is not necessarily adiabatic, it is accomplished so slowly that the system is in equilibrium at each stage, and no irreversible processes can occur. Since the probability of finding the system in a state of energy E is $f(E)$, the average energy imparted to the system is the sum $[\sum_L f(E)dE]$ over all states L of $f(E)dE$; this may be identified with the work done on the system.

The internal energy, which is defined as the average energy of the system, is $\sum_L f(E)E$; in the course of the process which is being considered, this will change by the amount $\sum_L [d\{f(E)\}E + f(E)dE]$ — which exceeds the work done by an amount $\sum_L d\{f(E)\}E$. According to the first law of thermodynamics, the latter therefore represents the 'amount of heat' δQ gained by the system in the course of the change. One may write, therefore,

$$(5.\,16) \qquad \delta Q = \sum_L d\,\{f(E)\}\,E.$$

Now, taking logarithms of the appropriate equation of (5. 15), one has

$$\beta E = \ln a - \ln \{f(E)\};$$

hence (5. 16) gives

$$\beta\,\delta Q = -\sum_L [1 + \ln \{f(E)\}]\,d\,\{f(E)\} + (1 + \ln a) \sum_L d\,\{f(E)\}.$$

Since $\sum_L f(E) = 1$, $\sum_L d\{f(E)\} = 0$, and the second part of this expression disappears. The first part is simply the differential of the quantity S defined in (5. 11); for, in equilibrium, the probability $f(L)$ of finding the system in the state L is identical with $f(E)$. The above formula may therefore finally be written in the form

$$\delta Q = (k\beta)^{-1}\,dS.$$

Now, according to a well-known consequence of the second law of thermodynamics, if the heat gained by a system is expressed in the form TdS, where dS is a perfect differential, the quantity S may be identified with the entropy, and T is the absolute temperature. It follows that

$$(5.\,17) \qquad \beta = 1/(kT)$$

where T is the thermodynamic temperature on the absolute scale.

Comparing this last result with (5. 6), it is obvious that T is identical with the quantity T_1 first defined in (3. 4). It has thus been established, on the basis of classical mechanics, that *the thermodynamical temperature of a fluid is a measure of the mean kinetic energy of the molecules.* A further important result is that, under conditions of thermal and mechanical equilibrium, the quantity S defined in (5. 11) is the entropy.

It is natural to enquire whether S continues to represent the entropy when the system is not in equilibrium. This has not been proved; however, the fact, discovered in the previous subsection, that S spontaneously increases with time under non-equilibrium conditions strongly suggests that it is true. One might agree to define the entropy under all circumstances by (5. 11), and it would then have all the properties required by thermodynamics. There is, however, a consequence of quantum mechanics, that since the state of a completely isolated system is invariable, its entropy, according to this definition, would also be invariable. It cannot be said that this conflicts with the physical conception of entropy, as an isolated system in a stationary state is not in principle susceptible to observation; but the idea—elaborated by DAVYDOV [1947]—that the spontaneous increase of entropy is a consequence of observability, is undeniably strange.

Another thermodynamical quantity which plays an important role in the molecular theory of fluids is the free energy (F) of Helmholtz. If U is the internal energy $\sum_L f(E)E$, this is expressed by

$$F = U - TS = \sum_L f(E) [E + kT \ln \{f(E)\}].$$

Using (5. 15) and (5. 17), it follows that

(5. 18) $$F = kT \ln a \sum_L f(E) = kT \ln a.$$

Hence the constant a is simply $\exp(\beta F)$, where F is the free energy of the system under consideration.

6. Statistical Mechanics of Fluids

The results (5. 8), (5. 9), (5. 17) and (5. 18) of the previous section, in their application to a simple fluid, consisting of N similar molecules, in thermal and mechanical equilibrium, can be summarized as follows. The probability of finding the volume elements $d\mathbf{x}^{(1)} \ldots d\mathbf{x}^{(N)}$ all occupied, by molecules with velocities in the ranges $\boldsymbol{\xi}^{(1)}$, $d\boldsymbol{\xi}^{(1)}$; ... $\boldsymbol{\xi}^{(N)}$, $d\boldsymbol{\xi}^{(N)}$ respectively, is

$$f_N \, d\boldsymbol{\xi}^{(1)} \ldots d\boldsymbol{\xi}^{(N)} \, d\mathbf{x}^{(1)} \ldots d\mathbf{x}^{(N)},$$

where

(6. 1) $$\begin{cases} f_N = \exp \beta (F - E), \\ E = \tfrac{1}{2} m \sum_{i=1}^{N} \boldsymbol{\xi}^{(i)2} + \varPhi \quad , \quad \beta = 1/kT, \end{cases}$$

and F and T represent the free energy and thermodynamic temperature.

The particular molecule which occupies each volume element is not specified; in fact, there are $N(N-1) \ldots 2 \cdot 1 = N!$ ways of distributing the N molecules among the N volume elements, and as the molecules are similar, these are all equally probable. The probability that each volume element is filled by a *particular* molecule in the specified velocity range is therefore

(6. 2) $$(N!)^{-1} \exp \beta (F - E) \, d\boldsymbol{\xi}^{(1)} \ldots d\boldsymbol{\xi}^{(N)} \, d\mathbf{x}^{(1)} \ldots d\mathbf{x}^{(N)}.$$

Since every molecule must have *some* velocity, and be located *somewhere* in the space occupied by the fluid, if this expression is summed for each molecule over all possible positions of the volume elements and all velocities, the result must be unity:

(6. 3) $$\frac{1}{N!} \int \overset{(2N)}{\cdots} \int \exp \beta (F - E) \, d\boldsymbol{\xi}^{(1)} \ldots d\boldsymbol{\xi}^{(N)} \, d\mathbf{x}^{(1)} \ldots d\mathbf{x}^{(N)} = 1.$$

The integrations with respect to $\mathbf{x}^{(1)} \ldots \mathbf{x}^{(N)}$ are over the whole space occupied by the fluid, which may be assumed to have a definite volume V. The integrations with respect to $\boldsymbol{\xi}^{(1)} \ldots \boldsymbol{\xi}^{(N)}$ are over all velocities, and can be performed without difficulty; on substituting for E one obtains [1])

$$\frac{1}{N!} \left(\frac{2\pi}{\beta m} \right)^{\frac{3N}{2}} \int \overset{(N)}{\cdots} \int \exp \beta (F - \varPhi) \, d\mathbf{x}^{(1)} \ldots d\mathbf{x}^{(N)} = 1.$$

The potential energy $\varPhi$ is, however, a rather complicated function of the positions of the molecules, and the remaining integrations are difficult to perform. Removing the constant factor $\exp \beta F$ from the integral to the opposite side of (6. 3), one obtains, however, the

---

[1]) See footnote to p. 41.

formal equation

(6. 4) $\exp(-\beta F) = \frac{\lambda^N}{N!} \int \overset{(N)}{\ldots} \int \exp(-\beta \Phi) \, d\mathbf{x}^{(1)} \ldots d\mathbf{x}^{(N)}, \qquad \lambda = \left(\frac{2\pi}{\beta m}\right)^{3/2}$

to determine the free energy

The result just obtained is very useful, as, given the free energy $F(V, T)$ expressed as a function of volume and temperature, all other thermodynamical quantities can be derived. The internal energy U of the fluid and the pressure p, for example, are given by

(6. 5) $\qquad\qquad\qquad U = \frac{\partial (\beta F)}{\partial \beta} \quad ; \quad p = -\frac{\partial F}{\partial V}.$

The first of these relations can be applied immediately by differentiating (6. 4) with respect to β; the expression for the internal energy obtained in this way, after simplification, is

(6. 6) $\qquad U = \frac{3N}{2\beta} + \frac{\lambda^N}{N!} \int \overset{(N)}{\ldots} \int \Phi \exp \beta (F - \Phi) \, d\mathbf{x}^{(1)} \ldots d\mathbf{x}^{(N)}.$

The first term in this expression, written in the form $3NkT/2$, is obviously the mean kinetic energy of the molecular motion in the fluid, being N times the mean kinetic energy $3kT/2$ of a single molecule. The second term represents the average value of the potential energy Φ of all the molecules in the fluid; for, it is readily seen from the expression (6. 2) for the probability distribution that the average value of any function $\Psi(\mathbf{x}^{(1)}, \ldots \mathbf{x}^{(N)})$ of the molecular co-ordinates is

(6. 7) $\qquad \langle \Psi \rangle = \frac{\lambda^N}{N!} \int \overset{(N)}{\ldots} \int \Psi \exp \beta (F - \Phi) \, d\mathbf{x}^{(1)} \ldots d\mathbf{x}^{(N)}.$

To evaluate the pressure by using the second of the relations (6. 5), it is necessary to express the right-hand side of (6. 4) explicitly as a function of the volume V. It may be supposed that the fluid is contained in a cubic vessel with side l, such that $l = V^{\frac{1}{3}}$. Each of the N 3-dimensional integrals in the formula (6. 4) is then explicitly of the form $\int_0^l \int_0^l \int_0^l$, and to remove the length $l = V^{\frac{1}{3}}$ from the integral sign it is necessary to change the variables of integration from $\mathbf{x}^{(i)}$ to $\mathbf{y}^{(i)} = V^{-\frac{1}{3}} \mathbf{x}^{(i)}$. If this is done, the volume elements $d\mathbf{x}^{(i)}$ become $V d\mathbf{y}^{(i)}$, and the function $\Phi(x^{(1)} \ldots \mathbf{x}^{(N)})$ becomes $\Phi(V^{\frac{1}{3}} \mathbf{y}^{(1)} \ldots V^{\frac{1}{3}} \mathbf{y}^{(N)})$. Now

$$\frac{\partial}{\partial V} (V \, d\mathbf{y}^{(1)} \ldots V \, d\mathbf{y}^{(N)}) = NV^{-1}(V \, d\mathbf{y}^{(1)} \ldots V \, d\mathbf{y}^{(N)})$$
$$= NV^{-1} \, d\mathbf{x}^{(1)} \ldots d\mathbf{x}^{(N)}$$

and

$$\frac{\partial}{\partial V} \Phi(V^{\frac{1}{3}} \mathbf{y}^{(1)} \ldots V^{\frac{1}{3}} \mathbf{y}^{(N)}) = \tfrac{1}{3} V^{-1} \sum_{i=1}^{N} \mathbf{y}^{(i)} \cdot \frac{\partial}{\partial \mathbf{y}^{(i)}} \Phi(V^{\frac{1}{3}} \mathbf{y}^{(1)} \ldots V^{\frac{1}{3}} \mathbf{y}^{(N)})$$

$$= \tfrac{1}{3} V^{-1} \sum_{i=1}^{N} \mathbf{x}^{(i)} \cdot \frac{\partial}{\partial \mathbf{x}^{(i)}} \Phi(\mathbf{x}^{(1)} \ldots \mathbf{x}^{(N)})$$

so

$$\frac{\partial}{\partial V} \int \overset{(N)}{\ldots} \int \exp(-\beta \Phi)\, d\mathbf{x}^{(1)} \ldots d\mathbf{x}^{(N)}$$

$$= \int \overset{(N)}{\ldots} \int \left\{ \frac{N}{V} - \frac{\beta}{3V} \sum_{i=1}^{N} \mathbf{x}^{(i)} \cdot \frac{\partial \Phi}{\partial \mathbf{x}^{(i)}} \right\} \exp(-\beta \Phi)\, d\mathbf{x}^{(1)} \ldots d\mathbf{x}^{(N)}.$$

By differentiating (6.4) with respect to volume one therefore obtains

(6.8) $\quad p = \dfrac{N}{\beta V} - \dfrac{\lambda^N}{3V N!} \int \overset{(N)}{\ldots} \int \sum_{i=1}^{N} \mathbf{x}^{(i)} \cdot \dfrac{\partial \Phi}{\partial \mathbf{x}^{(i)}} \exp \beta(F - \Phi)\, d\mathbf{x}^{(1)} \ldots d\mathbf{x}^{(N)}.$

The first term in this expression may be written in the form nkT, where n is the number density, and represents the pressure of an ideal gas, which arises entirely from the motion of the molecules. The second term, according to the formula (6.7), is the average value of

$$\tfrac{1}{3} \sum_{i=1}^{N} \mathbf{x}^{(i)} \cdot \frac{\partial \Phi}{\partial \mathbf{x}^{(i)}},$$

divided by the volume; this is the contribution to the pressure which arises from the attractions and repulsions between the molecules.

The results (6.6) and (6.8) simplify a little when the total potential energy Φ is expressed as the sum of the mutual potential energies of individual pairs of molecules in the fluid, thus:

(6.9) $\quad\quad\quad\quad\quad\quad \Phi = \sum_{i > j} \phi(r^{(ij)}),$

where $r^{(ij)}$ is the distance between the molecules at $\mathbf{x}^{(i)}$ and $\mathbf{x}^{(j)}$. The mean value of $\phi(r^{(ij)})$ does not depend on i or j, so the mean value of Φ is $\tfrac{1}{2}N(N-1)$ times the mean value of $\phi(r^{(12)})$. The expression (6.6) for the internal energy therefore reduces to

(6.10) $\quad U = \dfrac{3N}{2\beta} + \dfrac{\lambda^N}{2(N-2)!} \int \overset{(N)}{\ldots} \int \phi(r^{12}) \exp \beta(F - \Phi)\, d\mathbf{x}^{(1)} \ldots d\mathbf{x}^{(N)}.$

Also, since

$$\mathbf{x}^{(1)} \cdot \frac{\partial \phi(r^{(12)})}{\partial \mathbf{x}^{(1)}} + \mathbf{x}^{(2)} \cdot \frac{\partial \phi(r^{(12)})}{\partial \mathbf{x}^{(2)}} = r^{(12)} \phi'(r^{(12)}),$$

the mean value of

$$\sum_{i=1}^{N} \mathbf{x}^{(i)} \cdot \frac{\partial \Phi}{\partial \mathbf{x}^{(i)}}$$

is $\frac{1}{2}N(N-1)$ times the mean value of $r^{(12)}\phi'(r^{(12)})$, and the expression (6. 8) for the pressure reduces to

(6. 11) $p = nkT - \dfrac{\lambda^N}{6(N-2)!\,V} \int_{\cdots}^{(N)}\!\!\int r^{(12)} \phi'(r^{(12)}) \exp \beta(F-\Phi)\, d\mathbf{x}^{(1)}\ldots d\mathbf{x}^{(N)}.$

These formulae can be compared with those obtained in § 4, which were not restricted to equilibrium conditions, and were therefore more general. It can be seen that they agree, provided one substitutes in (4. 8) and (4. 11) the formula

(6. 12) $n_2(r^{(12)}) = \dfrac{\lambda^N}{(N-2)!} \int_{\cdots}^{(N-2)}\!\!\int \exp \beta(F-\Phi)\, d\mathbf{x}^{(3)}\ldots d\mathbf{x}^{(N)}$

for the molecular distribution function relating to two molecules. As a matter of fact, this formula is easily deduced from the probability distribution (6. 2): since there are $N(N-1)$ different ways of allotting two molecules to the volume elements $d\mathbf{x}^{(1)}$ and $d\mathbf{x}^{(2)}$, it is necessary only to multiply (6. 2) by this factor, and to sum over all velocities and all positions of the volume elements $d\mathbf{x}^{(3)} \ldots d\mathbf{x}^{(N)}$, to obtain the desired result.

The fact that the values of the internal energy and pressure calculated by means of statistical thermodynamics agree with those obtained by more direct methods is a necessary confirmation of the theory. At the same time, statistical thermodynamics provides the explicit formula (6. 12) for the radial distribution function which, if it could be evaluated, would supply all the information required concerning simple fluids in equilibrium. For systems not in equilibrium, (6. 12) does not apply, and other methods have to be devised.

7. The Theory of Partitions

The integral appearing on the right-hand side of (6. 4) is often referred to as the partition function. With its aid one can readily state the probability of any partition of the N molecules occupying the volume V. For example, if the total volume is divided into two parts V_A and V_B, so that $V_A + V_B = V$, the probability of finding N_A *particular* molecules in V_A and the remaining N_B molecules in V_B is simply

(7. 1) $\dfrac{\lambda^N}{N!} \int \underset{V_A}{\overset{(N_A)}{\cdots}} \int \int \underset{V_B}{\overset{(N_B)}{\cdots}} \int \exp \beta (F-\Phi)\, d\mathbf{x}^{(1)}\ldots d\mathbf{x}^{(N)},$

where the suffixes V_A and V_B signify that the integrations are to be carried out over the volumes thus indicated. There are $N!/(N_A!N_B!)$ ways of choosing the N_A from the total number N to occupy the volume V_A, so the probability of finding *any* N_A molecules in V_A and the rest in V_B is

$$(7.2) \quad p_{AB} = \frac{\lambda^N}{N_A!N_B!} \int \overset{(N_A)}{\cdots} \int_{V_A} \int \overset{(N_B)}{\cdots} \int_{V_B} \exp \beta (F - \Phi) \, d\mathbf{x}^{(1)} \ldots d\mathbf{x}^{(N)}.$$

The interaction of the molecules in V_A with those in V_B is merely a surface effect, and may therefore be neglected when the total number of molecules is large. Under such circumstances, it is permissible to write

$$(7.3) \quad \Phi = \Phi_A + \Phi_B$$

where Φ_A and Φ_B are the potential energies of the corresponding groups of molecules. Using the formula (6.4), one then sees that (7.2) reduces to

$$(7.4) \quad p_{AB} = \exp \beta (F - F_A - F_B)$$

where F_A is the free energy of N_A molecules in volume V_A, and F_B the free energy of N_B molecules in volume V_B.

Now the free energy of a system of identical molecules is proportional to the number of molecules, and otherwise depends only on the density when the temperature is fixed. Hence

$$(7.5) \quad \begin{cases} F_A = N_A\, F_1(N_A/V_A) = V_A\, n_A\, F_1(n_A), & n_A = N_A/V_A; \\ F_B = N_B\, F_1(N_B/V_B) = V_B\, n_B\, F_1(n_B), & n_B = N_B/V_B; \end{cases}$$

where $F_1(n)$ is the free energy per molecule at the density n. The number densities n_A, n_B in the two regions are related by

$$(7.6) \quad V_A n_A + V_B n_B = N.$$

Also, if the volumes V_A and V_B are kept unaltered, but the proportion of the N molecules in each supposed to vary, p_{AB} has its maximum value when $F_A + F_B$ has its minimum value, which is given by

$$\frac{\partial}{\partial n_A}(F_A + F_B) = 0.$$

Since, according to (7.6)

$$\frac{dn_B}{dn_A} = -\frac{V_A}{V_B},$$

this condition reduces to

$$(7.7) \quad \frac{\partial}{\partial n_A}\{n_A F_1(n_A)\} - \frac{\partial}{\partial n_B}\{n_B F_1(n_B)\} = 0.$$

It is obviously satisfied when $n_A = n_B$, showing that the state with uniform density $n = N/V$ is always the most probable one. The quantity $\frac{\partial}{\partial n}\{nF_1(n)\}$ is actually the thermodynamic potential per molecule, so that (7.7) may be interpreted as the condition of equilibrium at constant temperature and pressure.

To gain some idea of the relative probabilities in the neighbourhood of the equilibrium state, p_{AB} can be expressed in terms of $\delta n_A = n_A - n$, on the assumption that this quantity is small. Since $V_A \delta n_A + V_B \delta n_B = 0$, it follows from (7.5) that

$$(7.8) \quad F_A + F_B = VnF_1(n) + \tfrac{1}{2}\frac{V_A V}{V_B}\frac{\partial^2}{\partial n^2}\{n F_1(n)\}\,\delta n_A^2,$$

with neglect of terms of order $V_A^2 \delta n_A^3/V_B^2$. Now the isothermal compressibility ζ_T is given by

$$(7.9) \quad \zeta_T^{-1} = \frac{\partial p}{\partial n} = \frac{\partial}{\partial n}\left(n^2 \frac{\partial F_1}{\partial n}\right) = n\frac{\partial^2}{\partial n^2}(nF_1);$$

hence $\quad F_A + F_B = F + \tfrac{1}{2} N V_A\, \delta n_A^2/(\zeta_T V_B n^2).$

Substituting this result in (7.4), one finds that the probability of a fluctuation δn_A in the density of the region V_A is

$$(7.10) \quad p_{AB} = \exp\left(-\frac{\tfrac{1}{2}\beta V_A N}{\zeta_T V_B n^2}\delta n_A^2\right).$$

On account of the factor N in the exponent, this is very small unless either δn_A or the ratio of the two volumes V_A and V_B is very small — a fact reflected in the circumstance that p_{AB} is 1 when $\delta n_A = 0$. The probability of a configuration in which the density ratio n_A/n differs appreciably from 1 is in fact completely negligible when, as is always assumed, the number of molecules is very large. An exception arises when the compressibility ζ_T becomes very large or infinite; this is so, for example, in the region of condensation. The thermodynamic potential of a liquid is the same as that of its saturated vapour, so that the condition (7.7) can be satisfied in the condensation region without requiring that the densities n_A and n_B should be the same. The latter may then assume any values intermediate between the number densities of the saturated vapour and boiling liquid without infringing the condition of equilibrium.

Another interesting question can be answered with the help of the theory of partitions. Suppose one has a macroscopic region of volume V_A which is contained in a very much larger region, and wishes to know the probability of finding just N_A molecules with definite positions and velocities within V_A. The number N_A will, of course, be small compared with the total number of molecules N. If $V = V_A + V_B$ is again the total volume of the region, the probability of finding the volume elements $d\mathbf{x}^{(1)} \ldots d\mathbf{x}^{(N_A)}$ in the region V_A all occupied by molecules, and the other N_B molecules outside the region, is $p_{N_A} d\mathbf{x}^{(1)} \ldots d\mathbf{x}^{(N_A)}$, where

$$(7.11) \qquad p_{N_A} = \frac{\lambda^N}{N_B!} \int \overset{(N_B)}{\ldots} \int_{V_B} \exp \beta (F - \Phi) \, d\mathbf{x}^{(N_A+1)} \ldots d\mathbf{x}^{(N)}.$$

Again one can neglect the interaction between the two groups of molecules, so that (7.3) remains correct, and it follows from (6.4) that

$$(7.12) \qquad p_{N_A} = \lambda^{N_A} \exp \beta (F - F_B - \Phi_A).$$

Now, according to (7.5),

$$F_B = (N - N_A) \, F_1 (\overline{N - N_A} / \overline{V - V_A})$$

and, since N_A and V_A have been supposed to be very small compared with N and V, one has, very nearly,

$$F_B = (N - N_A) \left\{ F_1(n) + n \left(\frac{nV_A - N_A}{N - N_A} \right) F_1'(n) \right\}$$

$$F - F_B = N_A \{ F_1(n) + n F_1'(n) \} - n^2 F_1'(n) \, V_A.$$

Hence

$$(7.13) \qquad p_{N_A} = z^{N_A} \exp \{ -\beta (p V_A + \Phi_A) \}$$

where

$$(7.14) \qquad z = \lambda \exp \beta \{ F_1(n) + n F_1'(n) \}$$

is the thermodynamical quantity commonly referred to as the *activity*, or *fugacity*.

If N_A/V_A is the same as n, which is the mean density of the fluid, the formula (7.13) for the probability distribution of the N_A molecules within the volume V_A becomes the same as if there were no other molecules outside the region. It is, of course, only correct if the volume V_A is sufficiently large for the interactions of the molecules across the surface separating the two regions to be neglected; this is true of any macroscopic region, but not of microscopic domains.

CHAPTER III

THE STRUCTURE OF FLUIDS AT REST

1. The Scattering of X-Rays by Fluids

It has been seen already that, in order to specify many of the equilibrium properties of fluids from the molecular standpoint, it is necessary to have some knowledge of their molecular structure. The theoretical determination of molecular structure is possible in principle, but remains one of the most difficult problems of statistical physics. Fortunately there is another way of obtaining information concerning the structure of matter, which has been extensively exploited for crystalline solids, and to a lesser extent for liquids and gases. This is the experimental technique of X-ray scattering.

The experimental method generally adopted consists in the separation of a monochromatic constituent of a primary beam of X-radiation by means of the Bragg reflections from a crystal plane. The reflected beam is allowed to fall on a sample of the fluid, and the intensity of the scattered radiation can then be recorded photographically. The scattering by a fluid at rest is symmetrical about the incident beam, but shows a well-marked angular distribution from which the internal structure of the fluid, as described by the radial distribution function, can be directly inferred. A precisely similar phenomenon is observed in connection with the scattering of neutrons by liquids.

Some of the earliest quantitative work directed towards the determination of the structure of liquids by the study of X-ray scattering was conducted by TRILLAT [1930] and MENKE [1932], who, with Debije, developed the theory of the scattering in almost its present form [1]). More recently, the technique of the scattering of a monochromatic beam of neutrons has been developed by CHAMBERLAIN [1950].

Some of the theoretical aspects of the scattering of X-rays will now

---

[1]) Important qualitative work was conducted previously by KEESOM and DE SMEDT [1922, 1923], and ZERNIKE and PRINS [1927].

be considered. The scattering is due almost entirely to the scattering by individual electrons which enter into the constitution of the molecules of the fluid; the scattering by the atomic nuclei is negligible, on account of their greater mass. It is convenient therefore to begin by considering the scattering due to a single electron.

Radiation of any frequency is scattered by an electron in accordance with a law derived by Klein and Nishina. In quantitative terms, this law can be stated as follows. The average intensity (I) of the radiation scattered by a single electron, initially at rest in an incident beam of intensity I_0, is

(1. 1) $$I = \tfrac{1}{2} I_0 r_0^2 \lambda^2 (\lambda^2 - \lambda \sin^2 \theta + 1)/R^2$$

for a point at distance R from the electron and subtending an angle θ with the incident beam. Here r_0 is the classical 'radius' $e^2/\mu c^2$ of the electron, and λ is the ratio of the frequencies of the scattered and incident radiation. If h ($= 2\pi \hbar$) is Planck's constant, ν the frequency of the incident radiation, c the velocity of light, and μ the electronic mass, λ is given by the Compton formula

(1. 2) $$\lambda^{-1} = 1 + h\nu(1 - \cos \theta)/\mu c^2.$$

This is obviously not much different from unity for wave-lengths down to 1 Å, and for practical purposes Thomson's formula

(1. 3) $$I = \tfrac{1}{2} I_0 r_0^2 (1 + \cos^2 \theta)/R^2$$

is therefore a fair approximation which becomes exact at low frequencies. At low frequencies, however, the formula is modified if the electron is not at rest in its initial state; and owing to the thermal motion this is generally the case. Experimentally, the wave-length chosen for the study of the structure of fluids has to be comparable with the intermolecular distances, which allows one to use the approximation (1. 3), and at the same time to disregard the effect of the thermal motion on the scattering.

Thomson's formula (1. 3) is for unpolarized primary radiation. For radiation polarized in directions perpendicular and parallel to the plane of scattering the corresponding intensities are $I_0(r_0/R)^2$ and $I_0(r_0 \cos \theta/R)^2$ respectively.

The scattering of X-rays by a liquid is due to the scattering by the individual electrons which, together with the atomic nuclei, constitute the molecules of the liquid. It is well known, however, that the intensity of radiation due to two or more sources (electrons) is not in general the sum of the intensities which would result from the sources separately. Radiation has in this respect a wave-like character, and if there is any difference in phase between waves arriving by different paths, the result will be that the waves annihilate one another in some places and reinforce one another elsewhere. In this way the characteristic interference patterns are produced. The effect of the interference of the radiation scattered by the electrons associated with the various molecules of a fluid remains to be investigated.

Assuming that one has a large number of electrons situated at the points $\mathbf{x}_1, \mathbf{x}_2, \ldots \mathbf{x}_M$ in the fluid, the radiation scattered by an electron situated at the point $\mathbf{x}_\nu$ will have the phase factor $\exp i(\mathbf{k} - \mathbf{k}_0) \cdot \mathbf{x}_\nu$, where $\mathbf{k}_0$ and $\mathbf{k}$ are the wave vectors corresponding to the incident and scattered radiation respectively. Neglecting the change of frequency on scattering, it may be assumed that $\mathbf{k}$, as well as $\mathbf{k}_0$, has the magnitude $2\pi\nu/c$, so that the magnitude of $\mathbf{k} - \mathbf{k}_0$ is

$$(1.4) \qquad s = |\mathbf{k} - \mathbf{k}_0| = (4\pi\nu \sin \tfrac{1}{2}\theta)/c$$

where θ is again the scattering angle. The amplitude of the scattered radiation arriving at the distant point considered from the electron at $\mathbf{x}_\nu$ is $I_0^{\frac{1}{2}}(r_0/R) \exp i(\mathbf{k} - \mathbf{k}_0) \cdot \mathbf{x}_\nu$, with or without a factor $\cos \theta$ according to the direction of polarization of the incident beam. The resultant amplitude at a distant point is therefore

$$I_0^{\frac{1}{2}}(r_0/R) \sum_\nu \exp i(\mathbf{k} - \mathbf{k}_0) \cdot \mathbf{x}_\nu [\cos \theta]$$

and the resultant intensity, which is the square of the modulus of the resultant amplitude, is

$$(1.5) \qquad I = \tfrac{1}{2}(1 + \cos^2 \theta) I_0 (r_0/R)^2 \sum_{\nu,z} \exp i(\mathbf{k} - \mathbf{k}_0) \cdot (\mathbf{x}_\nu - \mathbf{x}_z)$$

for unpolarized radiation.

The task of calculating the intensity of the X-radiation scattered by a liquid now reduces to the evaluation of the average value of the double sum $\sum_{\nu,z} \exp i(\mathbf{k} - \mathbf{k}_0) \cdot (\mathbf{x}_\nu - \mathbf{x}_z)$. It is convenient to consider a division of the electrons into groups and sub-groups associated with the individual molecules and atoms respectively. One can write

$$(1.6) \qquad \sum_\nu \exp i(\mathbf{k} - \mathbf{k}_0) \cdot \mathbf{x}_\nu = \sum_j f_{(j)} \exp i(\mathbf{k} - \mathbf{k}_0) \cdot \mathbf{x}^{(j)}$$

where the $\mathbf{x}^{(j)}$ represent the mass-centres of the systems of electrons of the molecules, and

(1.7) $$f_{(j)} = \sum_{\nu(j)} \exp i (\mathbf{k} - \mathbf{k}_0) \cdot (\mathbf{x}_\nu - \mathbf{x}^{(j)})$$

the latter summation being carried out only over the electrons associated with a given molecule (j). When the atomic and electronic structure of the molecule is known, the average value of the right-hand side of (1.7) is easily evaluated; in view of the relation (1.4) it is obviously a function of $\nu \sin \frac{1}{2}\theta$ only. For monatomic molecules, f is called the atomic scattering factor and can be found tabulated for every kind of atom in most works on crystallography (e.g. BRAGG & BRAGG [1933]). Molecular scattering factors are also tabulated, but may be calculated directly from a knowledge of the structure of the molecule (cf. MENKE [1932]).

As a result of this simplification, one requires only to calculate the average value of

$$f^2 \sum_{i,j} \exp i (\mathbf{k} - \mathbf{k}_0) \cdot (\mathbf{x}^{(i)} - \mathbf{x}^{(j)}),$$

where the summations refer now only to the molecules of the fluid. Now, if $n_2(\mathbf{x}^{(i)}, \mathbf{x}^{(j)}) d\mathbf{x}^{(i)} d\mathbf{x}^{(j)}$ is the probability of finding molecules in each of two distinct volume elements $d\mathbf{x}^{(i)}$ and $d\mathbf{x}^{(j)}$, the average value of $\sum_{i,j} \exp i (\mathbf{k} - \mathbf{k}_0) \cdot (\mathbf{x}^{(i)} - \mathbf{x}^{(j)})$ is

$$N + \iint n_2(\mathbf{x}^{(i)}, \mathbf{x}^{(j)}) \exp i (\mathbf{k} - \mathbf{k}_0) \cdot (\mathbf{x}^{(i)} - \mathbf{x}^{(j)}) d\mathbf{x}^{(i)} d\mathbf{x}^{(j)}.$$

The first term N, representing the total number of molecules, comes from the terms with $i = j$, and is associated with the scattering which would result if there were no interference between the radiation scattered by different molecules. The second term represents the effect of this interference.

If the fluid is in equilibrium, n_2 depends only on the distance r between $\mathbf{x}^{(i)}$ and $\mathbf{x}^{(j)}$; also one has $(\mathbf{k} - \mathbf{k}_0) \cdot (\mathbf{x}^{(i)} - \mathbf{x}^{(j)}) = rs \cos \varphi$ where φ is the angle between the vectors $\mathbf{r} = \mathbf{x}^{(i)} - \mathbf{x}^{(j)}$ and $\mathbf{s} = \mathbf{k} - \mathbf{k}_0$. Hence the average value of the double sum in (1.5) is, finally [1],

$$Nf^2 \left\{ 1 + n^{-1} \int n_2(r) \exp (irs \cos \varphi) d\mathbf{r} \right\}$$

where n is the number density of the molecules; and (1.5) is reduced to

(1.8) $I = \frac{1}{2} (1 + \cos^2 \theta) I_0 (r_0/R)^2 Nf^2 \left\{ 1 + n^{-1} \int n_2(r) (\sin rs/rs) 4\pi r^2 dr \right\}.$

[1] In the following integral, the vector $\mathbf{r}$ must be restricted to the confines of the liquid to secure convergence.

The parameter s depends on the scattering angle in the way indicated by (1.4); hence, if I is known as a function of θ, the last equation may be regarded as an integral equation to determine the radial distribution function $n_2(r)$.

It is of interest to examine the behaviour of the scattering formula when the positions of the molecules are supposed nearly independent of one another, as they would be if there were no interactions between them. Then $n_2(r) = n^2$, and the integral expression in (1.8) is nearly zero everywhere, except near $s = 0$, where it becomes very large. This means that the only effect of the interference is to produce a large amount of scattering in very small scattering angles, where it is indistinguishable from the incident beam. It is desirable to subtract this contribution from the total scattering, which can be done quite generally by replacing $n_2(r)$ by $n_2(r) - n^2$ in the integral expression of (1.8). Then, if one defines the function $i(s)$ by

$$(1.9) \qquad s\,i(s) = 4\pi n^{-1} \int_0^\infty r\{n_2(r) - n^2\} \sin(rs)\,dr.$$

the equation (1.8) becomes

$$(1.10) \qquad \tfrac{1}{2}(1 + \cos^2\theta)\,I_0\,(r_0/R)^2\,Nf^2\,\{1 + i(s)\} = I.$$

The integral equation (1.9) is easily solved by a Fourier transformation; the solution is, in fact,

$$(1.11) \qquad 2\pi^2 r\,\{n_2(r) - n^2\} = n\int_0^\infty s\,i(s) \sin(rs)\,ds.$$

From measurements of the scattered intensity I for angles between 0 and π, the function $i(s)$ is determined by the formula (1.10) for values of s between 0 and $4\pi\nu/c$. Since, with a proper choice of the frequency ν of the primary radiation, the function $i(s)$ becomes very small for values of s in excess of or even in the neighbourhood of $4\pi\nu/c$, the formula (1.11) then allows one to calculate the radial distribution function with very small error. The harmonic integral is in practice easily evaluated by making a sine Fourier analysis of the function $s\,i(s)$ measured for 12 equally spaced values of $\sin\tfrac{1}{2}\theta$ between 0 and 1, i.e., for $\theta = 2\sin^{-1}(k/12)$, $k = 1, \ldots, 12$.

The measurements have been made and the analysis carried through for a large number of liquids, and, in a few instances, over a satisfactory range of pressure and temperature. The results up to a few years ago are summarized in an excellent review by GINGRICH [1943].

Some examples for liquid argon, based on the work of EISENSTEIN & GINGRICH [1942], are shown in fig. 1 of chapter II (p. 29).

2. Optical Scattering and Density Fluctuations

The pattern exhibited by the X-radiation scattered by fluids is an interference effect, and it might therefore be expected to disappear if the wave-length of the radiation is increased so far as to considerably exceed the intermolecular distances. This is in fact true of scattering through large angles; but at small angles it is possible to have an appreciable intensity of scattering even at optical frequencies. From the formulae (1.9) and (1.10) it can be seen immediately that the function $1 + i(s)$ which gives the scattering intensity at an angle $\theta = 2 \sin^{-1} (cs/4\pi\nu)$ approaches the limit

$$(2.1) \qquad 1 + i(0) = 1 + 4\pi n^{-1} \int_0^\infty \{n_2(r) - n^2\} r^2 \, dr$$

at small angles, independently of the frequency. The magnitude of this quantity can be shown to be closely related to fluctuations in density within the fluid. The relation between the scattering of light by fluids and density fluctuations was first examined in detail by ORNSTEIN & ZERNIKE [1914], and has been given a systematic interpretation in terms of molecular theory by YVON [1937a, 1937b].

Let attention be restricted to a macroscopic volume V_A within the fluid. If N is the total number of molecules and V the total volume of the fluid, the average number of molecules to be found in V_A is obviously NV_A/V, or nV_A, where n is the number density. It is required to calculate now the mean square deviation from this average number.

If $m(\mathbf{x})$ is a function defined to be equal to unity when $\mathbf{x}$ lies within the volume V_A, and zero otherwise, the number of molecules within V_A is always

$$(2.2) \qquad N_A = \sum_{i=1}^{N} m(\mathbf{x}^{(i)})$$

where the $\mathbf{x}^{(i)}$, $i = 1 \ldots N$, represent the positions of the molecules. The average value of $m(\mathbf{x}^{(i)})$ is just the probability of finding the *particular* molecule (i) in the volume V_A and is therefore

$$(2.3) \qquad \overline{m(\mathbf{x}^{(i)})} = \frac{1}{N} \int_{V_A} n_1 \, d\mathbf{x} = nV_A/N.$$

Similarly the average value of $m(\mathbf{x}^{(i)})m(\mathbf{x}^{(j)})$ is the probability of finding the two molecules (i) and (j) in the volume V_A, i.e.,

(2.4) $$\overline{m(\mathbf{x}^{(i)})\,m(\mathbf{x}^{(j)})} = \frac{1}{N(N-1)} \iint_{V_A} n_2\, d\mathbf{x}^{(1)}\, d\mathbf{x}^{(2)},$$

assuming that i and j are not the same. Now, it is clear from (2.2) that the average value $\overline{N}_A$ of N_A is N times the average value of any one of the $m(\mathbf{x}^{(i)})$:

(2.5) $$\overline{N_A} = N\,\overline{m(\mathbf{x}^{(i)})}.$$

Also, since $\{m(\mathbf{x}^{(i)})\}^2 = m(\mathbf{x}^{(i)})$, the average value of N_A^2 is

(2.6) $$\overline{N_A^2} = N\,\overline{m(\mathbf{x}^{(i)})} + N(N-1)\,\overline{m(\mathbf{x}^{(i)})\,m(\mathbf{x}^{(j)})}.$$

Combining these results, one finds that the mean square fluctuation of the number of molecules in the volume V_A is

(2.7) $$\begin{cases} \overline{(N_A - \overline{N_A})^2} = \overline{N_A^2} - (\overline{N_A})^2 \\ \qquad = \{1 + n^{-1} \int (n_2 - n^2)\, d\mathbf{x}^{(2)}\}\,\overline{N_A}. \end{cases}$$

The quantity bracketed is exactly that which appears in (2.1); it can be seen, therefore, that the ratio of the mean square fluctuation of the number of molecules to the mean number in a small volume is exactly proportional to the small-angle scattering of light.

The same quantity can be calculated directly with the help of the theory of partitions, which was explained in § 7 of chapter II. There it was shown that the probability of finding a partition of the N molecules, such that the mean density in V_A exceeds the mean density n of the whole fluid by the amount δn_A, is proportional to

$$p_{AB} = \exp\left\{ -\frac{\tfrac{1}{2}\beta V_A N}{\zeta_T V_B n^2}\, \delta n_A^2 \right\}$$

where ζ_T is the isothermal compressibility, and $\beta = 1/(kT)$. The mean square of the fluctuation $V_A \delta n_A$ in the number of molecules in the volume V_A is therefore

$$\frac{\int_{-\infty}^{\infty} \exp\left\{ -\frac{\tfrac{1}{2}\beta V_A N}{\zeta_T V_B n^2}\, \delta n_A^2 \right\} V_A^2\, \delta n_A^2\, d(\delta n_A)}{\int_{-\infty}^{\infty} \exp\left\{ -\frac{\tfrac{1}{2}\beta V_A N}{\zeta_T V_B n^2}\, \delta n_A^2 \right\} d(\delta n_A)} = \frac{\zeta_T V_A V_B n^2}{\beta N}.$$

If the volume V_A is very small compared with the total volume, V_B/N will be almost equal to $V/N = n^{-1}$, so that the above expression reduces to

(2. 8) $\qquad \zeta_T V_A n/\beta = \{n + \int (n_2 - n^2)\, d\mathbf{x}^{(2)}\} V_A$

by comparison with (2. 7).

This shows that the intensity of the radiation scattered by a fluid at small angles to the direction of the incident beam is proportional to the isothermal compressibility, to the number density, and to the absolute temperature. Since in a gas at high temperatures the ratio ζ_T/β is almost unity, but in a liquid, especially near the freezing point, the compressibility ζ_T is very small, it can be seen that the small-angle scattering per molecule will be greatest in gases and smallest in liquids far removed from the critical point. This conclusion is confirmed not only by observations on the scattering of light by fluids, but also by the X-ray scattering experiments. Due to the very large value of the compressibility, a particularly large amount of small angle scattering is observed in the neighbourhood of the critical point. The scattering of light in this region is in fact great enough to be visually observable, leading to the phenomenon of opalescence.

3. The Structure of Fluids

From the information derived from the X-ray scattering experiments, it is possible to form a good picture of the configuration of the molecules in fluids. The radial distribution function, which gives the relative probability of finding a molecule at any distance from a given molecule is necessarily zero for very small distances, on account of the strong repulsion which prevents the overlapping of the outer electron shells of the molecules. From a distance, which may be identified with the molecular diameter, however, the probability distribution rises rapidly to a pronounced maximum. In liquids the height of this maximum indicates the presence of anything from 3 or 4 to 10 or 11 other molecules at a distance somewhat greater than a molecular diameter from the one considered; these molecules constitute what is known as the first co-ordination shell. Beyond this shell the probability distribution falls to a subsidiary minimum, but in liquids there are usually a number of further maxima, corresponding to the presence of other co-ordination shells beyond the first. EISENSTEIN & GINGRICH [1942] report the following peaks (co-ordination shells) and co-ordination numbers (numbers of molecules in the corresponding co-ordination shells) in liquid argon for points distributed along the saturated vapour line.

Temperature ° K.	First peak distance	First co-ordination number	Second peak distance	Second co-ordination number
(crystal	3.82 A	12	5.4 A	6)
84.4	3.79 A	10.5	5.3 A	
91.8	3.79 A	7	4.7 A	4
144.1	3.8 A	4.2	5.4 A	
149.3	4.5 A	6		
(vapour	4.1 A	2		)

The observations in the liquid range from the triple point to the critical point; and adjacent observations for the crystalline solid and saturated vapour are included. It will be seen that the distance of the first maximum tends to increase, and the number of molecules in the first co-ordination shell to decrease, as the temperature rises and the density is diminished. The second peak tends to disappear as the critical point is approached, and is non-existent in the vapour.

It may be concluded from these observations, and many similar observations for other fluids, that, even near the melting curve, the structure of a liquid is very different from that of a solid, where there is a long-range ordering of the molecules and a fixed integral number of molecules in the various co-ordination shells. There is no evidence to suggest in general the existence of an ordered structure of the molecules, even over microscopic domains of the liquid, of the kind associated with the crystalline state. Such a structure is, in fact, clearly incompatible with those observations showing a first co-ordination shell not more than half complete. Only near the freezing point might one suppose that a degree of short-range order exists, with some sort of lattice structure broken up by the incompleteness of many co-ordination shells. This has been emphasized because, in attempts to justify the adoption of quasi-crystalline models of a liquid, misleading descriptions of the molecular structure of fluids have been given currency in the past. It is certain that cybotactic groups of the order of 1,000 molecules are of rare occurrence in most liquids.

Nevertheless the idea of cybotactic groups, developed originally by STEWART [1930] has considerable value in describing the behaviour of liquids not too far removed from their freezing points. There can be no question that such liquids have a rather close-packed structure; the density does not differ much from the density of the crystal; and the immediate environment of a molecule in a liquid near the freezing point is not very different from that which it experiences in the crystal. Those properties of a crystal which depend on the existence of only a short-range order will not be much changed on melting. This applies, for example, to the specific heat and electrical conductivity of metals, and the dielectric constant of polar substances.

The X-ray diffraction observations all reveal a considerable amount of scattering at small angles in the neighbourhood of the condensation line. This is especially marked in the region surrounding the critical point, and tends to diminish somewhat at lower temperatures. An example showing considerable small-angle scattering is provided by

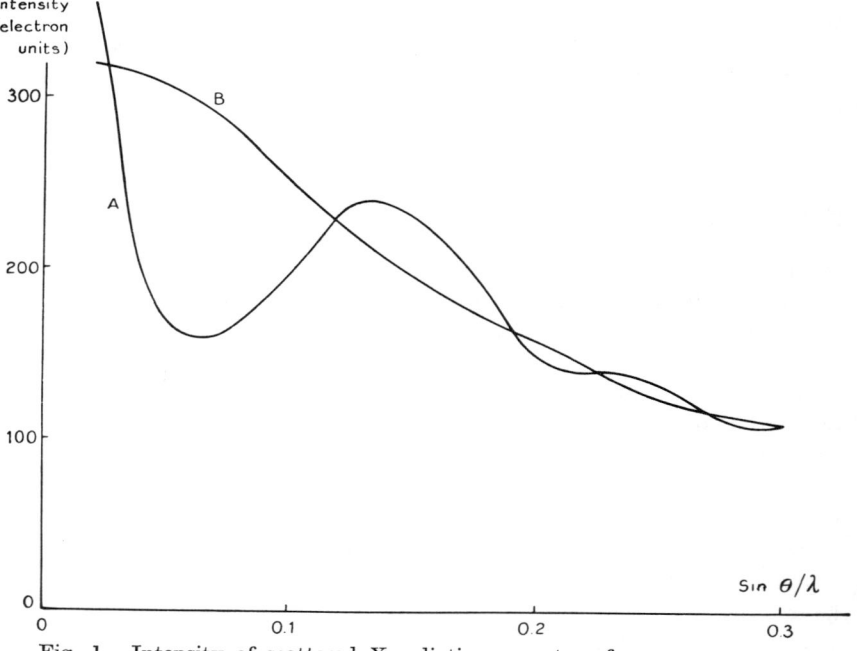

Fig. 1. Intensity of scattered X-radiation per atom for argon.
Curve A in liquid argon at 154° K. and 51.1 atm. (near critical point).
Curve B for an isolated atom. The difference is due to interference.

the curve for liquid argon at a temperature of 154° K., and a pressure of 51.1 atm. (very near the critical point), which is shown in fig. 1.

Such a phenomenon is in agreement with the relation between small-angle scattering and the compressibility established in the previous section; an exactly similar effect is found in the scattering of light by fluids. It also indicates the existence of large density fluctuations in the fluid, and cautions against a too static conception of the molecular structure. It is evident that, on account of their thermal motion, the relative configuration of the molecules is always changing, and the radial distribution function provides only a statistical picture, in which fluctuations both in time and from place to place in the fluid have been smoothed out. A more direct reminder of this is of course found in the phenomenon of the Brownian motion of particles suspended in a fluid. However, in their effect on most macroscopic properties, such fluctuations can be disregarded. The radial distribution alone is sufficient to determine almost every macroscopic property of a fluid in equilibrium.

4. Distribution Functions in General

Thus far there has been occasion to consider probability distributions, such as the velocity distribution function, and the radial distribution, referring to only one or two molecules in the fluid, and the distribution function defined for equilibrium in § 6 of chapter II, which refers to the entire molecular assembly. Between these two extremes one can interpolate a sequence of distribution functions referring to an arbitrary number of molecules in the fluid. These will now be defined.

Consider a set of q distinct volume elements $d\mathbf{x}^{(1)}, d\mathbf{x}^{(2)}, \ldots, d\mathbf{x}^{(q)}$, situated at the positions $\mathbf{x}^{(1)}, \mathbf{x}^{(2)}, \ldots, \mathbf{x}^{(q)}$ in the fluid. The probability that, at a given time t, each of these is occupied by a molecule, will be proportional to $d\mathbf{x}^{(1)} d\mathbf{x}^{(2)} \ldots d\mathbf{x}^{(q)}$, and can therefore be represented by $n_q d\mathbf{x}^{(1)} d\mathbf{x}^{(2)} \ldots d\mathbf{x}^{(q)}$, where the coefficient n_q will in general depend on the time t as well as the positions $\mathbf{x}^{(1)}, \mathbf{x}^{(2)}, \ldots, \mathbf{x}^{(q)}$ of the volume elements. When $q = 1$, n_q is simply the number density of the molecules as already defined, and when $q = 2$, n_q is the molecular distribution function which reduces to the radial distribution function in equilibrium.

As a simple consequence of the assumption that the molecules are all alike, the probability of finding a molecule in a given volume element cannot depend on the particular molecule involved; n_q must therefore be a symmetrical function of the positions $\mathbf{x}^{(1)}, \mathbf{x}^{(2)}, \ldots, \mathbf{x}^{(q)}$, and remain unchanged under any permutation of them. In fact the definition of n_q does not specify the particular molecule which occupies each of the q volume elements. If there are, in all, N similar molecules, the probability that a *particular one* of these will be found in each of the q volume elements is

$$\{(N-q)!/N!\}\, n_q\, d\mathbf{x}^{(1)} \ldots d\mathbf{x}^{(q)}.$$

This result enables one to establish a connection between n_q and n_{q-1}. For, since the qth molecule must be somewhere in the volume V occupied by the fluid, if one sums the above expression over all positions of the qth volume element, the result must be the probability of finding each of the remaining $q-1$ volume elements occupied by a particular molecule. Therefore

$$\int \{(N-q)!/N!\}\, n_q\, d\mathbf{x}^{(q)} = \{(N-q+1)!/N!\}\, n_{q-1}.$$

Increasing the value of q by 1, and cancelling a redundant factor, one has

(4.1) $$\int n_{q+1}\, d\mathbf{x}^{(q+1)} = (N-q)\, n_q.$$

This relation is correct even when $q = N$, since n_{N+1} vanishes identically by virtue of its definition and the fact that $N+1$ molecules can never be found within the volume V.

It is possible, in a similar way, to define a set of velocity distribution functions of which a typical member f_q relates to a set of q molecules at the positions $\mathbf{x}^{(1)}, \mathbf{x}^{(2)}, \ldots, \mathbf{x}^{(q)}$ in the fluid. Let the probability of finding the volume elements $d\mathbf{x}^{(1)}, d\mathbf{x}^{(2)}, \ldots, d\mathbf{x}^{(q)}$ occupied at time t by molecules with velocities in the ranges $\boldsymbol{\xi}^{(1)}, d\boldsymbol{\xi}^{(1)}; \ldots; \boldsymbol{\xi}^{(q)}, d\boldsymbol{\xi}^{(q)}$ respectively be $f_q\, d\boldsymbol{\xi}^{(1)} \ldots d\boldsymbol{\xi}^{(q)}\, d\mathbf{x}^{(1)} \ldots d\mathbf{x}^{(q)}$. Then, if φ is the velocity distribution function defined in § 3 of chapter II, one has

$$f_1 = n\varphi.$$

This relation may be generalized. For, since each molecule must have a velocity somewhere in the range from zero to infinity, one must have

(4.2) $$\int \overset{(q)}{\ldots} \int f_q\, d\boldsymbol{\xi}^{(1)} \ldots d\boldsymbol{\xi}^{(q)} = n_q.$$

Hence one may write

(4.3) $$f_q = n_q\, \varphi_q,$$

where the function φ_q satisfies

(4.4) $$\int \overset{(q)}{\cdots} \int \varphi_q \, d\boldsymbol{\xi}^{(1)} \ldots d\boldsymbol{\xi}^{(q)} = 1.$$

This is the generalization of (3.1) of chapter II. Most of the other formulae of § 3, chapter II also have an easy generalization; for example, the mean velocity of a molecule (i) in a cluster of molecules situated at the positions $\mathbf{x}^{(1)}, \mathbf{x}^{(2)}, \ldots, \mathbf{x}^{(q)}$ at time t is

(4.5) $$\mathbf{u}_q^{(i)} = \int \overset{(q)}{\cdots} \int \boldsymbol{\xi}^{(i)} \, \varphi_q \, d\boldsymbol{\xi}^{(1)} \ldots d\boldsymbol{\xi}^{(q)}.$$

It is obvious that a relation exists, completely analogous to (4.1), connecting the distribution functions f_q and f_{q+1}; this is

(4.6) $$\iint f_{q+1} \, d\mathbf{x}^{(q+1)} \, d\boldsymbol{\xi}^{(q+1)} = (N-q) f_q.$$

In equilibrium, it is known from (6.1) of chapter II that the distribution function f_N, relating to the entire molecular assembly, is

$$f_N = \exp \beta (F - E), \quad E = \tfrac{1}{2} m \sum \boldsymbol{\xi}^{(i)2} + \Phi,$$

where $\beta = 1/(kT)$, T is the absolute temperature, F is the free energy of the system, and Φ is the total potential energy, expressed as a function of the positions of the N molecules. From this, the equilibrium values of the distribution functions f_q and n_q are easily deduced. With the help of (4.6), one has

(4.7) $$(N-q)! f_q = \int \overset{(2N-2q)}{\cdots} \int \exp \beta (F-E) \, d\boldsymbol{\xi}^{(q+1)} \ldots d\boldsymbol{\xi}^{(N)} \, d\mathbf{x}^{(q+1)} \ldots d\mathbf{x}^{(N)}.$$

Also, with the help of (4.2), one obtains

$$n_N = \lambda^N \exp \beta (F - \Phi), \qquad \lambda = (2\pi/\beta m)^{3/2},$$

and

(4.8) $$(N-q)! n_q = \lambda^N \int \overset{(N-q)}{\cdots} \int \exp \beta (F - \Phi) \, d\mathbf{x}^{(q+1)} \ldots d\mathbf{x}^{(N)}.$$

These expressions are, of course, only formal. It has so far proved impracticable to calculate the radial distribution function of a liquid, for example, directly from the formula (4.8). In order to do this, approximative methods have had to be devised.

The most useful approximate methods are based on a differential equation which can be obtained from (4.8) by differentiating with respect to $\mathbf{x}^{(i)}$. The result is

(4.9) $$\begin{cases} \dfrac{(N-q)!}{(-\beta)} \dfrac{\partial n_q}{\partial \mathbf{x}^{(i)}} = \lambda^N \int \overset{(N-q)}{\cdots} \int \dfrac{\partial \Phi}{\partial \mathbf{x}^{(i)}} \exp \beta (F - \Phi) \, d\mathbf{x}^{(q+1)} \ldots d\mathbf{x}^{(N)} \\ \qquad = \int \overset{(N-q)}{\cdots} \int \sum_{j=1}^{N} \dfrac{\partial \phi^{(ij)}}{\partial \mathbf{x}^{(i)}} n_N \, d\mathbf{x}^{(q+1)} \ldots d\mathbf{x}^{(N)}. \end{cases}$$

The summation appearing in the last integral can be split into two parts $\sum_{j=1}^{q}$ and $\sum_{j=q+1}^{N}$, of which the second gives $N-q$ terms, all equal to another for reasons of symmetry:

$$\int {}^{(N-q)}\!\!\ldots \int \sum_{j=q+1}^{N} \frac{\partial \phi^{(ij)}}{\partial \mathbf{x}^{(i)}} n_N \, d\mathbf{x}^{(q+1)} \, d\mathbf{x}^{(N)} =$$
$$= (N-q) \int {}^{(N-q)}\!\!\ldots \int \frac{\partial \phi^{(i,\,q+1)}}{\partial \mathbf{x}^{(i)}} n_N \, d\mathbf{x}^{(q+1)} \ldots d\mathbf{x}^{(N)}$$
$$= (N-q)(N-q-1)! \int \frac{\partial \phi^{(i,\,q+1)}}{\partial \mathbf{x}^{(i)}} n_{q+1} \, d\mathbf{x}^{(q+1)},$$

with the help of (4.1). Hence (4.9) reduces to

(4.10) $\qquad \dfrac{1}{(-\beta)} \dfrac{\partial n_q}{\partial \mathbf{x}^{(i)}} = \sum_{j=1}^{q} n_q \dfrac{\partial \phi^{(ij)}}{\partial \mathbf{x}^{(i)}} + \int n_{q+1} \dfrac{\partial \phi^{(i,\,q+1)}}{\partial \mathbf{x}^{(i)}} \, d\mathbf{x}^{(q+1)}.$

This equation, though it is valid only in equilibrium, can be used, in conjunction with suitable approximations, to determine the radial distribution function n_2 and higher distribution functions if required.

5. Some Properties of the Entropy

In this section some combinations of the distribution functions will be introduced which are of particular importance in describing the entropy of a fluid. They are functions of the positions and velocities of a group of molecules which are almost zero except when the molecules are very close together. Consider for example the quantity

$$\exp z_2^{(ij)} = f_2^{(ij)}/f_1^{(i)} f_1^{(j)}$$

where $f_2^{(ij)}$ is the second velocity distribution function relating to the pair of molecules i and j, and $f_1^{(i)}$ and $f_1^{(j)}$ are the first velocity distribution functions for the molecules i and j separately. When the molecules are far apart, the probability that one has a given position or velocity is practically independent of the position or velocity of the other; hence the above expression is almost unity, and $z_2^{(ij)}$ is almost zero. In general $\exp z_2^{(ij)}$ measures the probability of finding the two molecules with given positions and velocities, relative to a situation in which they are considered not to influence one another.

A similar consideration applies to the expression

$$\exp z_3^{(ijk)} = f_3^{(ijk)} f_1^{(i)} f_1^{(j)} f_1^{(k)} / f_2^{(jk)} f_2^{(ki)} f_2^{(ij)}.$$

An argument can be given to make it appear plausible that this quantity is not very different from unity even when the molecules

i, j, k are close together. For $f_3^{(ijk)}/f_1^{(i)}f_1^{(j)}f_1^{(k)}$ is the relative probability of finding three molecules with given positions and velocities, and

$$f_2^{(jk)}/f_1^{(j)}f_1^{(k)}, \quad f_2^{(ki)}/f_1^{(k)}f_1^{(i)}, \quad f_2^{(ij)}/f_1^{(i)}f_1^{(j)}$$

are the relative probabilities of finding the molecules with the same positions and velocities *in pairs*. It seems reasonable to expect that the first quantity is approximately the product of the latter three; then $z_3^{(ijk)}$ would be approximately zero even when the three molecules are not far removed from one another. This approximation has been used by KIRKWOOD [1942] and others to simplify the difficult calculations which arise in connection with the statistical mechanics of liquids. A better approximation still would be obtained by assuming that the corresponding function z_4, relating to a group of 4 molecules, is zero.

The general definition of these functions is by means of the set of equations

(5. 1) $\begin{cases} \ln f_1^{(i)} = z_1^{(i)} \\ \ln f_2^{(ij)} = z_2^{(ij)} + z_1^{(i)} + z_2^{(j)} \\ \ln f_3^{(ijk)} = z_3^{(ijk)} + z_2^{(jk)} + z_2^{(ki)} + z_2^{(ij)} + z_1^{(i)} + z_1^{(j)} + z_1^{(k)}, \end{cases}$

etc., where in general the z's corresponding to all different groupings of the molecules considered appear on the right-hand side. By solving these equations, one has

(5. 2) $\begin{cases} z_1^{(i)} = \ln f_1^{(i)} \\ z_2^{(ij)} = \ln f_2^{(ij)} - \ln f_1^{(i)} - \ln f_1^{(j)} \\ z_3^{(ijk)} = \ln f_3^{(ijk)} - \ln f_2^{(jk)} - \ln f_2^{(ki)} - \ln f_2^{(ij)} \\ \qquad\quad + \ln f_1^{(i)} + \ln f_1^{(j)} + \ln f_1^{(k)}, \end{cases}$

where, again, terms corresponding to all different groupings appear on the right-hand side, with a factor $(-1)^q$, where q is the number of molecules left over from the group.

To see the connection of the z-functions with the entropy, one has first to recall the definition suggested by the equation (5. 11) of chapter II. When the system consists of N identical molecules, this reduces to

(5. 3) $\qquad S = -\dfrac{k}{N!} \int \overset{(2N)}{\cdots} \int f_N \ln f_N \, d\mathbf{x}^{(1)} \ldots d\mathbf{x}^{(N)} \, d\boldsymbol{\xi}^{(1)} \ldots d\boldsymbol{\xi}^{(N)}.$

To support this definition, it will be shown that (i) it is correct when the fluid is in thermal and mechanical equilibrium, and (ii) when the fluid is not in thermal and mechanical equilibrium, S is always less than the value S^0 which it assumes in an equilibrium state with the same internal energy and average density.

The internal energy U is the average value of the energy E of the system of molecules, and is therefore given by

(5. 4)
$$\begin{cases} U = \dfrac{1}{N!} \int \overset{(2N)}{\cdots} \int f_N E \, d\mathbf{x}^{(1)} \ldots d\mathbf{x}^{(N)} \, d\boldsymbol{\xi}^{(1)} \ldots d\boldsymbol{\xi}^{(N)} \\ E = \sum_{i=1}^{N} \tfrac{1}{2} m \boldsymbol{\xi}^{(i)2} + \boldsymbol{\Phi}. \end{cases}$$

Also the value of f_N in equlibrium is given by

(5. 5) $$\ln f_N^0 = \beta (F - E), \quad (\beta = 1/kT),$$

where F is the free energy, acccording to § 6 of chapter II. In general, f_N can be expressed in the form

(5. 6) $$f_N = a f_N^0,$$

where a is a function of positions and velocities satisfying the equation

(5. 7)
$$\begin{cases} \int \overset{(2N)}{\cdots} \int f_N^0 (a - 1) \, d\mathbf{x}^{(1)} \ldots d\mathbf{x}^{(N)} \, d\boldsymbol{\xi}^{(1)} \ldots d\boldsymbol{\xi}^{(N)} \\ = \int \overset{(2N)}{\cdots} \int (f_N - f_N^0) \, d\mathbf{x}^{(1)} \ldots d\mathbf{x}^{(N)} \, d\boldsymbol{\xi}^{(1)} \ldots d\boldsymbol{\xi}^{(N)} = 0. \end{cases}$$

Now, if (5. 6) and (5. 5) are substituted into (5. 3), the result is

(5. 8)
$$\begin{cases} S = -\dfrac{k}{N!} \int \overset{(2N)}{\cdots} \int f_N \{\beta (F - E) + \ln a\} \, d\mathbf{x}^{(1)} \ldots d\mathbf{x}^{(N)} \, d\boldsymbol{\xi}^{(1)} \ldots d\boldsymbol{\xi}^{(N)} \\ = - k\beta (F - U) - \dfrac{k}{N!} \int \overset{(2N)}{\cdots} \int f_N^0 \, a \ln a \, d\mathbf{x}^{(1)} \ldots d\mathbf{x}^{(N)} \, d\boldsymbol{\xi}^{(1)} \ldots d\boldsymbol{\xi}^{(N)} \end{cases}$$

according to (5. 4). The first term reduces to $(U - F)/T$, which is an accepted thermodynamical expression for the entropy of a system in thermal and mechanical equilibrium. The second term, with the help of (5. 7), can be written in the form

$$- k \int \overset{(2N)}{\cdots} \int f_N^0 \, (a \ln a - a + 1) \, d\mathbf{x}^{(1)} \ldots d\mathbf{x}^{(N)} \, d\boldsymbol{\xi}^{(1)} \ldots d\boldsymbol{\xi}^{(N)}$$

where f_N^0 is positive, and $a \ln a - a + 1$ is also an essentially positive function of a for the positive values of a which are required by (5. 6). Hence S is never greater than the entropy of the equilibrium state with the same mean density and internal energy, which it approaches as the system approaches equilibrium. This completes the proof of the assertions (i) and (ii) above.

Now the entropy will be expressed in terms of the distribution functions z_1, z_2, etc. This can be done by substituting the expression for $\ln f_N$, provided by (5. 1), into (5. 3). Since the distribution functions are symmetrical in the co-ordinates and velocities, one obtains

(5. 9)
$$\begin{aligned} S &= -\frac{k}{N!} \int \overset{(2N)}{\cdots} \int f_N \left\{ \binom{N}{1} z_1^{(1)} + \binom{N}{2} z_2^{(12)} + \ldots + z_N \right\} d\mathbf{x}^{(1)} \ldots d\boldsymbol{\xi}^{(N)} \\ &= -k \left\{ \frac{1}{1!} \iint f_1 z_1 \, d\mathbf{x}^{(1)} d\boldsymbol{\xi}^{(1)} + \frac{1}{2!} \iiiint f_2 z_2 \, d\mathbf{x}^{(1)} \ldots d\boldsymbol{\xi}^{(2)} + \ldots \right\}. \end{aligned}$$

Hence $-k z_1(\boldsymbol{\xi}^{(1)}, \mathbf{x}^{(1)})$ can be regarded as the contribution to the entropy of a molecule with velocity $\boldsymbol{\xi}^{(1)}$ at the point $\mathbf{x}^{(1)}$;

$$-k z_2(\boldsymbol{\xi}^{(1)}, \boldsymbol{\xi}^{(2)}, \mathbf{x}^{(1)}, \mathbf{x}^{(2)})$$

is an additional contribution due to the interaction of a pair of molecules with given positions and velocities; and so on.

The functions z_q evidently play a rather important role in the theory of fluids. In equilibrium one has, according to (4. 7) and (4. 8),

$$\lambda^q f_q = \exp\left(-\tfrac{1}{2}\beta m \sum_{i=1}^{q} \boldsymbol{\xi}^{(i)2}\right) n_q$$

so that

(5. 10)
$$\begin{cases} z_2 = \ln(n_2/n^2), \\ z_3 = \ln(n_3 n^3 / n_2^{(23)} n_2^{(31)} n_2^{(12)}), \end{cases}$$

etc., and

(5. 11) $S = -kV \left\{ \dfrac{3}{2} \ln \dfrac{m\beta}{2\pi e} + \tfrac{1}{2} \int n_2 \ln(n_2/n^2) \, d\mathbf{x}^{(2)} + \ldots \right\}.$

If one writes, for small q,

(5. 12) $n_q = n^q \exp(-\beta \psi_q),$

ψ_q is a mean potential energy of the group of q molecules. The equations (5. 10) reduce to

(5. 13)
$$\begin{cases} z_2 = -\beta \psi_2, \\ z_3 = -\beta (\psi_3 - \psi_2^{(23)} - \psi_2^{(31)} - \psi_2^{(12)}). \end{cases}$$

Hence, while, in equilibrium, z_2 is proportional to the mean potential energy of a pair of molecules, z_3 varies as the difference between the mean potential energy of a group of three molecules and the sum of the mutual potential energies of the pairs which it contains. The other z_q's may be interpreted in a similar way.

6. Radial Distribution

The behaviour of any fluid is a consequence of its molecular structure, which, in the equilibrium state, is most conveniently described by the radial distribution function. Although one has in the formula (4. 8) [with $q = 2$, or (6. 11) of chapter II] an explicit expression for this function, it has so far proved impossible to evaluate the multiple integrals involved, except in the form of a power series in the density, valid only for gases. DE BOER [1940] has used this power series, which begins thus:

$$(6.\ 1) \quad \begin{cases} n_2(r) = n^2 \exp\{-\beta\phi(r)\}\{1 + n \int u(\mathbf{s})\, u(\mathbf{s}+\mathbf{r})\, d\mathbf{s} + \ldots\}, \\ u(\mathbf{r}) = \exp\{-\beta\phi(r)\} - 1 \end{cases}$$

to determine the radial distribution function in compressed gases. He assumed (cf. § 7) a potential function of the form

$$\phi(r) = 4\varepsilon\{(\sigma/r)^{12} - (\sigma/r)^6\},$$

and calculated the integral

$$I(r) = 3/(2\pi\sigma^3) \int u(\mathbf{s})\, u(\mathbf{s}+\mathbf{r})\, d\mathbf{s}$$

for various values of $r^* = r/\sigma$ and $T^* = kT/\varepsilon$. The expression in the brackets in (6. 1) can be written in the form

$$1 + \tfrac{2}{3}\pi\sigma^3 n\, I(r).$$

De Boer's results were as follows:

TABLE I
Values of $I(r)$

r^* \ T^*	1.000	1.141	1.333	1.666	2.000	2.666	4.000	6.666
0.9	+ 0.789	+ 0.510	+ 0.340	+ 0.262	+ 0.239	+ 0.270	+ 0.315	+ 0.348
1.0	+ 0.575	+ 0.314	+ 0.174	+ 0.117	+ 0.113	+ 0.149	+ 0.212	+ 0.257
1.2	+ 0.229	+ 0.089	− 0.054	− 0.084	− 0.070	− 0.018	+ 0.048	+ 0.102
1.4	+ 0.087	− 0.059	− 0.165	− 0.187	− 0.173	− 0.126	− 0.062	− 0.006
1.6	+ 0.085	− 0.056	− 0.176	− 0.210	− 0.203	− 0.170	− 0.116	− 0.065
1.8	+ 0.202	+ 0.051	− 0.097	− 0.137	− 0.157	− 0.148	− 0.112	− 0.070
2.0	+ 0.423	+ 0.246	+ 0.063	− 0.002	− 0.049	− 0.070	− 0.059	− 0.040
2.2	+ 0.578	+ 0.354	+ 0.194	+ 0.107	+ 0.038	+ 0.004	− 0.018	− 0.017
2.4	+ 0.477	+ 0.309	+ 0.190	+ 0.114	+ 0.064	+ 0.016	− 0.003	− 0.007
2.6		+ 0.222			+ 0.044	+ 0.017		
2.8		+ 0.146			+ 0.031			
3.0	+ 0.134		+ 0.062		+ 0.022			

It will be noticed that although, at very high temperatures, there is only one maximum of $I(r)$, at distances somewhat less than σ, at lower temperatures a second maximum appears at distances rather greater than σ — an indication of the existence of a second coordination shell. This happens even above the critical temperature, which is given by $T^* = 1.26$ in the units employed. The curves derived from de Boer's calculations agree very well with the experimentally determined curves for the gas.

6.1 Radial Distribution in Liquids

To determine the radial distribution function for a liquid other methods have to be used. These are most conveniently based on a direct application of the principles of statistical mechanics, according to which (cf. § 5 of chapter II) the probability that a system of any kind will be found in a state of energy E must be proportional to $\exp(-\beta E)$. Thus, the probability that a pair of molecules will be found separated by a distance r in a fluid must be proportional to $\exp\{-\beta\psi_2(r)\}$, where $\psi_2(r)$ is a mean potential energy of the molecules in this configuration. It follows that the radial distribution function is given by the formula

$$(6.2) \qquad n_2(r) = \text{const.} \cdot \exp\{-\beta\psi_2(r)\}$$

the constant being determined by the requirement that $n_2(r)$ should approach the value n^2 as r becomes very large. To determine the mean potential energy $\psi_2(r)$, only approximate methods are at present available. There exists, however, an exact relation between $\psi_2(r)$ and the mutual potential energy of a pair of molecules represented by $\phi(r)$; this is derived in the following way.

The mean force acting on a molecule at the point $\mathbf{x}^{(1)}$, when it is known that a second molecule is situated at the point $\mathbf{x}^{(2)}$, is $-\frac{\partial \psi_2(r)}{\partial \mathbf{x}^{(1)}}$. This force is the resultant of that due to the second molecule alone, and the mean force due to the other molecules. Since the probability of finding a molecule in a volume element $d\mathbf{x}^{(3)}$ at $\mathbf{x}^{(3)}$, when it is already known that other molecules are located at $\mathbf{x}^{(1)}$ and $\mathbf{x}^{(2)}$, is $n_3 d\mathbf{x}^{(3)}/n_2$, it follows that the mean resultant force on the molecule at $\mathbf{x}^{(1)}$, due to molecules other than the one at $\mathbf{x}^{(2)}$, is

$$-\int \frac{n_3}{n_2} \frac{\partial \phi(s)}{\partial \mathbf{x}^{(1)}} d\mathbf{x}^{(3)},$$

where s is the distance between the points $\mathbf{x}^{(1)}$ and $\mathbf{x}^{(3)}$. Hence

(6.3) $$\frac{\partial \psi_2(r)}{\partial \mathbf{x}^{(1)}} = \frac{\partial \phi(r)}{\partial \mathbf{x}^{(1)}} + \int \frac{n_3}{n_2} \frac{\partial \phi(s)}{\partial \mathbf{x}^{(1)}} d\mathbf{x}^{(3)}.$$

This formula is exact, and would enable one to determine ψ_2 if the distribution function n_3/n_2 for the third molecule were known. By combining the equations (6.2) and (6.3), one obtains further

(6.4) $$\frac{\partial n_2}{\partial \mathbf{x}^{(1)}} + \beta \left\{ n_2 \frac{\partial \phi}{\partial \mathbf{x}^{(1)}} + \int n_3 \frac{\partial \phi(s)}{\partial \mathbf{x}^{(1)}} d\mathbf{x}^{(3)} \right\} = 0.$$

Now, if n_3 can be expressed in terms of n_2, the last result will provide an integro-differential equation for the determination of n_2. No exact expression exists; however, an approximate relation is suggested by the considerations of the previous section. If z_3 is assumed to be negigible, one will have

(6.5) $$n^3 n_3(\mathbf{x}^{(1)}, \mathbf{x}^{(2)}, \mathbf{x}^{(3)}) = n_2(\mathbf{x}^{(2)}, \mathbf{x}^{(3)}) \, n_2(\mathbf{x}^{(3)}, \mathbf{x}^{(1)}) \, n_2(\mathbf{x}^{(1)}, \mathbf{x}^{(2)})$$

— the so-called superposition approximation. Physically it amounts to assuming that the relative probability $n_3/nn_2^{(12)}$ of finding a third molecule at $\mathbf{x}^{(3)}$ in the presence of two other molecules at $\mathbf{x}^{(1)}$ and $\mathbf{x}^{(2)}$ respectively is the same as the product of the relative probabilities $n_2^{(13)}/n^2$ and $n^{(23)}/n^2$ of finding it in the same position, in the presence of the other two molecules separately. The approximation has the merit that it correctly describes the behaviour of the distribution function n_3 both when any one of the three molecules is far from the other two, and when any two of the three are very close; further, it is known from other considerations to be very nearly correct both for rare gases, and the solid state. Of course, none of these considerations fully justifies its use in the theory of liquids; it can be judged finally only in the light of its consequences.

The following solution of the equation (6.4) requires what is essentially the generalization of some analysis due to KIRKWOOD & BOGGS [1942]. After substitution from (6.5), (6.4) assumes the form

(6.6) $$\frac{\mathbf{r}}{r} \frac{d}{dr} \{\ln n_2(r) + \beta \phi(r)\} + \beta n^{-3} \int n_2(s) \, n_2(t) \, \phi'(s) \, (\mathbf{s}/s) \, d\mathbf{s} = 0,$$

where $\mathbf{s} = \mathbf{x}^{(3)} - \mathbf{x}^{(1)}$ and $\mathbf{t} = \mathbf{x}^{(2)} - \mathbf{x}^{(3)} = \mathbf{r} - \mathbf{s}$; the variable t is not independent, but given by

$$t^2 = r^2 + s^2 - 2 \, rs \cos \theta,$$

if θ is the angle between the vectors **r** and **s**. The vector equation (6.6) reduces to its component in the direction of **r**, namely

(6.7)
$$\frac{d}{dr}\{\ln n_2(r) + \beta\phi(r)\} = -2\pi\beta n^{-3} \int_0^\infty \int_0^\pi n_2(t)\, n_2(s)\, \phi'(s) \cos\theta \sin\theta\, d\theta\, s^2\, ds$$
$$= \frac{\pi\beta}{n^3} \int_0^\infty \int_{|r-s|}^{|r+s|} n_2(t)\, t \left(\frac{t^2-s^2}{r^2} - 1\right) dt\, n_2(s)\, \phi'(s)\, ds.$$

This equation can be integrated with respect to r; after a somewhat tedious integration by parts, it yields

(6.8)
$$\ln\{n_2(r)/n^2\} + \beta\phi(r)$$
$$= \frac{\pi\beta}{n^3 r} \int_0^\infty \int_{-s}^s (s^2 - t^2)(t+r)\{n_2(|t+r|) - n^2\}\, dt\, n_2(s)\, \phi'(s)\, ds.$$

This is an integral equation for the determination of $n_2(r)$; it is, however, non-linear, and therefore impossible to solve exactly by existing analytical methods. One may, however, make use of the circumstance that $n^2 \exp\{-\beta\phi(r)\}$ is a fairly good first approximation to $n_2(r)$, and substitute

(6.9) $$n_2(|r|) = n^2 \exp\{-\beta\phi(r) + f(r)\},$$

assuming that $f(r)$ is small, so that squares and higher powers of $f(r)$ can be neglected. This approximation reduces (6.8) to

(6.10)
$$-rf(r) = \pi n \int_0^\infty \int_{-s}^s (s^2 - t^2)(t+r)\{f(t+r) + u(t+r) +$$
$$+ f(t+r)\, u(t+r)\}\, dt\, u'(s)\{1 + f(s)\}\, ds,$$

where

(6.11) $$u(r) = \exp\{-\beta\phi(r)\} - 1.$$

Even now the equation is not easily soluble, and it is necessary to make a further approximation by replacing $f(r)$ by an average value

$$\overline{f(r)} = \varepsilon - 1$$

wherever it occurs multiplied by $u'(r)$ or $u(r)$. This approximation is the last, and probably the most drastic of those required for the analytic solution of the integral equation; it has the consequence that some of the subsequent developments are only semi-quantitatively correct. It gives

$$(6.12) \begin{cases} rf(r) = -\pi n \int_0^\infty \int_{-s}^s (s^2 - t^2)(t+r)\{f(t+r) + \varepsilon u(t+r)\}\,dt\,\varepsilon u'(s)\,ds \\ = 2\pi n \int_0^\infty \int_{-s}^s (t+r)\{f(t+r) + \varepsilon u(t+r)\}\,dt\,\varepsilon u(s)\,s\,ds, \end{cases}$$

with the help of an integration by parts.

One can now solve the equation quite easily by Fourier transformations. One uses the fact that, if $f(r)$ is *any* even function of r, and

$$(6.13) \qquad i s g(s) = \int_{-\infty}^\infty rf(r)\exp(2\pi irs)\,dr,$$

then

$$(6.14) \qquad i r f(r) = \int_{-\infty}^\infty sg(s)\exp(2\pi irs)\,ds.$$

If the function $v(s)$ is defined by

$$(6.15) \qquad i s v(s) = \int_{-\infty}^\infty ru(r)\exp(2\pi irs)\,dr,$$

one finds, by taking the Fourier transform of (6.12),

$$(6.16) \qquad sg(s) = ns\{g(s) + \varepsilon v(s)\}\,\varepsilon v(s).$$

Substituting the value of $g(s)$ thus obtained in (6.14), one has

$$(6.17) \qquad i r f(r) = \int_{-\infty}^\infty \frac{sn\varepsilon v(s)\exp(2\pi irs)\,ds}{1 - n\varepsilon v(s)}.$$

Since, by virtue of (6.15) and (6.11), $v(s)$ can be regarded as a known function, this, in conjunction with (6.9), is the required solution of the integral equation.

For sufficiently low densities, the function $n\varepsilon v(s)$ never exceeds unity; the integrand in (6.17) is then everywhere finite, and may in fact be developed in powers of the density. The solution then evidently corresponds to the gaseous phase. At higher densities, $n\varepsilon v(s)$ attains the value unity and a singularity appears in the integrand; then the integral is indeterminate to the extent of an arbitrary multiple of $\sin(2\pi r s_0)$, where s_0 is the positive root of the equation

$$(6.18) \qquad n\varepsilon v(s_0) = 1.$$

It is, in fact, easily verified that if $f(r)$ is one solution of the integral equation (6.12), so is

$$f(r) + A_0 r^{-1} \sin(2\pi r s_0)$$

where A_0 is any constant. However, only one of the infinitely many solutions is physically admissible. The requisite solution is determined by the condition that the integral

$$\int \{n_2(r) - n^2\} \, dr$$

should be convergent; this integral, as was seen in § 2, is connected with the density fluctuations in the fluid, and is therefore necessarily finite. Such a condition can only be satisfied if $f(r)$ decreases at least as rapidly as r^{-3} for large values of r. This will be so if, and only if, the path of integration associated with the integral in (6.17) is chosen so as to enter the upper half of the complex plane in the neighbourhood of the singularities at $s = \pm s_0$, thus:

The solution so obtained, when a real root exists of the equation (6.18), corresponds to the liquid state. Thus, the simplifications required to effect a solution of the integral equation have not obscured the fundamental distinction between the two phases. Though the results are not exact, it has seemed worth while to derive them, since they mirror the nature of their exact counterparts.

The parameter ε has not yet been determined. One can see from some work of MONTROLL & MAYER [1941] that the special choice $\varepsilon = 1$ corresponds to the assumption that the molecules of the fluid are associated only in chains and rings. This could not possibly be correct for the liquid, and in fact from its definition as an average value of $1 + f(r)$ weighted by the function $u(r)$, it is clear that ε must be given by

$$(6.19) \qquad (\varepsilon - 1) \int_{-\infty}^{\infty} u(r) \, r^2 \, dr = \int_{-\infty}^{\infty} f(r) \, u(r) \, r^2 \, dr.$$

Substituting from (6.17) for $f(r)$, this equation reduces to

$$(6.20) \qquad 2\pi (\varepsilon - 1) v(0) = \int_{-\infty}^{\infty} \frac{n\varepsilon \{s \, v(s)\}^2 \, ds}{1 - n\varepsilon v(s)}$$

which is sufficient to determine ε as a function of temperature and density. Only for a rare gas is it permissible to assume that $\varepsilon = 1$.

6.2 More Exact Methods

If it is required to obtain more accurate values of the radial distribution function, it is better not to try to linearize the non-linear integral equation (6. 8), but to solve it directly. An excellent method for the accomplishment of this task has been devised by A. G. McLellan, and will now be described. One assumes a solution of the type

$$(6.21) \quad n_2(r) = n^2 \exp\{-\beta\phi(r)\} \left(1 + \sum_{m=1}^{\infty} c_m r^m\right), \qquad (r < 3r_0/2),$$

for values of r less than $3r_0/2$, r_0 being the value of r for which $\phi(r)$ has its minimum. For values of r greater than r_0, it is sufficient to assume

$$(6.21a) \quad n_2(r) = n^2 \exp\{-\beta\phi(r)\} \qquad (r > 3r_0/2).$$

When the trial solution (6. 21) is substituted into (6. 8), the latter reduces to

$$(6.22) \quad \ln\left(1 + \sum_m c_m r^m\right) = (n/n_0)\left\{\gamma(r) + \sum_m c_m A_m(r) + \sum_{m,n} c_m c_n B_{mn}(r)\right\}$$

where n_0 is the density of the crystalline solid, and $\gamma(r)$, $A_m(r)$ and $B_{mn}(r)$ involve integrals of the type

$$\int_0^\infty \int_{-s}^s (s^2 - t^2)(t+r)^m \exp\{-\beta\phi(t+r)\} \, dt \, s^n \exp\{-\beta\phi(s)\} \, \phi'(s) \, ds.$$

The latter are best evaluated by replacing $\exp{-\beta\phi(t+r)}$ by an equivalent polynomial function of $t + r$ within a suitable range depending on the temperature, and by the values 0 and 1 below and above this range respectively.

The remaining task is then to determine the values of the coefficients c_m which satisfy the equation (6. 22) for all values of r. This can be accomplished by solving the linearized equation

$$\sum_m c_m r^m = (n/n_0)\left\{\gamma(r) + \sum_m c_m A_m(r)\right\}$$

to obtain a first approximation, and then applying an iterative process to (6. 22) to obtain more exact solutions. In this way one can find solutions valid for all densities up to that of the crystal.

The values of $\gamma(r)$ calculated by McLellan for argon are given in the following table.

TABLE II

Values of $\gamma(r)$

r/r_0	0.9	1.0	1.1	1.2	1.3	1.4	1.5
120° K.	+ 0.75	+ 0.38	+ 0.14	+ 0.01	− 0.03	+ 0.01	+ 0.01
150° K.	+ 0.69	− 0.03	− 0.23	− 0.30	− 0.37	− 0.33	+ 0.18

There is a marked change in the character of this function below the critical point, corresponding to the development of the co-ordination shells beyond the first. This change is reflected in the character of the radial distribution functions, whose computed values for argon are given in table III below. The first set is for the temperature 120° K. and the density 0.685 n_0 of the liquid at its saturated vapour pressure; the second is for the temperature 150° K. and a density 0.322 n_0 corresponding roughly to the critical point.

TABLE III

Values of $n_2(r)/n^2$

r/r_0	0.9	1.0	1.1	1.2	1.3	1.4	1.5
120° K.	1.89	3.43	2.46	1.75	1.42	1.29	1.29
150° K.	1.46	2.18	1.77	1.41	1.18	1.09	1.21

These curves agree well with the experimental curves of Eisenstein & Gingrich (*vide* fig. 1 of chapter II), except that, particularly at the lower temperature, the theory predicts somewhat *more coherent* coordination shells. The agreement might be improved by increasing the accuracy of the calculations described. It is, however, quite probable that under the experimental conditions the sharpness of the peaks of the radial distribution function has been blurred. This would account for the rather noticeable discrepancy between the experimental radial distribution curve and the requirements of the experimental equation of state. The discrepancy is fortunately not very important, as most macroscopic quantities depend mainly on the position and amplitude of the first coordination shell, and only to a minor extent on the subsequent fluctuations of the radial distribution function. It will be seen (§ 5 of chapter IV) that the distribution functions obtained by the method of this section leads to an equation of state which is in excellent agreement with the experimental data.

7. Intermolecular Forces

The idea that two molecules should interact with one another, so that their mutual potential energy can be represented by some function $[\phi(r)]$ of the distance between them, has become familiar already through the earlier considerations of this book. So far, however, the origin of the intermolecular forces, or the way in which they can be determined either theoretically or experimentally, has not been considered in detail. To remedy this, it will be necessary to set aside the intricacies of the molecular structure of the fluid, and direct attention to the structure of the molecules themselves.

The various kinds of intermolecular forces can be classified roughly as chemical, or electrostatic, or of the van der Waals' type. These divisions are mutually exclusive only if they are effected in rather a special way. For intermolecular forces are all, in some sense electrostatic in origin. A complete understanding of their nature, however, cannot be attained on the basis of classical mechanics: especially in relation to the forces of a chemical nature, the concepts of quantum mechanics are essential.

The existence of chemical forces, between molecules not permanently bound together, considerably complicates the theory of what are known as *associated* liquids. These liquids, which include all the glasses, are characterized macroscopically by an abnormally high viscosity. From the molecular point of view, the high viscosity is explained as due to an *association* of the molecules, which are more or less loosely bound together by weak chemical forces. The association is not permanent, but can be broken up by the thermal motion, and therefore decreases with increasing temperature. There is no rigid distinction between associated fluids and other substances whose molecules dissociate at high temperatures, except that one does not normally speak of associated gases or solids, especially where the chemical composition is well defined.

An understanding of the phenomenon of association requires some study of the nature of chemical forces. These are due to the interactions between the electrons and nuclei of different atoms, but cannot be accounted for simply in terms of the Coulombian forces between these charged particles. For example, the phenomenon of saturation — the inability of the chemical forces to bind more than a certain fixed number of atoms to a given atom — could never be explained by purely classical concepts. Without entering into a detailed description of atomic structure, it may be stated that an atom consists of a central positively charged nucleus, and a cloud of electrons whose motion is restricted to certain well-defined orbits relative to the nucleus. Moreover, not more than two electrons may occupy any orbit. Each electron has a spin (↑), and the spins of two electrons may be either parallel (↑ ↑) or antiparallel (↑ ↓). If two electrons occupy the same orbit, their spins must be antiparallel; for, according to a principle due to Pauli, it is *impossible* for two electrons to be found in exactly the same state; in particular, they cannot have the same spin and the same orbit in the same atom. Generally speaking, two electrons with parallel spin will tend to repel one another; those with antiparallel spin, to attract one another. Thus, if two atoms possess unpaired electrons with opposite spins, they will attract one another, quite independently of the ordinary Coulombian forces which may exist. Such attractions are mainly responsible for chemical bonds.

Often the associations of atoms produced by chemical bonds are rather permanent, and complex molecules are formed by the chemical combination of the atoms. It may happen, however, that the potential energy released in the formation of a bond is of the same order as the energy fluctuations arising from the thermal motion, so that the association is broken up from time to time if the temperature is high enough. It is this type of bond which is operative in associated liquids.

A pair of electrons with antiparallel spins cannot attract a third electron. Therefore the only type of molecule which can form an associated liquid is one in which the electrons are not already associated in pairs, so that there is a resultant electronic spin. The unpaired electrons are generally localized on definite atoms or groups of atoms in the molecule, and the bonds are therefore also localized in the molecule. The nature of the association between the molecules thus obviously depends on the number of unpaired electrons in each molecule. If there is only one, a molecule is able to bind only one other molecule, and the molecules must be associated in pairs. If there are two, on the other hand, the molecules can form chains and rings; and, if there are several, quite complicated configurations may result. Clearly, whatever the number of unpaired electrons, the phenomenon of association considerably adds to the complication of the molecular theory. A formal generalization of the theory which might be applied to the study of associated liquids will be described in chapter VI, but, up to the present, little more than qualitative conclusions have been adduced. Obviously phenomena which are independent of the existence of association between the molecules are most conveniently studied in simple fluids where no association exists, and for this reason attention will generally be restricted to such fluids. The form of interaction which best represents the action of the chemical bond has been given by Morse. It is expressed by the formula

$$(7.1) \qquad \phi(r) = -Me^{-r/\sigma} + Ne^{-r/\varrho}$$

where the constants σ and ϱ vary from substance to substance. For example, the interaction between two hydrogen atoms is very well described by the potential function

$$109.46 \{-2\exp(-x) + \exp(-2x)\} \ k \ \text{cal/mole},$$
$$x = 1.0298 \{(r/a_0) - 1.401\}$$

where a_0 is the distance of minimum potential energy. It should be remembered, however, that chemical forces have the property of saturation. It would not be correct to suppose, for example, that a third hydrogen atom might be attracted to the pair already considered; in fact, it is much more likely to be repelled.

It is convenient to consider next the purely electrostatic forces which may be exerted between the molecules of a liquid. These are easily understood as due to the Coulombian forces between the various charged particles which enter into the constitution of the individual molecules. It is possible for a molecule to be ionized; it then carries a resultant positive or negative charge which exerts a relatively powerful influence on other charged particles for a long distance. More frequently, a molecule has no resultant charge, but possesses a dipole moment, which may be imagined as arising from equal and opposite charges at a short distance from one another in the molecule. Polar fluids are made up of such molecules; they are characterized by abnormally high viscosity and surface tension, and anisotropy near the surface. Some of the more important effects have been described by HENNIKER [1949]. If a molecule has no dipole moment, a quadrupole moment may produce the same effects to a lesser degree.

There is little difficulty in predicting electrostatic forces from a knowledge of the electronic structure of the molecules. Like chemical forces, however, they have the effect of considerably complicating the molecular theory. One difficulty is that the potential energy between molecules with dipole or quadrupole moments depends on the orientation of the molecules as well as the distance between them. The generalization of the theory presented in chapter VI is well adapted to overcome this difficulty. Nevertheless, in the study of phenomena which do not require the presence of molecules with electric moments, it is better to a confine attention to non-polar fluids in which such molecules do not exist.

It has been seen that, in non-associated fluids, the electrons which enter into the structure of the molecules are, in general, distributed in pairs with opposite spins; and further that, in non-polar fluids, the molecules have no electric moment. Both of these conditions are realized, for example, in the inert gases, and the monatomic liquids which they form in condensation. The inert gases offer the additional simplicity of possessing spherically symmetrical molecules. Diatomic substances do not, of course, have spherically symmetrical molecules, but as, in a fluid, the molecules rotate freely and are distributed statistically over all orientations, it often happens that the deviation from spherical symmetry can be ignored for most purposes. Molecules of more complicated structure are also frequently either roughly spherical in shape, or else distributed over all orientations in the fluid so as to produce the same effect.

If the molecules of a non-polar and non-associated liquid have, or may be supposed to have, spherical symmetry, the mutual potential energy of any pair of molecules may be represented as a function $\phi(r)$ of the distance between their mean centres. This potential energy is a result of the van der Waals' forces, which are compounded of a comparatively long-range attraction, and a much more intense short-range repulsion. The short-range repulsion is easily understood as a consequence of the electrostatic repulsion between the electrons which constitute the periphery of both molecules. It can be calculated without great difficulty from a knowledge of the electronic structure of the molecules; the quantum-mechanical treatment which can be given in simple examples shows that its potential is a product of a function of the form $e^{-r/\varrho}$ and another more slowly varying factor $N(r)$ which can usually be treated as a constant.

The attractive forces can also be understood on electrostatic grounds, provided that it is borne in mind that the structure of a molecule is not rigid, but easily deformed by external electric fields. Thus, although the molecules are both without resultant electric moment, they will tend to polarize one another when brought close together; and the effect of the mutual polarization is, on the average, an attractive one. An exact treatment of this effect requires detailed quantum-mechanical considerations which were first made by LONDON [1930]; these will not be reproduced here, but it will be mentioned simply that the mutual potential energy of two molecules arising from London's forces is practically proportional to r^{-6}, though there are much smaller contributions proportional to r^{-8} and higher even powers of r^{-1}.

The simplest adequate representation of the van der Waals' forces is by means of the formula

$$(7.2) \qquad \phi(r) = -\mu r^{-6} + N e^{-r/\varrho}$$

with constants μ, N and ϱ varying from substance to substance. This is not, however, a very convenient expression to use for analytical purposes. Many physical properties of fluids are not very sensitive to the precise nature of the repulsive field, and one may therefore use a formula of the type

$$(7.3) \qquad \phi(r) = -\mu r^{-6} + \nu r^{-m}$$

without risk of great error. Indeed, if one compares this formula with the previous one, it will be found that the difference between them can be made very small except at very short distances—where neither is properly valid—by the adjustment of the parameters μ and ν. The value of m may be taken to be any between 8 and 14; perhaps the best fit is obtained with a value of m nearly equal to 12.

The numerical constants involved in the law of force between two molecules may be determined theoretically for simple molecules, from a knowledge of their structure. The interaction between two helium atoms has been calculated by SLATER & KIRKWOOD [1931]. To determine the forces between molecules of more complicated structure, however, it is necessary to make use of some kind of experimental data. Measurements of the first virial coefficient are very well adapted to this purpose, and have been used extensively (cf. § 4 of chapter VII). Observations of the intensity of scattered X-radiation furnish another potentially very valuable source of information concerning the intermolecular forces. Some information concerning results already obtained will be found in § 4 of chapter VII.

CHAPTER IV

CONDENSATION AND THE LIQUID STATE

1. The Nature of Condensation

All fluids have certain properties in common, the most spectacular of which are manifest in the phenomena of freezing and condensation. The latter — the phenomenon of condensation — presents a very interesting challenge to the molecular theory of the condensed phase. Obviously no theory could claim to be complete in any sense which did not elucidate the nature of the passage between the liquid and the gas. The fact is, however, that up to recent years the molecular mechanism which leads to a sudden change in the state of a substance has been understood only very imperfectly.

It is, of course, easy to describe the process from a phenomenological point of view. If the temperature of a gas is kept constant, and the pressure is increased indefinitely, one of two things may happen. If the temperature exceeds the critical temperature, the density increases in a uniform way up to a point where the substance solidifies; no process of condensation is observed, although the density in the final stages is comparable with that of the liquid. If the temperature is lower than the critical temperature, on the other hand, at a certain pressure condensation is possible, in which the density of the fluid undergoes a discontinuous change, together with the internal energy and many other thermodynamical quantities.

From the study of the X-ray scattering data it has been seen that this change is associated with a change in the molecular structure, from a state in which the molecules spend most of their time at a relatively large distance, beyond the range of the intermolecular forces, from their nearest neighbours, to a fairly close-packed state in which there are always several other molecules within the sphere of interaction of any given molecule. The latent heat evolved in condensation may be understood as the potential energy lost by the molecules when they associate in this way. In the condensed state a molecule is always in a potential 'well' associated with the neighbouring molecules, and a certain amount of energy is required to extract the molecule from the fluid; this is the heat of vaporization per molecule. The fact that evaporation occurs spontaneously at a free surface of the liquid can be attributed to fluctuations in the liquid which allow the more energetic of the molecules to escape. All this can be given a quantitative description in molecular terms without much difficulty. A less easy task is to account for the *suddenness* with which the transition from gas to liquid is achieved when the energy of the fluid is progressively reduced.

The key to the situation is the existence, in a certain region of the phase diagram, of two states of the fluid—the liquid and the gas—with different densities at the same temperature and pressure. This can, in principle, be inferred from the equation of state, which furnishes the relation between pressure, temperature and density. Only the difficulties in the evaluation of the partition function have prevented an exact quantitative treatment of the equation of state. An important advance in this direction has, however, been led by Mayer, who analysed the partition function in terms of the various ways in which the molecules may associate under the influence of their mutual attractions.

In any state of the fluid, there will be a large number of groups of molecules which are called clusters. In any cluster, each molecule lies within the sphere of attraction of at least one other molecule of the cluster. Only in a rare gas are most of the molecules free, and clusters of two or more molecules are comparatively rare. As the density is increased, clusters become very much more numerous, and, as the point of condensation is reached, clusters containing a comparatively large number of molecules may occur. It should be recognized that, since the molecules are in rapid motion, a cluster in the gas is always a rather temporary association of molecules, and the number of molecules in a cluster is subject to rapid fluctuations in time.

The concept of clusters in a gas should be distinguished from that of cybotaxis (cf. § 3 of chapter III) in a liquid. There is some analogy between the two, since the development of large clusters in a gas is characteristic of incipient condensation in much the same way as the development of large cybotactic groups is a necessary preliminary to crystallization. The van der Waals' forces are normally responsible for both. The difference arises from the fact that the molecules in a cluster are able to move freely relative to one another, whereas those in a cybotactic group are bound in an imperfect sort of lattice structure.

A liquid drop should be regarded as a giant cluster containing a very large number of molecules; and, from this point of view, the distinction between gas and liquid lies in whether the clusters are of macroscopic or microscopic dimensions. The phenomenon of condensation may then be understood in terms of a certain instability of the vapour, after a certain density is reached. Then, if a cluster of macroscopic dimensions is formed, it will tend to grow owing to the fact that the molecules of the gas are captured more rapidly than the faster moving molecules of the cluster can escape. The point of equilibrium of such a macroscopic cluster with the vapour is reached when the number of molecules entering and leaving the cluster in a short interval is, on the average, the same. The density of the vapour when this point is reached may be identified with the density of saturation.

The density of saturation is not, however, independent of the size of the liquid drops formed by the growth of the clusters. A small drop has a convex periphery through which the molecules inside have an appreciably greater chance to escape than through a flat or nearly flat interface. In consequence, the density of saturation will be larger in the neighbourhood of small drops than large ones. This circumstance favours the eventual development of a few large drops of liquid, instead of many small ones, in a condensing gas. For, by a process of diffusion, molecules will migrate in the vapour from the vicinity of small drops to the neighbourhood of large drops, where the density of saturation is lower. In this way, the small drops tend to evaporate and the large ones to grow.

A point of interest is that, in the absence of a cluster of macroscopic dimensions, a considerable degree of supersaturation may exist in the vapour without the incidence of condensation. Owing to the tendency of small clusters to evaporate, the flunctuations in the size of the microscopic clusters in the vapour are not generally great enough to ensure that a macroscopic cluster will appear, except at extremely long intervals of time. This makes possible the existence of a range of metastable states of the (supercooled) vapour up to a point where the compressibility becomes so large that the latent instability can no longer be checked. It is also possible for metastable states of the (superheated) liquid to arise in a very large volume of the fluid, where the free surface is relatively so small that the normal equilibrium with the vapour cannot be attained.

The idea of molecular clusters has proved a very fruitful one in the study of the thermodynamics of imperfect gases, where it is found that every kind of cluster makes a distinctive and easily isolated contribution to the thermodynamical properties. For example, the ideal gas law $p = nkT$ can be shown to be correct only so long as no clusters of two or more molecules are present in the fluid. The presence of clusters, except at very high temperatures, has the effect of reducing the pressure below that predicted by the ideal gas law, because the clusters have some of the attributes of complex molecules, and their formation involves a reduction in the effective number density. Quantitatively, the effect of clusters of a given number (s) of molecules on the thermodynamical quantities is proportional to the corresponding power of the density. The effect of clusters containing even a fairly small number of molecules is therefore very small at low densities, but increases very rapidly once a sufficiently high density is attained. When one considers the giant clusters of which liquid drops are composed, it is easy to understand that their influence will become effective within a density range which is very short indeed: the almost discontinuous nature of the process of condensation is an obvious consequence of the cluster theory [1]).

Mayer has made the distinction between two types of cluster. The reducible type consists of two or more groups of molecules, bound together by only one molecule which is common to each group. The irreducible type, on the other hand, has the property that any part of it is connected to the whole by at least two molecules. A reducible cluster can be constructed from its irreducible components stage by stage, by selecting any one of them and attaching the others one by one, at each stage identifying one molecule of the attached irreducible cluster with one of the molecules of the incomplete cluster already formed. Each type of cluster is represented by a characteristic integral,

---

[1]) The mathematical discussion, however, presents numerous pitfalls which few authors have been able to avoid completely. In § 4 below will be found a correct formulation of the theory so far as the author has been able to determine on the basis of his own work and communication with the leading authorities in the field. He is particularly indebted to Prof. J. E. Mayer for illuminating correspondence on this subject.

known as a cluster integral, and an integral representing a reducible cluster is a simple product of the integrals representing the component irreducible clusters.

It turns out that the pressure (p) can be expressed as a power series, either in the activity (z) with coefficients which are proportional to the reducible cluster integrals, or in the density (n) with coefficients which are proportional to the irreducible cluster integrals, cf. (3. 13) and (3. 26) below. These series, as has been remarked by YVON [1949], are always convergent so long as the volume (V) occupied by the fluid is finite. However, the convergence is non-uniform, for the liquid state, in passing from a fluid of finite to one of infinite extent. The phenomenon of condensation is therefore closely associated with the divergence of the series already mentioned, for an infinite fluid. This is a transparent consequence of the fact that, in passing from the vapour to the liquid, the individual clusters lose their identity and a single giant cluster is formed.

2. The Cluster Integrals

The starting point of Mayer's investigation of the phenomenon of condensation is the formula for the free energy (F)

(2. 1) $\begin{cases} \exp(-\beta F) = \lambda^N Q_N/N!, \quad \lambda = (2\pi/\beta m)^{3/2}, \\ Q_N = \int \overset{(N)}{\cdots} \int \exp\left\{-\beta \sum_{i<j} \phi(\mathbf{x}^{(i)}, \mathbf{x}^{(j)})\right\} d\mathbf{x}^{(1)} \ldots d\mathbf{x}^{(N)}, \end{cases}$

which is also the equation (6. 4) of chapter II, with the total potential energy Φ of the fluid expressed as the sum of the mutual potential energies of all pairs of molecules in the fluid. Mayer (vide MAYER [1937], MAYER & ACKERMANN [1937], and MAYER & HARRISON [1938]) in extending an earlier development of URSELL [1927], succeeded in evaluating the phase integral Q_N in terms of a series of irreducible integrals the first few of which are easily calculated; the result proved to throw very valuable light on the phenomenon of condensation. The present treatment incorporates some improvements of Mayer's method due to BORN & FUCHS [1938] and UHLENBECK & KAHN [1938].

It is convenient to begin by introducing the function

(2. 2) $\qquad u^{(ij)} = \exp\left\{-\beta\phi(\mathbf{x}^{(i)}, \mathbf{x}^{(j)})\right\} - 1$

which is almost zero when the positions of the two molecules i and j are far apart. Then

$$(2.3) \quad \begin{cases} Q_N = \int \overset{(N)}{\ldots} \int \prod_{i<j} \{1 + u^{(ij)}\} \, d\mathbf{x}^{(1)} \ldots d\mathbf{x}^{(N)} \\ = \int \overset{(N)}{\ldots} \int \sum u^{(ij)} \, u^{(kl)} \ldots u^{(yz)} \, d\mathbf{x}^{(1)} \ldots d\mathbf{x}^{(N)}, \end{cases}$$

where the summation extends over all different products of the $u^{(ij)}$ for which $i < j$ and $j \leqslant N$. The number of factors in the products ranges from 0 to $\tfrac{1}{2}N(N-1)$. Consider a typical product,

$$u^{(24)} \, u^{(35)} \, u^{(56)} \, u^{(57)} \, u^{(58)} \, u^{(68)},$$

say. This can be represented by a bond-figure diagram in which every molecule is represented by a point, and bonds are drawn connecting those molecules which are associated in each factor of the product.

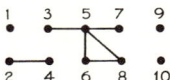

Figure 1. A typical bond-figure.

Those groups of molecules, for example [1], [2, 4], [3, 5, 6, 7, 8], [9], [10] in figure 1, which are mutually connected but have no connection with the other molecules, are called clusters. Thus in figure 1 are shown 3 'clusters' of one molecule, one cluster of 2 molecules, and one cluster of 5 molecules.

The number of ways of subdividing the N molecules into k_1 clusters of one molecule, k_2 clusters of 2 molecules, ..., k_q clusters of q molecules, is the number of partitions $[N!/\{(1!)^{k_1}(2!)^{k_2} \ldots (q!)^{k_q}\}]$ of the N molecules in this way, divided by the number of ways $[k_1!k_2! \ldots k_q!]$ of permuting clusters with the same number of molecules among themselves. Hence Q_N, expressed in the form (2.3), reduces to

$$(2.4) \qquad Q_N = N! \sum \frac{(a_1/1!)^{k_1}}{k_1!} \frac{(a_2/2!)^{k_2}}{k_2!} \ldots,$$

where the summation extends over all values of k_1, k_2, k_3 etc. which satisfy the condition

$$(2.5) \qquad \sum q k_q = k_1 + 2\,k_2 + 3\,k_3 + \ldots = N,$$

and the quantitites a_1, a_2, a_3 etc. represent the integrals

$$(2.6) \quad \begin{cases} a_1 = \int d\mathbf{x}^{(1)} \\ a_2 = \iint u^{(12)}\, d\mathbf{x}^{(1)}\, d\mathbf{x}^{(2)} \\ a_3 = \iiint (u^{(12)}\, u^{(13)}\, u^{(23)} + u^{(12)}\, u^{(13)} + u^{(12)}\, u^{(23)} + \\ \qquad\qquad + u^{(13)}\, u^{(23)})\, d\mathbf{x}^{(1)} d\mathbf{x}^{(2)}\, d\mathbf{x}^{(3)}. \end{cases}$$

In the integral expression for a_q one has to include all different products of the $u^{(ij)}$ for $i < j \leqslant q$ which correspond to a single cluster. Remembering that $u^{(ij)}$ depends only on the relative position $\mathbf{x}^{(j)} - \mathbf{x}^{(i)}$ of the molecules i, j, one finds from (2. 6) that

$$(2.7) \quad a_1 = V,\ a_2 = V\alpha_2,\ a_3 = V(\alpha_3 + 3\alpha_2^2),\ a_4 = V(\alpha_4 + 6\alpha_3\alpha_2 + 16\alpha_2^3),$$

where α_2, α_3, α_4 etc. are the irreducible integrals

$$(2.8) \quad \begin{cases} \alpha_2 = \int u^{(12)}\, d\mathbf{x}^{(1)} \\ \alpha_3 = \iint u^{(12)}\, u^{(13)}\, u^{(23)}\, d\mathbf{x}^{(1)}\, d\mathbf{x}^{(2)} \\ \alpha_4 = \iiint (u^{(12)}\, u^{(13)}\, u^{(14)}\, u^{(23)}\, u^{(24)}\, u^{(34)} + 6\, u^{(12)}\, u^{(13)}\, u^{(14)}\, u^{(23)}\, u^{(24)} \\ \qquad\qquad + 3\, u^{(12)}\, u^{(13)}\, u^{(24)}\, u^{(34)})\, d\mathbf{x}^{(1)}\, d\mathbf{x}^{(2)}\, d\mathbf{x}^{(3)}. \end{cases}$$

These integrals are derived from the irreducible clusters which cannot be split into two clusters joined by a single molecule.

Figure 2. The irreducible clusters.

The cluster of 5 molecules shown in figure 1 is reducible, being constructed from the irreducible clusters [35], [57], and [568], which are connected together only by the molecule 5.

It is important to have the general formula for a_q in terms of the α_r, which reduces to (2. 7) for $q = 1, 2, 3$. It is required to determine the number of different ways of constructing reducible clusters of q molecules from k_2 irreducible clusters of 2 molecules, k_3 irreducible clusters of 3 molecules, ..., and k_r irreducible clusters of r molecules. To obtain this result, notice that any reducible cluster can be formed from the irreducible clusters by selecting any one of the latter and connecting the others one by one. The number of connections is thus always one less than the number of irreducible clusters (Σk_s). Also the number of molecules (q) in the resultant reducible cluster must be equal to the total number ($\Sigma s k_s$) in the component irreducible clusters, less the number of connections ($\Sigma k_s - 1$). Hence the k_s are subject to the restriction

$$(2.9) \qquad q - 1 = k_2 + 2k_3 + \ldots = \sum_s (s-1) k_s$$

but are otherwise arbitrary. The number of ways in which the number $q - 1$ can be subdivided in the way required by the restriction (2.9) is the number of partitions

$$[(q-1)!/\{(1!)^{k_2} \ldots (\overline{r-1}!)^{k_r}\}]$$

of $q - 1$, divided by the number of ways $[k_1! \ldots k_r!]$ of permuting similar components. Also, since there are $\Sigma k_s - 1$ connections between the irreducible clusters, and each of these connections can be identified with any one of the q molecules of the final reducible cluster, there are $q^{(\Sigma k_s - 1)}$ ways of making the required connections. It follows that there are

$$\frac{(q-1)!}{(1!)^{k_2} \ldots (r-1!)^{k_r}} \cdot \frac{q^{(\Sigma k_s - 1)}}{k_2! \ldots k_r!}$$

ways of forming reducible clusters of q molecules from the irreducible clusters at one's disposal; and the general formula for a_q corresponding to (2.7) is

$$(2.10) \qquad a_q = V(q-1)! \, q^{-1} \sum \frac{(qa_2/1!)^{k_2}}{k_2!} \frac{(qa_3/2!)^{k_3}}{k_3!} \ldots$$

where the summation extends over all values of k_2, k_3, etc. which satisfy (2.9).

By means of the formulae (2.4) and (2.10), the partition integral Q_N has been expressed in terms of the irreducible cluster integrals a_r. It is rather remarkable that these two formulae bear such a close resemblance to one another.

3. The Virial Series

Proceeding from the formulae (2.4) and (2.10), Mayer found it possible to derive expressions for all thermodynamical quantities in terms of the 'cluster integrals' a_q and a_r. The derivation which follows has been adopted because it is the only one which makes clear the full range of validity of the results obtained. The volume V of the fluid is kept fixed at some very large value, so that to consider changes in the mean density of the fluid the number of molecules must be supposed to vary.

To begin with, it is convenient to adopt a customary notation by re-writing the equations (2.4) and (2.10) in the form

(3.1) $$\begin{cases} Q_N = N! \sum \frac{(Vb_1)^{k_1}}{k_1!} \frac{(Vb_2)^{k_2}}{k_2!} \ldots, \\ b_q = a_q/(Vq!), \quad (\sum q k_q = N), \end{cases}$$

and

(3.2) $$\begin{cases} b_q = q^{-2} \sum \frac{(q\beta_1)^{k_1}}{k_1!} \frac{(q\beta_2)^{k_2}}{k_2!} \ldots, \\ \beta_s = a_{s+1}/s!, \quad (\sum s k_s = q - 1). \end{cases}$$

It is clear from the definitions (2.6) and (2.8) that b_q and β_s so defined will be independent of the volume V of integration for small values of q and s; however, the same is not necessarily true for values comparable with the number of molecules in the condensed fluid.

From (3.1) it can be seen that $Q_N/N!$ is the coefficient of z^N in the series expansion of the function $\exp(V \Sigma b_q z^q)$, so

(3.3) $$\sum_N Q_N z^N/N! = \exp(V \sum b_q z^q).$$

The coefficients Q_N of the series on the left hand side are defined by (2.1); it follows, indeed, from this equation that, if $F(N, V)$ is the free energy of N molecules occupying the volume V, then

(3.4) $$\sum_N Q_N z^N/N! = \sum_N \exp\{-\beta F(N, V)\} (z/\lambda)^N.$$

The equations (3.3) and (3.4) are true for all values of z; in particular, they are true if z is the thermodynamical quantity known as the *activity*, which is a function of density defined by

(3.5) $$z = \lambda \exp \beta \{F_1(n) + n F_1'(n)\},$$

where $F_1(n)$ is the free energy per molecule at some fixed value of the number density n, and $F_1'(n)$ is the derivative with respect to n at the same density. The pressure p and the isothermal compressibility ζ_T at this density are given by

(3.6) $$\begin{cases} p = n^2 F_1'(n), \\ \zeta_T^{-1} = 2 n F_1'(n) + n^2 F_1''(n). \end{cases}$$

When (3.5) is substituted into (3.4), the result is

(3.7) $$\sum_N Q_N z^N/N! = \sum_N \exp N\beta \{F_1(n) + n F_1'(n) - F_1(N/V)\}$$

since $F(N, V)$, the free energy of N molecules at the density N/V, can be written in the form $NF_1(N/V)$.

It will now be shown that the maximum term of the series (3.7) is the one with the value of N given most nearly by $N = nV$, and further, that the terms in the neighbourhood of this maximum are so large in comparison with the rest that they account for practically the entire value of the sum, when V is large. First, to find the maximum term, one needs only to differentiate the exponent in (3.7) with respect to N, equating the result to zero, thus:

(3.8) $\quad F_1(n) + nF'_1(n) - F_1(N/V) - (N/V)\,F'_1(N/V) = 0.$

This is obviously satisfied by $N = nV$. Next, the value of the terms in the neighbourhood of the maximum can be obtained by expanding the exponent of (3.7) in powers of $\varDelta = N - nV$. Writing

(3.9) $\qquad\qquad\qquad N = nV + \varDelta,$

one has, with the help of (3.6),

$$(3.10)\ \begin{cases} N\,\{F_1(n) + nF'_1(n) - F_1(N/V)\} = \\ = (nV + \varDelta)\,\{nF'_1(n) - \varDelta V^{-1}\,F'_1(n) - \tfrac{1}{2}\varDelta^2\,V^{-2}\,F''_1(n)\} \\ \qquad\qquad\qquad = pV - \varDelta^2/(2\,nV\zeta_T), \end{cases}$$

very nearly, so long as $\varDelta$ is small compared with nV. Under the same condition, the exponential factor of (3.7) is therefore

$$\exp\,(\beta pV)\cdot\exp\,\{-\beta\varDelta^2/(2\,nV\zeta_T)\}.$$

Now, when $\varDelta$ is of the order $(nV)^{2/3}$ — which is still small compared with nV — the second factor of the above expression is already negligible compared with unity, when V is large. Hence there is negligible error in writing, instead of (3.7),

(3.11) $\quad \sum_N Q_N\,z^N/N! = \exp\,(\beta pV)\sum_N \exp\,\{-\beta(N - nV)^2/(2nV\zeta_T)\}.$

Again because V is large, the sum on the right-hand side can be replaced by the integral

$$\int \exp\,\{-\beta Vv^2/2\,n\zeta_T\}\cdot(Vdv) = (2\pi n\zeta_T V/\beta)^{\frac{1}{2}},$$

wherein $v = \varDelta/V$. By comparison of this result with (3.3), one has

(3.12) $\qquad \beta pV + \tfrac{1}{2}\ln\,(2\pi n\zeta_T\,V/\beta) = V\sum b_q\,z^q.$

The term $\tfrac{1}{2}\ln\,(2\pi n\zeta_T V/\beta)/V$ which occurs on dividing by the volume

V is negligible when V is large; hence

(3. 13) $$\beta p = \sum b_q z^q.$$

The result just obtained gives the pressure in terms of the activity z, when the cluster integrals b_q have been evaluated. It is correct for both liquid and gas. It is more convenient, however, for many purposes, to have the pressure expressed in terms of the density. To achieve this end, differentiate (3. 13) with respect to the density, still keeping the volume V fixed, thus:

(3. 14) $$\beta \frac{dp}{dn} = \frac{1}{z} \frac{dz}{dn} \sum q b_q z^q;$$

then since, according to (3. 5) and (3. 6),

(3. 15) $$\frac{n}{z} \frac{dz}{dn} = \beta \frac{dp}{dn},$$

it follows that

(3. 16) $$\sum q b_q z^q = n,$$

unless $dp/dn = 0$. This equation makes it possible in principle to solve for z in terms of n, though to do it directly would be rather tedious. Fortunately, with the help of (3. 2), an exact solution can be obtained by a method which will now be explained.

It can be seen at once from (3. 2) that $q^2 b_q$ is the coefficient of ζ^{q-1} in the series expansion of the function $\exp(q \sum \beta_s \zeta^s)$ in powers of ζ; hence, the expression $\sum q b_q z^q$ which occurs on the left-hand side of (3. 16) must be the coefficient of ζ^{-1} in the expansion of

(3. 17) $$f(\zeta) = \sum_q q^{-1} (z/\zeta)^q \exp(q \sum \beta_s \zeta^s)$$

in ascending and descending powers of ζ.

Two things will now be assumed provisionally. The first is that, for a given value of z, a root ζ_0 exists of the equation

(3. 18) $$z = \zeta_0 \exp(-\sum \beta_s \zeta_0^s).$$

Secondly, it will be supposed that, for this value of ζ_0, z is an increasing function of ζ_0; one finds, by differentiating (3. 18) with respect to ζ_0, that the condition under which this is so is

(3. 19) $$1 - \sum s \beta_s \zeta_0^s > 0.$$

The significance of these two assumptions, which are obviously always correct for sufficiently small values of z, will be examined presently. Now, if ζ is given any value slightly in excess of ζ_0,

$$(z/\zeta) \exp \sum \beta_s \zeta^s = \{\zeta_0 \exp(-\sum \beta_s \zeta_0^s)\}/\{\zeta \exp(-\sum \beta_s \zeta^s)\} < 1,$$

and the series (3. 17) can therefore be summed, thus:

(3. 20) $$f(\zeta) = -\ln\{1 - (z/\zeta)\exp \sum \beta_s \zeta^s\}.$$

Substituting from (3. 18) into (3. 20), one has further

(3. 21) $$\begin{cases} f(\zeta) = -\ln(1 - \zeta_0/\zeta) - \\ \qquad -\ln[1 + \zeta_0\{1 - \exp \sum \beta_s(\zeta^s - \zeta_0^s)\}/(\zeta - \zeta_0)]. \end{cases}$$

Now ζ_0 is less than ζ, and, if the difference between them is small,

$$\zeta_0\{1 - \exp \sum \beta_s(\zeta^s - \zeta_0^s)\}/(\zeta - \zeta_0) \sim \sum s\beta_s \zeta_0^s,$$

which is less than 1, according to (3. 19). Hence both logarithms may be expanded to provide a series expression for $f(\zeta)$ in ascending and descending powers of ζ. Only the first logarithm yields inverse powers of ζ, and since $\Sigma q b_q z^q$ is the coefficient of ζ^{-1} in the series expansion of $f(\zeta)$, one has immediately

(3. 22) $$\sum q b_q z^q = \zeta_0$$

But, by comparison with (3. 16), it then follows that $\zeta_0 = n$; so, substituting this value in (3. 18), one obtains finally

(3. 23) $$z = n \exp(-\sum \beta_s n^s).$$

This is the required solution of the equation (3. 16). With its help, all thermodynamical quantities can be expressed in powers of the density n. For example, by differentiation of (3. 23) with respect to the density, one has

(3. 24) $$\frac{n}{z}\frac{dz}{dn} = 1 - \sum s\beta_s n^s$$

and by comparison with (3. 15) it follows that the isothermal compressibility ζ_T is given by

(3. 25) $$\beta \zeta_T^{-1} = \beta \frac{dp}{dn} = 1 - \sum s\beta_s n^s.$$

Integrating this equation with respect to n, and observing that the pressure must vanish at zero density, one obtains further

(3. 26) $$\beta p = n - \sum s\beta_s n^{s+1}/(s+1).$$

The two assumptions made in the course of this derivation are seen, *a posteriori*, always to be justified. For $\zeta_0 = n$, and there is always a value of n corresponding to a given value of the activity ζ, either in the liquid or the gas; and since the isothermal compressibility ζ_T in (3. 25) is a positive quantity, the condition (3. 19) will also be satisfied. Of course, this conclusion has been reached by a somewhat circular argument, but it is always possible to verify that (3. 23) is the solution of (3. 16) by a process of direct substitution, using (3. 2) to eliminate the reducible cluster integrals b_q.

The coefficients $s\beta_s/(s+1)$ of the powers of the density in the 'virial' series (3. 26) are called the virial coefficients. Supposing that the intermolecular potential is known, the first few of these can be determined fairly easily as a function of temperature, and compared with the values experimentally determined for the imperfect gas. Alternatively, as will be explained in § 4 of chapter VII, the intermolecular force can be inferred from a knowledge of the variation of the second virial coefficient $\frac{1}{2}\beta_1$ with temperature. The formulae of this section provide a practicable way of calculating thermodynamic quantities for the gas. They are also correct for the liquid, but for reasons which will presently appear, they do *not* provide a convenient basis for calculations relating to the condensed phase.

4. The Theory of Condensation

The theory of condensation which Mayer proposed on the basis of the developments of the previous section is properly founded on a single further assumption, derived from experience. It must be assumed that, below a certain critical temperature T_c, it is possible for the homogeneous fluid to exist in two stable states with different densities, at the same temperature and pressure. The two states, of course, correspond to the liquid and the vapour. They are represented, for a given temperature (T), by the points A and B respectively in fig. 3, where the pressure (p) is shown as a function of the specific volume (n^{-1}).

There are no homogeneous stable states for densities between A and B, but there are metastable states of the vapour represented by points, such as P, on the curve somewhat to the left of B, and metastable states of the liquid represented by points on the curve somewhat to the right of A. There are also inhomogeneous states, representing stable mixtures of the liquid and gas, represented by points on the straight line ACB; the latter will be referred to as the line of normal condensation. Figure 4 represents the free energy (F_1)

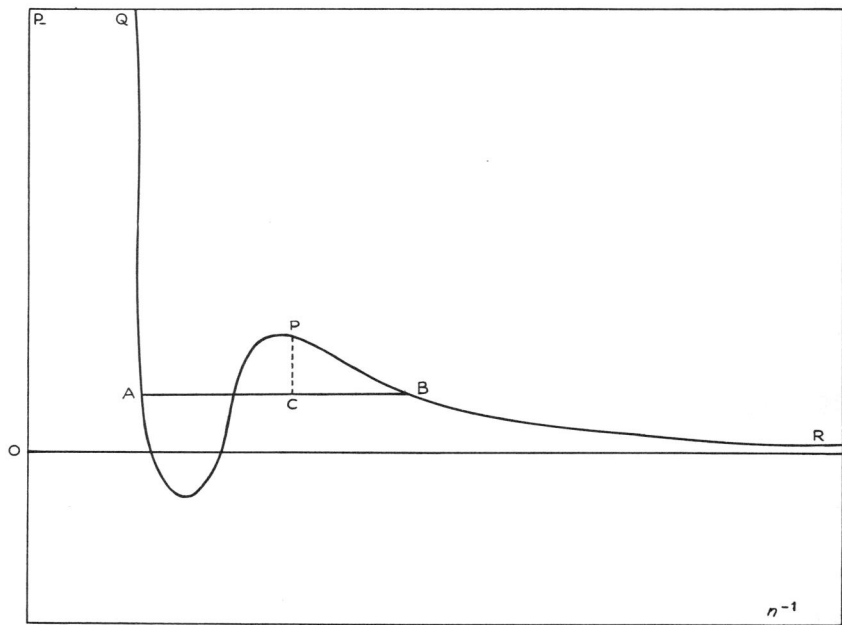

Fig. 3. The isotherm in the pressure-specific volume diagram.

per molecule at the same temperature (T) as a function of the specific volume (n^{-1}). The points A, B, C, P, etc., correspond roughly to those shown in fig. 3.

The free energy (F_1) per molecule derived from the formula (2.1) is a continuous, differentiable function of the density N/V. It can be shown, on the basis of the theory of partitions, that between A and B it must be represented by the straight line ACB which represents the stable mixtures of liquid and gas, rather than the curve APB which embraces the metastable states. The proof is as follows. If one divides the volume V into two parts V_A and V_B, one can show, as in § 7 of chapter II, that the probability of a configuration in which N_A molecules occupy the volume V_A, and $N_B = N - N_A$ molecules occupy the volume $V_B = V - V_A$, is

(4.1) $$p_{AB} = \exp \beta (F - F_A - F_B)$$

where F_A and F_B represent the free energies $F(N_A, V_A)$ and $F(N_B, V_B)$ of the two groups of molecules. Now the quantity p_{AB}, being a probability, must be less than 1, so

(4. 2) $\quad\quad\quad F(N, V) < F(N_A, V_A) + F(N_B, V_B).$

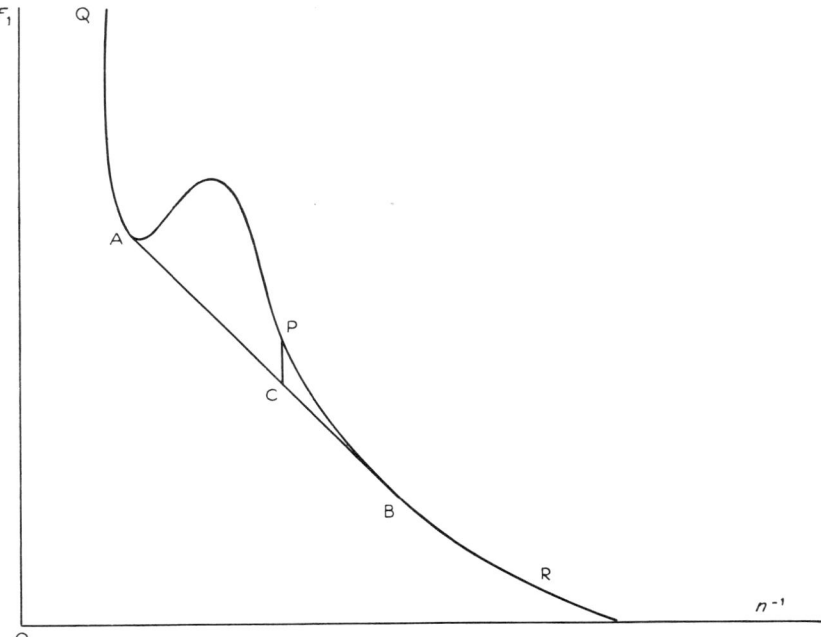

Fig. 4. The isotherm in the free energy-specific volume diagram.

On dividing by N, this inequality becomes

(4. 3) $\quad F_1(N/V) \leqslant (N_A/N) F_1(N_A/V_A) + (N_B/N) F_1(N_B/V_B).$

One cannot exclude the possibility that the two sides of (4. 3) are effectively equal to one another, because the difference between the two sides of (4. 2) could be negligibly small when divided by the number of molecules (N), which is always assumed to be very large. Indeed, *unless* the sign of equality in (4. 3) is possible, the fundamental hypothesis—that there exist two different states with equal pressure on the same isotherm—could not be true. For let N_A/V_A and N_B/V_B be the densities at A and B respectively, and let N_A, N_B be in the ratio AC : CB. Then (4. 3) asserts that the free energy per molecule for the specific volume

$$V/N = (N_A/N)(V_A/N_A) + (N_B/N)(V_B/N_B)$$

cannot be represented by a point *above* C. It can therefore be represented only by the point C itself. Thus, in the condensation region, the free energy per unit volume derived from the partition function must be represented by a point on the line of normal condensation AB. It follows further that, in the condensation region, all formulae derived from (2. 1) must apply to the stable mixtures, rather than the metastable inhomogeneous states. It will be seen presently, however, that formulae for the metastable states of the vapour can be derived by a simple limiting procedure.

It may be noticed in passing that the demonstrations given in the previous section do not suffice to establish the equations (3. 16) and (3. 23) with complete rigour on the condensation line, because (3. 16) follows from (3. 14) only when the compressibility is finite, and the proof given of (3. 23) also presupposes a finite positive value of the compressibility. However, direct substitution shows that (3. 23) must satisfy (3. 16) identically, provided that the series $\Sigma q b_q z^q$ and $\Sigma \beta_s n^s$ are convergent; and the convergence of these series is not in question so long as V is finite, because one can see from the definitions of b_q and β_s that $b_q \sim V b_{q-1}/q$ and $\beta_s \sim V \beta_{s-1}/s$ for very large values of q and s. It is, besides, clear on physical grounds that the terms representing clusters of more than nV molecules cannot make any effective contribution to the sums of the series. Then, since (3. 23) is always the solution of (3. 16), there is a one-to-one correspondence between the values of n and z; and both series are necessarily monotonic. It follows that the equation (3. 23) must be considered correct even on the singular condensation line.

Now let attention be given to the series

(4. 4) $$\begin{cases} \Sigma_q = \Sigma q b_q z^q & [= n], \\ \Sigma_s = \Sigma s \beta_s n^s & [= 1 - \beta(dp/dn)], \end{cases}$$

contained in the equations (3. 16) and (3. 25) respectively. The activity z, as one can readily infer from the definition (3. 5), has a constant value z_0 along the condensation line ACB. On the other hand, the sum of the series Σ_q increases from n_B to n_A between B and A; it is therefore discontinuous, or very nearly so, when z passes through the value z_0. Since the early terms of Σ_q are slowly varying functions of z, such an abrupt increase can be explained only on the supposition that the terms corresponding to large values of q suddenly make a contribution to the sum, when the activity passes the value z_0. This result may be interpreted as indicating that clusters containing a large number of molecules become suddenly important in passing from the gas to the liquid. Similarly the series Σ_s, whose sum is simply related to the compressibility, must suffer *two* discontinuities, one at the density n_B, and a second at the density n_A. It may be inferred that the terms representing the giant clusters first become important at the density n_B; and that their effect is modified at the density n_A, in a way which requires further consideration.

Though Σ_q and Σ_s depend only on the density n when the temperature is fixed, the cluster integrals b_q and β_s depend on the absolute volume V for large values of q and s respectively. More precisely, the b_q and β_s must become volume-dependent when q and s attain a value $n_0 V$, proportional to the volume V, where n_0 is some density appreciably less than that of the liquid. Let the series Σ_q and Σ_s each be divided into two parts, thus:

(4. 5) $$\begin{cases} \Sigma_q = \Sigma_q^{(1)} + \Sigma_q^{(2)}, \\ \Sigma_s = \Sigma_s^{(1)} + \Sigma_s^{(2)}, \end{cases}$$

in such a way that $\Sigma_q^{(1)}$ and $\Sigma_s^{(1)}$ include only the terms with q- and s-values less than $n_0 V$, which are volume-independent, and $\Sigma_q^{(2)}$ and $\Sigma_s^{(2)}$ the terms which are dependent on volume. For the gas, $\Sigma_q^{(2)}$ and $\Sigma_s^{(2)}$ both have a sum which is effectively zero. In passing the value z_0 of the activity, the value of $\Sigma_q^{(2)}$ rises abruptly from zero to some finite value. Similarly, in passing the density n_B, the value of $\Sigma_s^{(2)}$ rises abruptly from zero to some finite value. There are two possibilities, when this occurs, according as the terms at the end of the series $\Sigma_q^{(1)}$ and $\Sigma_s^{(1)}$ assume an appreciable value, or not. If these terms are negligible, as in the case of the series $\Sigma_s^{(1)}$, a further discontinuity in the sum of the series may be expected at some higher value (n_A) of the density, where they first become finite. If, on the other hand, as in the case of the series $\Sigma_q^{(1)}$, they are already finite, no further discontinuity need be anticipated.

It is a condition of the convergence of the series $\Sigma_q^{(1)}$ as V tends to infinity that the final terms should be negligible. This condition is satisfied for the gas, but not for the liquid; hence, although Σ_q is always finite, the infinite series $\lim_{V\to\infty}\Sigma_q^{(1)}$ becomes divergent when z exceeds the value z_0. This conclusion—that, in an infinite liquid, the series Σ_q is divergent—was first reached by Mayer. It may be shown in a similar way that the infinite series $\lim_{V\to\infty}\Sigma_s^{(1)}$ is convergent for the gas, but divergent for the liquid. The singularity is not necessarily at the density n_A, because it is not a *sufficient* condition for the convergence of a series that the terms 'at the end' of the series should be negligible. It is certainly true, however, that the series must diverge at some density n_0 intermediate between n_A and n_B. Assuming that n_0 is distinct from n_B, there will be a range of densities between n_0 and n_B where the series $\lim_{V\to\infty}\Sigma_s^{(1)}$ is convergent, but different from Σ_s. Below the density n_0, terms representing the giant clusters make no contribution to the sum $\lim_{V\to\infty}\Sigma_s^{(1)}$, which, accordingly, can describe the metastable states of the vapour. These conclusions are confirmed by an approximative method to be explained in the next section.

It will be noticed from (3.25) that when Σ_s attains the value 1, the pressure p is stationary, and the compressibility infinite. Among the metastable states of the vapour, this merely indicates the existence of a maximum of the pressure, with negative compressibilities at slightly higher densities which are, of course, never physically realized. Among the stable states, the stationary value occurs on the condensation line, below the critical temperature. In the vicinity of the critical temperature, it is possible for a small region to exist where the compressibility increases continuously to infinity, instead of discontinuously, as in the process of normal condensation. MAYER & HARRISON [1938] made the suggestion that this might be related to the observations of Maass and others (cf. MAASS & GEDDES [1937]) indicating that one could have a separation of two phases without the appearance of a meniscus. It is still not clear whether the experiments in this region are fully reproducible, however, and the theoretical position also requires further investigation.

5. Approximate Methods

Though the exact analysis of the partition function just described throws valuable light on the process of condensation and the nature of the liquid state, the formulae which it provides for the thermodynamical functions are not adapted to numerical computation, except at gaseous densities. Other methods have therefore been developed, which require only a knowledge of the radial distribution. It has been seen (§ 4 of chapter II) that the pressure and the internal energy per molecule are given by the formulae

(5. 1)
$$\begin{cases} p = n\beta^{-1} - \tfrac{1}{6} \int n_2(r) \, \phi'(r) \, r \, d\mathbf{r}, \\ U_1 = \tfrac{3}{2}\beta^{-1} + \tfrac{1}{2} n^{-1} \int n_2(r) \, \phi(r) \, d\mathbf{r}, \end{cases}$$

respectively. Also, the pressure and internal energy, expressed in terms of the number density (n) and inverse temperature (β), are sufficient for the determination of the free energy per molecule, by integration of the differential equation

$$d(\beta F_1) = \beta p \, dn/n^2 + U_1 \, d\beta.$$

Thus, all thermodynamical quantities can be derived if the radial distribution function $n_2(r)$ is known.

One source of information concerning this function is provided by the X-ray scattering experiments; unfortunately, the results are not yet sufficiently accurate for direct use in conjunction with the formulae (5. 1). The reason is that the latter are extremely sensitive to the precise value of $n_2(r)$ in the repulsive part of the intermolecular field of force, where the multipliers $\phi'(r)$ and $\phi(r)$ have very large values. In this region only the theoretical determinations of $n_2(r)$ are as yet sufficiently reliable. For numerical computation, one therefore turns to the solutions of the integral equation for $n_2(r)$ derived in § 6 of chapter III.

The analytical solution will be considered first. This [cf. (6. 9), (6. 17), (6. 15), (6. 11) and (6. 20) of chapter III] has the form

(5. 2)
$$\begin{cases} n_2(r) = n^2 \exp\{-\beta\phi(r) + f(r)\}, \\ i\,r f(r) = \int_{-\infty}^{\infty} \dfrac{sn\varepsilon v(s) \exp(2\pi i r s) ds}{1 - n\varepsilon v(s)}, \\ i\,s v(s) = \int_{-\infty}^{\infty} r u(r) \exp(2\pi i r s) dr, \\ u(r) = \exp\{-\beta\phi(r)\} - 1, \end{cases}$$

where ε is the solution of the equation

(5. 3)
$$\int_{-\infty}^{\infty} \frac{\{sv(s)\}^2 \, ds}{1 - n\varepsilon v(s)} = \frac{2\pi(\varepsilon - 1) v(0)}{n\varepsilon}.$$

The equation of state which results from the substitution of the solution (5. 2) into the formula (5. 1) for the pressure has been derived by RODRIGUEZ [1948].

To determine the function $v(s)$, Rodriguez first made a Fourier analysis of the function $\exp(2\gamma r/a) r u(r)$, thus expressing $r u(r)$ in the form

$$(5.4) \qquad r u(r) = \exp(-2\gamma r/a) \sum_{k=1}^{h} c_k \sin(2\pi k r/a)$$

within the interval $0 < r < a$. The values of the constants γ and a are so chosen that both sides of (5. 4) are small when $r > a$. Substituting (5. 4) in the appropriate formula of (5. 2), one obtains

$$(5.5) \qquad s v(s) = \sum_{k=1}^{h} \frac{c_k}{2\pi} \left\{ \frac{\gamma a}{\gamma^2 + (as-k)^2} - \frac{\gamma a}{\gamma^2 + (as+k)^2} \right\}.$$

The roots of the equation

$$(5.6) \qquad n \varepsilon v(s_m) = 1$$

may be supposed to have been determined by ordinary algebraic or numerical methods; including complex roots, there are $4h$ of them altogether, consisting of $2h$ pairs which differ only in sign.

The function $f(r)$, expressed as a definite integral in (5. 2), can be evaluated by contour integration. The appropriate contour consists of the real axis and the semi-circle of infinite radius in the upper half-plane. Since $s^3 v(s) \to 0$ uniformly on the infinite semicircle, one obtains by the theory of residues

$$(5.7) \qquad r f(r) = -2\pi \sum_{m}{}' \frac{s_m v(s_m)}{v'(s_m)} \exp 2\pi i r s_m$$

where the summation Σ' extends over all the complex roots in the upper half-plane. Similarly the equation (5. 3) for the determination of ε reduces to

$$(5.8) \qquad -\sum_{m}{}' \frac{s_m^2 \{v(s_m)\}^2}{v'(s_m)} = 2\pi (\varepsilon - 1) v(0).$$

Since the function $f(r)$ has been evaluated, it remains only to substitute the value of $n_2(r)$ given by (5. 2) into the equation of state (5. 1). In view of the fact that the solution is in any case a linear approximation, it is permissible to neglect squares and higher powers of $f(r)$, and set

$$n_2(r) = n^2 \exp\{-\beta \phi(r)\} \{1 + f(r)\}.$$

The resulting integral

$$\int f(r) \exp\{-\beta\phi(r)\}\, \phi'(r)\, d\mathbf{r} = 4\pi\beta^{-1} \int_0^\infty u(r)\,\{rf'(r) + 3f(r)\}\, r^2\, dr$$

can then be calculated quite easily, using the formula (5. 4) for $u(r)$ and (5. 7) for $f(r)$.

The isotherms which Rodriguez calculated by this method are in good numerical agreement with experimental values for the vapour, but, owing to the linearization approximation, the results for the condensed phase are only qualitative. They are nevertheless of interest, as they exhibit clearly the singularity which divides the liquid from the vapour. The way in which this singularity appears can be descibed as follows.

At sufficiently low densities the equation (5. 6) has no roots on the real axis; the $4h$ roots are in fact equally divided between the upper

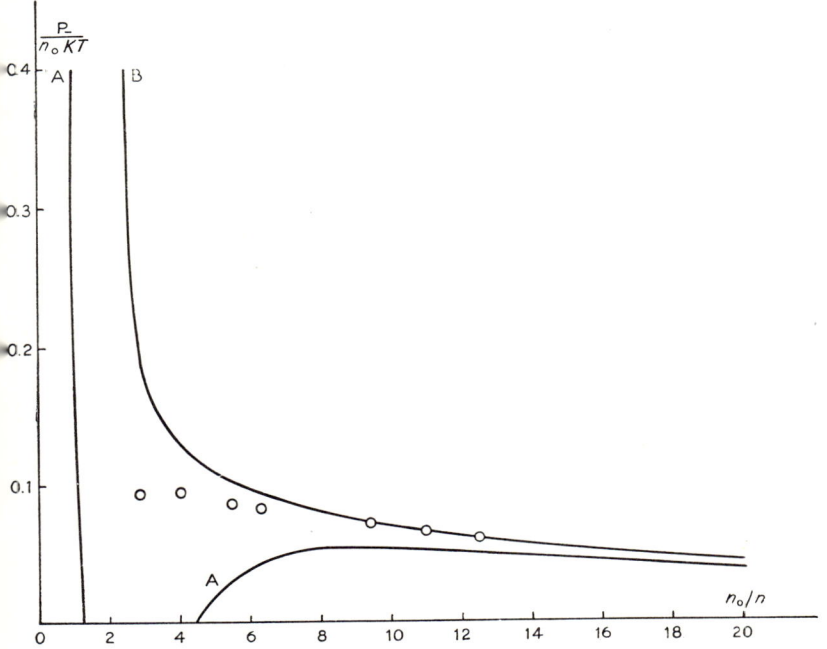

Fig. 5. Isotherms in the pressure-specific volume diagram. Curve A in argon at 120° K. Curve B at 150° K., near the critical point ⊙ Circles represent experimental observations on the critical isotherm. (Units in terms of the number density n_0 for the crystalline phase).

and lower halves of the complex plane. As the density is increased, however, a value n_0 of n will obviously be reached where (5. 6) has, for the first time, a real root—or rather two equal and opposite real roots, which may coincide at the origin. The density n_0 marks the transition from gas to liquid, and many physical quantities, such as the compressibility, have a discontinuity at this point. The reason for the discontinuity is that, whereas at gaseous densities there are $2h$ roots in the upper half-plane, on passing the density n_0 one or more of these is lost to the real axis, and have to be omitted from the summations of (5. 7) and (5. 8). The discontinuity lies below the horizontal axis in figure 5.

The isotherms shown in fig. 5 are actually those for 120° K. and 150° K. in argon which were computed by McLellan, using the more exact method for determining the radial distribution function described in § 6.2 of chapter III. Experimental points on the isotherm for 151.9° K. are plotted for comparison.

The critical temperature (T_c), density (mn_c) and pressure (p_c) derived by linear interpolation from these isotherms are compared with the experimentally determined values in the following table:

TABLE I
Critical Data for Argon

	T_c (° K.)	mn_c (gr/cm³)	P_c (atm.)
Calculated	140	0.40	46
Observed	150.7	0.531	48

The agreement is good, and could probably be improved by more exact calculations.

The isotherms shown in fig. 5 are both for the homogeneous fluid, and at the lower temperature metastable or unstable states are represented in the condensation region rather than stable mixtures of the two phases. This is a consequence of the use of the superposition approximation

$$n^3 n_3 = n_2^{(23)} n_2^{(31)} n_2^{(12)}$$

which is valid only for homogeneous states of the fluid.

6. The Solid State

Of the two states of aggregation of matter—the fluid state, which is characterized by *disorder* in the configuration of the molecules, and the crystalline state, in which the molecules remain for long periods in the vicinity of fixed lattice sites—only the first is directly related to the present considerations. As a matter of fact, the theory of crystalline solids is in many ways much simpler than that of fluids, mainly owing to the fact that by their nature such solids are incapable of flow. Though the molecules are not at rest, they may be imagined as performing continuous small oscillations about their lattice positions under the influence of the forces exerted by the other molecules. The lattice site is the position of minimum mean potential energy of each molecule, and, if the displacements of the molecules from their lattice positions are not too large, the total potential energy of the crystal will be a quadratic function of these displacements. As a result, the motion of each molecule is, very nearly, a superposition of a very large number of harmonic motions, with frequencies depending on the intermolecular forces. Even for a crystal of finite dimensions, the number of frequencies is so large that one has effectively a continuous spectrum. The determination of this spectrum is the principal task presented by the theory of solids (cf. BORN [1923]).

> Even at low temperatures, the picture of a solid based on a perfectly ordered system of lattice sites is probably an oversimplification which would make it impossible to explain a number of experimental findings, notably the low thermal conductivity and mechanical strength of crystalline media. It seems necessary to suppose that a perfect crystal of large dimensions is not only non-existent in nature, but for some reason impossible in principle. There is considerable evidence indicating that actual solids possess various faults which disrupt the regular lattice arrangement in such a way that the crystal is broken up into domains of microscopic, though not molecular, dimensions. It is possible that the reason for this is to be found in the province of the quantum theory.

As the temperature rises, the amplitude of the oscillations performed by the molecules also increases until, just below the melting point, individual molecules are able to leave their lattice sites and move, either into interstitial positions with a displacement of the surrounding molecules, or to sites vacated by other molecules. It is possible for this to happen fairly extensively without the destruction of the lattice arrangement as a whole. However, at the melting point, the agitation becomes sufficiently great for the majority of the molecules to leave their lattice sites more or less permanently, and the lattice disintegrates. This process has been elaborated by FRENKEL [1946] in his recent treatise.

The process of melting, then, is to be regarded as the transition of a substance from a state of order to one of disorder among the molecules. To understand why it occurs so abruptly, one has to advance considerations similar to those which arise in the theory of condensation. One knows already from experience that there is a range of metastable states below the normal freezing point which can be attained by supercooling the pure liquid in the absence of any nucleus to promote crystallization. Once a crystal exceeding a certain microscopic size has formed in the supercooled liquid, it will continue to grow, with evolution of heat due to the lower potential energy of the ordered configuration of the molecules. Equilibrium is reached when the faster moving molecules near the boundary of the lattice structure are lost at the same rate as the accrescence of the molecules from the liquid. The necessary and sufficient condition for the occurrence of freezing is that, over a certain range, two states, one ordered and the other disordered, should exist at the same temperature and pressure; and the freezing point is determined thermodynamically by the condition that at this temperature of normal freezing the activity of the two phases is the same.

At present the phenomenon of freezing need be considered only from the point of view of the disordered state. This is perhaps fortunate, since most attempts to consider the melting of a solid purely as a consequence of the instability of the lattice structure predict a melting point which is considerably too high. On the other hand, the difficulties inherent in a strictly quantitative theory of condensed fluids makes it impossible to give a perfectly exact treatment of freezing phenomena.

KIRKWOOD & MONROE [1941] are the authors of a theory which is qualitatively correct and also fairly satisfactory from the quantitative point of view; it is essentially their method which will be described in this and the following section. It is supposed that the lattice sites of the crystalline phase are known; then the principal distinction between the liquid and the crystal is exhibited by the behaviour of the number density in its variation from point to point in the medium. Recalling that the number density $n(\mathbf{x})$ measures the relative probability of finding a molecule at a given point **x**, one sees that, whereas in the fluid it has everywhere the same value, in the crystal this function must have a periodic character, with maxima at the lattice sites and minima in the interstitial positions. From the theoretical point of view, it is necessary to show that there are in fact solutions of the equations of equilibrium, corresponding to a periodic variation of the number density over macroscopic regions of space. The phase of these periodic variations will, of course, be undetermined; one cannot hope to predict the absolute positions of the lattice sites.

The sharpness of the maxima of the number density function in the crystalline phase depends on the degree of thermal motion of the molecules, and will be most accentuated at low temperatures. If there were no thermal motion at all—as one would expect, on the basis of classical theory, at absolute zero—the density function would be infinite at the lattice points and zero elsewhere. Near the melting point, on the other hand, there is a considerable degree of thermal motion, and the variation in the number density is not so great. Nevertheless, the density function of the ordered phase is sharply differentiated from that of the liquid, which is uniform in character.

The molecular distribution function $n_2(\mathbf{x}^{(1)}, \mathbf{x}^{(2)})$ also behaves differently in the two phases. In the fluid, it depends only on the distance r between the two positions $\mathbf{x}^{(1)}$ and $\mathbf{x}^{(2)}$ referred to; in the crystal, however, this is no longer so, because n_2 is proportional to the densities at both $\mathbf{x}^{(1)}$ and $\mathbf{x}^{(2)}$. It is, nevertheless, assumed by Kirkwood and Monroe that the *relative* distribution function g defined by

(6. 1) $$n_2(\mathbf{x}^{(1)}, \mathbf{x}^{(2)}) = g\, n(\mathbf{x}^{(1)})\, n(\mathbf{x}^{(2)})$$

depends only on the distance r; in fact, they adopt for both phases a value of g determined by extrapolation from the X-ray scattering experiments. This is the essence of their approximation; it enables one to determine the number density function, which is found to have, over the appropriate range of temperature and pressure, a periodic form corresponding to the ordered phase as well as a constant value corresponding to the fluid phase. Which of these—the crystal or the fluid—obtains in nature depends on the question of thermodynamic stability already considered.

7. The Theory of Freezing

The number density function $n(\mathbf{x})$, according to the general principles of statistical mechanics explained in chapter II, is expressed in terms of the mean potential energy $\phi^*(\mathbf{x})$ of a molecule situated at the point $\mathbf{x}$ by means of the relation

(7.1) $$n(\mathbf{x}) = \text{const.} \cdot \exp\{-\beta\phi^*(\mathbf{x})\},$$

where $\beta = 1/(kT)$ and T is the absolute temperature. By differentiation of this relation, one has

$$\frac{1}{\beta n}\frac{\partial n}{\partial \mathbf{x}} = -\frac{\partial \phi^*}{\partial \mathbf{x}}.$$

The right hand side of this equation represents the mean force acting on a molecule at the point $\mathbf{x}$, which, if one disregards gravity, is simply the mean resultant force due to all the other molecules (cf. chapter II, § 4),

$$-\frac{1}{n}\int n_2 \frac{\partial \phi}{\partial \mathbf{x}^{(1)}} d\mathbf{x}^{(2)},$$

where ϕ is the mutual potential energy of a pair of molecules. Hence

(7.2) $$\begin{cases} \dfrac{\partial \phi^*}{\partial \mathbf{x}^{(1)}} = \int \dfrac{n_2(\mathbf{x}^{(1)}, \mathbf{x}^{(2)})}{n(\mathbf{x}^{(1)})} \dfrac{\partial \phi}{\partial \mathbf{x}^{(1)}} d\mathbf{x}^{(2)} \\ \qquad = \int g(r) \dfrac{\partial \phi(r)}{\partial \mathbf{x}^{(1)}} n(\mathbf{x}^{(2)}) d\mathbf{x}^{(2)}, \end{cases}$$

with the help of (6.1). This equation can be integrated to determine the mean potential energy ϕ^*, thus:

(7.3) $$\phi^*(\mathbf{x}^{(1)}) = \int \{\int_\infty^r g(s) \phi'(s) ds\} n(\mathbf{x}^{(2)}) d\mathbf{x}^{(2)}.$$

Substituting in (7.1), one has, therefore,

(7.4) $$\ln\{\chi n(\mathbf{x}^{(1)})\} = \beta \int \{\int_\infty^r g(s) \phi'(s) ds\} n(\mathbf{x}^{(2)}) d\mathbf{x}^{(2)},$$

where χ is some parameter independent of $\mathbf{x}^{(1)}$.

This is an integral equation to determine the density function n, somewhat analogous to that used to determine the radial distribution function in § 6 of chapter III. One may attempt to solve the present equation also by Fourier transformations. Guided by the expectation that the solution will have a periodic character, one writes

(7.5) $$n(\mathbf{x}) = \sum_{\mathbf{h}} s(\mathbf{h}) \exp(2\pi i \mathbf{h} \cdot \mathbf{x}),$$

where the complete set of vectors **h** are the co-ordinate vectors of the points of some lattice structure, reciprocal to the expected space lattice. The method of determining the coefficients $s(\mathbf{h})$ from (7. 4) is described in the next paragraph, which may be omitted by those interested only in the result. If

(7. 6) $\qquad \beta \int_{r}^{\infty} g(s) \, \phi'(s) \, ds = \int a(k) \exp(2\pi i \mathbf{k} \cdot \mathbf{r}) \, d\mathbf{k},$

one has, by Fourier transformation,

(7. 7) $\begin{cases} a(k) = \beta \int \{ \int_{r}^{\infty} g(\varrho) \, \phi'(\varrho) \, d\varrho \} \exp(2\pi i \mathbf{k} \cdot \mathbf{r}) \, d\mathbf{r} \\ \qquad = -\dfrac{4\pi\beta}{(2\pi k)^3} \int_{0}^{\infty} g(r) \, \phi'(r) \, (2\pi k r \cos 2\pi k r - \sin 2\pi k r) \, dr. \end{cases}$

Also, substituting (7. 5) and (7. 6) into the right-hand side of (7. 4), one obtains

(7. 8) $\begin{cases} \ln \{\chi n(\mathbf{x}^{(1)})\} = \sum_{\mathbf{h}} s(\mathbf{h}) \exp(2\pi i \mathbf{h} \cdot \mathbf{x}^{(1)}) \iint a(k) \exp 2\pi i \, (\mathbf{k}+\mathbf{h}) \cdot \mathbf{r} \, d\mathbf{k} \, d\mathbf{r} \\ \qquad = \sum_{\mathbf{h}} a(h) \, s(\mathbf{h}) \exp(2\pi i \mathbf{h} \cdot \mathbf{x}^{(1)}). \end{cases}$

The parameter χ can be determined by integrating this equation over the volume of the unit cell $(\varDelta)$ of the space lattice, which is assumed to contain just one molecule, thus:

(7. 9) $\qquad \chi = \int_{\varDelta} \exp \{ \sum_{\mathbf{h}} a(h) \, s(\mathbf{h}) \exp(2\pi i \mathbf{h} \cdot \mathbf{x})\} \, d\mathbf{x}.$

By virtue of this result, χ can be regarded as a function of the Fourier coefficients $s(\mathbf{h})$, and by taking the partial derivative with respect to $s(-\mathbf{h})$, one has

$$\frac{\partial \chi}{\partial s(-\mathbf{h})} = a(h) \int_{\varDelta} \exp(-2\pi i \mathbf{h} \cdot \mathbf{x}) \exp \{ \sum a(h) \, s(\mathbf{h}) \exp 2\pi i \mathbf{h} \cdot \mathbf{x}\} \, d\mathbf{x}.$$

According to (7. 8), this reduces to

(7. 10) $\begin{cases} \dfrac{\partial (\ln \chi)}{\partial s(-\mathbf{h})} = a(h) \int_{\varDelta} \exp(-2\pi i \mathbf{h} \cdot \mathbf{x}) \, n(\mathbf{x}) \, d\mathbf{x} \\ \qquad = a(h) \, s(\mathbf{h}), \end{cases}$

with the help of (7. 5). As χ is a known function of the $s(\mathbf{h})$, by giving **h** each of its possible values in this last result, one obtains a set of simultaneous equations which suffice to determine the Fourier coefficients $s(\mathbf{h})$.

There is always a solution such that $s(\mathbf{h})$ vanishes when $\mathbf{h}$ is different from zero; this corresponds to the disordered fluid state. For suitable densities and temperatures, however, there may be in addition a non-trivial solution corresponding to the ordered crystalline state.

In practice it is probable that only the $s(\mathbf{h})$ with quite small values of h are large enough to influence the thermodynamical quantities to any important extent, even in the ordered state. Kirkwood and Monroe have treated the application to the face-centred cubic lattice —the only stable cubic lattice—by neglecting all the $s(\mathbf{h})$ except those corresponding to $\mathbf{h} = 0$ and $\mathbf{h} = (\pm a, \pm a, \pm a)$, where a is the side of the unit cell. For $\mathbf{h} = 0$, one has, by integrating (7. 5) over the volume of the unit cell (Δ),

$$(7.\,11) \qquad s_0 = s(0, 0, 0) = a^{-3};$$

also from (7. 7) it follows that

$$(7.\,12) \qquad a_0 = a(0) = -\tfrac{1}{3} \beta \int g(r)\,\phi'(r)\,r\,d\mathbf{r}.$$

The values of $s(\mathbf{h})$ and $a(h)$ for $\mathbf{h} = (\pm a, \pm a, \pm a)$ will be denoted by s and a respectively; the equation (7. 8) then reduces to

$$(7.\,13) \quad \chi n(\mathbf{x}) = \exp\{a_0 s_0 + 8as \cos(2\pi x_1/a) \cos(2\pi x_2/a) \cos(2\pi x_3/a)\}.$$

Hence

$$(7.\,14) \quad \begin{cases} \chi = a^3 \int_0^1 \int_0^1 \int_0^1 \exp\{a_0 s_0 + 8as \cos 2\pi y_1 \cos 2\pi y_2 \cos 2\pi y_3\}\, d\mathbf{y} \\ = a^3 \exp(a_0 s_0) \sum_{n=0}^{\infty} \{(2n)!\}^2 (as)^{2n}/(n!)^3 \end{cases}$$

and s is determined at the root of the equation

$$(7.\,15) \qquad \frac{\partial (\ln \chi)}{\partial s} = 8as,$$

which, in conjuction with (7. 14), can be solved numerically for any given value of a.

For $a < 0.973$, it is found that the only solution is that $(s = 0)$ which corresponds to the disordered phase. For $a > 0.973$, there are, in addition, two non-vanishing solutions, of which the larger corresponds to the more stable state with lower internal energy and is tabulated below.

8 a s	a	s	$\ln \chi - a_0 s_0$
2.82	0.9730	0.362	0.506
3.00	0.9734	0.3853	0.5738
3.50	0.9783	0.4472	0.7819
4.00	0.9897	0.5052	1.0203
4.50	1.0082	0.5579	1.287
5.00	1.0325	0.6053	1.578
5.50	1.0633	0.6466	1.892
6.00	1.0991	0.6824	2.223
6.50	1.1400	0.7127	2.572
7.00	1.1848	0.7385	2.936
7.50	1.232	0.7609	3.312
8.00	1.283	0.7794	3.696

This is the solution for the ordered phase; it is applicable to any solid with a face-centred crystalline structure. It remains to examine the thermodynamical properties and equilibrium of the two phases. The pressure (p) and the internal energy per molecule (U_1) are given quite generally by the formulae

$$(7.16) \begin{cases} \beta p = a^{-3} \int_\Delta n(\mathbf{x}^{(1)}) \left\{1 - \tfrac{1}{6} \beta \int g(r) \phi'(r) rn(\mathbf{x}^{(2)}) d\mathbf{x}^{(2)}\right\} d\mathbf{x}^{(1)}; \\ \beta U_1 = \int_\Delta n(\mathbf{x}^{(1)}) \left\{\tfrac{3}{2} + \tfrac{1}{2} \beta \int g(r) \phi(r) n(\mathbf{x}^{(2)}) d\mathbf{x}^{(2)}\right\} d\mathbf{x}^{(1)}, \end{cases}$$

where the integration $\int_\Delta \ldots d\mathbf{x}^{(1)}$ is taken over the unit cell, whose volume is a^3. By substitution from (7.5) these expressions can be evaluated in terms of the Fourier coefficients $s(\mathbf{h})$, and the principal thermodynamical quantities are therefore determined for both phases. One of the conditions of the equilibrium of the phases—that the pressure should be the same for each—can be applied immediately. The other—that the activity of the two phases should be the same —is more difficult to apply, since one requires the absolute difference ΔF_1 between the free energies per molecule of the crystal and the fluid at the same temperature and pressure. This should strictly be obtained by integrating the differential equation

$$d(\beta F_1) = U_1 \, d\beta/\beta + \beta p \, dn/n^2$$

between absolute zero, where $\Delta F_1 = \Delta U_1$, and the temperature considered [1]. A more practical procedure, however, is to make use of some experimental datum to fix the freezing point, for example,

---

[1] Thus making use of Nernst's heat theorem in thermodynamics.

the density of the liquid phase at this temperature. Kirkwood and Monroe were able in this way to calculate the freezing parameters for argon and make a direct comparison with experiment. In the following table, T_m is the melting point, ΔS represents the difference in entropy between the two phases and R the gas constant; ΔV is the change in volume on freezing in cm^3 per mole, and p_m the pressure in kg/cm^2; V_l represents the observed specific volume in cm^3/gm of the liquid at the melting point.

T_m	$\Delta S/\mathrm{R}$ calculated	$\Delta S/\mathrm{R}$ observed	ΔV calculated	ΔV observed	P_m	V_l
83.9	1.74	1.68	3.25	3.53	1	0.709
119.7	0.70	1.10	0.62	1.88	1650	0.661
183.2	0.48	0.71	0.33	0.92	5350	0.631

In view of the approximations made, the agreement between the observed and calculated values is very satisfactory at ordinary pressures; at very high pressures, the agreement is only semi-quantitative, probably as a result of the use of an inexact expression for the relative distribution function.

8. The Cell Model for Liquids

One of the earliest successful models of a liquid was due essentially to EYRING & HIRSCHFELDER [1937] and LENNARD–JONES & DEVONSHIRE [1938]. The idea of these authors was based on the fact that, especially near the melting point, a molecule in a liquid has not very much freedom of movement. It is surrounded by the molecules of the first coordination shell, and though it is in continual motion due to the thermal agitation, its translational motion is so inhibited by this environment that it cannot do much more than perform oscillations about a mean position, at any rate over short intervals of time. Such oscillations may be compared with those performed by a particle in a potential well, i.e., a field of force which becomes strongly repulsive at a short distance from the centre. Again, one might compare the motion to that executed by a molecule in a crystal, consisting of perpetual small oscillations about a lattice site. Such a model must obviously be treated with reserve, not merely because there is no long-range order in a liquid, but because the first co-ordination shell is never much more than half complete except very near the melting point, so that a molecule, like a Brownian particle, always possesses some degree of translational motion. For example, Lennard–Jones & Devonshire assumed that the number of nearest neighbours of a given molecule was constant and not much less than that obtaining in the crystal, whereas an examination of the experimental data discussed in § 3 of chapter III shows that such an assumption is far from correct. The agreement of the numerical results, and especially the critical constants, obtained by these authors, with experiment must therefore be regarded as largely coincidental.

The principal defect of most model-theories, of which the cell model theory is an example, is that it is impossible to form a reliable estimate of the errors to which their numerical predictions are subject, coupled with the circumstance that it is very difficult to see how they can be improved. The theory of Lennard–Jones & Devonshire, on the other hand, is exceptional, since KIRKWOOD [1950] has shown recently how it is connected with the exact statistical theory. As a result of Kirkwood's work, various possible developments of the cell model are revealed, which may in the future make it the most convenient approach to the molecular theory of liquids.

According to (6. 3) of chapter II, or (2. 1) of this chapter, the free energy F of a system of N molecules, confined within a volume V at the temperature $T = 1/(\beta k)$, is given by

(8. 1) $\begin{cases} \exp(-\beta F) = \lambda^N Q/N! \\ \lambda = (2\pi/\beta m)^{3/2}, \; Q = \int \overset{(N)}{\ldots} \int \exp(-\beta \Phi) \, d\mathbf{x}^{(1)} \ldots d\mathbf{x}^{(N)}. \end{cases}$

To introduce the cell method, the volume V is divided into N cells $\varDelta_1, \varDelta_2, \ldots \varDelta_N$, each of volume V/N. Then

(8. 2) $$Q = \sum_{i_1=1}^{N} \sum_{i_2=1}^{N} \ldots \sum_{i_N=1}^{N} \int_{i_1} \int_{i_2} \ldots \int_{i_N} \exp(-\beta \Phi) \, d\mathbf{x}^{(1)} \ldots d\mathbf{x}^{(N)}$$

where a suffix (i) to a sign of integration indicates that the integration is to be carried out only over the corresponding cell $(\varDelta_i)$. All those configurations in which $m_1, m_2, \ldots m_N$ molecules occupy the cells $\varDelta_1, \varDelta_2, \ldots \varDelta_N$ respectively give the same contribution $Q(m_1, \ldots m_N)$ to the sum (8. 2); and, as there are $N!/(m_1! \ldots m_N!)$ such configurations one has

(8. 3) $Q = \sum N!(m_1! \ldots m_N!)^{-1} \int_N \overset{(m_N)}{\ldots} \int_N \ldots \int_1 \overset{(m_1)}{\ldots} \int_1 \exp(-\beta \Phi) \, d\mathbf{x}^{(1)} \ldots d\mathbf{x}^{(N)}.$

Of all the possible configurations, that in which precisely one molecule occupies each cell is the most probable. It would therefore probably be a fairly good approximation to replace Q by the quantity

$$(8.\,4) \quad \begin{cases} Q^* = N!\,Q(1,\,1,\,\ldots\,1) \\ = N!\displaystyle\int_N \ldots \int_1 \exp \beta \Phi \, d\mathbf{x}^{(1)} \ldots d\mathbf{x}^{(N)}, \end{cases}$$

where the suffixes $N, \ldots 1$ again indicate that the integrations are restricted to the Nth ... 1st cells respectively. This formula, indeed, would be exact for the crystalline state; to allow for deviations in the liquid, one writes in general

$$(8.\,5) \quad Q = \sigma^N Q^*$$

where σ is a quantity which varies from 1 in the crystalline state to $e \doteqdot 2.7$ in the rare gas. In Eyring's terminology, $k \ln \sigma$ is the 'communal entropy' per molecule.

A set of thermodynamical quantities F^*, U^*, S^* can be defined by the equations

$$(8.\,6) \quad \begin{cases} \exp(-\beta F^*) = \lambda^N Q^*/N!, \\ f_N^* = \exp \beta(F^* - E), \\ U^* = \displaystyle\int_N \ldots \int_1 \int \int \overset{(N)}{\ldots} \int f_N^* E \, d\boldsymbol{\xi}^{(1)} \ldots d\boldsymbol{\xi}^{(N)} \, d\mathbf{x}^{(1)} \ldots d\mathbf{x}^{(N)}, \\ S^* = -k \displaystyle\int_N \ldots \int_1 \int \int \overset{(N)}{\ldots} \int f_N^* \ln f_N^* \, d\boldsymbol{\xi}^{(1)} \ldots d\boldsymbol{\xi}^{(N)} \, d\mathbf{x}^{(1)} \ldots d\mathbf{x}^{(N)}, \end{cases}$$

where E is the total energy of the system, defined as in (6. 1) of chapter II. It follows easily from (8. 1), (8. 5), (8. 4) and (8. 6) that

$$(8.\,7) \quad \begin{cases} \beta F = \beta F^* - N \ln \sigma, \\ F^* = U^* - TS^*. \end{cases}$$

According to the second of these relations, the partial free energy is given by

$$(8.\,8) \quad \beta F^* = \int_N \ldots \int_1 \int \int \overset{(N)}{\ldots} \int f_N^* (\beta E + \ln f_N^*) \, d\boldsymbol{\xi}^{(1)} \ldots d\boldsymbol{\xi}^{(N)} \, d\mathbf{x}^{(1)} \ldots d\mathbf{x}^{(N)}.$$

At this stage, the fundamental approximation is made: one writes

$$(8.\,9) \quad f_N^* = \prod_{i=1}^{N} \{ n^*(\mathbf{x}^{(i)}) \, \lambda \exp(-\tfrac{1}{2} m \beta \boldsymbol{\xi}^{(i)2}) \}.$$

Thus, one assumes that each molecule has a certain probability distribution $n^*(\mathbf{x})$ in its cell, but neglects the correlation between the positions of molecules in neighbouring cells. It is important to notice that this approximation might be improved on, by doubling the size of the cell, and allowing two molecules to occupy each cell [1]); however, only the first approximation, represented by (8. 9), will here be considered. It may be assumed that

$$(8.10) \qquad \int_i n^*(\mathbf{x}^{(i)})\, d\mathbf{x}^{(i)} = 1$$

since this leads to

$$(8.11) \qquad \int_N \ldots \int \int_1^{(N)} \int f_N^* \, d\boldsymbol{\xi}^{(1)} \ldots d\boldsymbol{\xi}^{(N)} d\mathbf{x}^{(1)} \ldots d\mathbf{x}^{(N)} = 1,$$

as required by (8. 6) and (8. 4)

By substituting the formula (8. 9) into the expression (8. 8) for the partial free energy, one obtains

$$(8.12) \quad \begin{cases} \beta F^* = N \{\ln \lambda + \tfrac{1}{2}\beta \sum_{j \neq i} \int_j \int_i n^*(\mathbf{x}^{(i)})\, n^*(\mathbf{x}^{(j)})\, \phi^{(ij)}\, d\mathbf{x}^{(i)}\, d\mathbf{x}^{(j)} + \\ \qquad\qquad + \int_i n^*(\mathbf{x}^{(i)}) \ln n^*(\mathbf{x}^{(i)})\, d\mathbf{x}^{(i)}\}. \end{cases}$$

Since this formula depends on an approximation, it can never be exact; however, the *best possible* choice of the distribution function $n(\mathbf{x})$ will be that which makes the free energy assume its minimum value for the temperature and density considered. To determine this minimum value, one contemplates a small variation $\delta n(\mathbf{x})$ in $n(\mathbf{x})$. The corresponding change in F^*, given by

$$(8.13) \quad \beta\, \delta F^* = N! \int_i \delta n^*(\mathbf{x}^{(i)}) \{\beta \sum_{j \neq i} \int_j n^*(\mathbf{x}^{(j)})\, \phi^{(ij)}\, d\mathbf{x}^{(j)} + \ln n^*(\mathbf{x}^{(i)}) + 1\}\, d\mathbf{x}^{(i)}$$

must vanish at the minimum. Now, according to (8. 10), $\delta n(\mathbf{x})$ must satisfy

$$\int_i \delta n^*(\mathbf{x}^{(i)})\, d\mathbf{x}^{(i)} = 0,$$

but is otherwise completely arbitrary. Consequently, one may infer from (8. 13) that the best value of $n^*(\mathbf{x})$ satisfies

$$(8.14) \qquad \ln n^*(\mathbf{x}^{(i)}) + \beta \sum_{j \neq i} \int_j n^*(\mathbf{x}^{(j)})\, \phi^{(ij)}\, d\mathbf{x}^{(j)} = \beta a$$

where a is some constant. This constant is not arbitrary, but is determined by the condition (8. 10).

The result (8. 14) is an integral equation for the determination of the probability distribution function $n^*(\mathbf{x})$; it can be solved by

---
[1]) This was pointed out to the author by Prof. J. E. Mayer.

successive approximations. In first approximation, one may assume

$$(8.15) \qquad n^*(\mathbf{x}^{(i)}) = \delta(\mathbf{x}^{(i)} - \mathbf{x}_0^{(i)})$$

where $\mathbf{x}_0^{(i)}$ is the centre of the cell $\varDelta_i$, and $\delta(\mathbf{x} - \mathbf{x}_0)$ is the δ-function which vanishes when $\mathbf{x}$ is appreciably different from $\mathbf{x}_0$, but is so large near $\mathbf{x}_0$ that $\int \delta(\mathbf{x} - \mathbf{x}_0) d\mathbf{x} = 1$. Substituting this first approximation into the integral of (8. 14), one obtains

$$(8.16) \qquad n^*(\mathbf{x}^{(i)}) = \exp \beta \{a - \sum_{j \neq i} \phi(|\mathbf{x}_0^{(j)} - \mathbf{x}^{(i)}|)\}$$

as the second approximation to $n^*(\mathbf{x})$. The next approximation would be obtained by substituting this second approximation into the integral of (8. 14); and the whole process could be repeated indefinitely to obtain the solution of (8. 14) to any desired degree of accuracy.

All thermodynamical quantities may be derived from the free energy, which results from combining (8. 7), (8. 12) and (8. 14):

$$(8.17) \quad \beta F = N \{\ln(\lambda/\sigma) + \beta a - \tfrac{1}{2} \beta \sum_{j \neq i} \int_i \int_j n(\mathbf{x}^{(i)}) n(\mathbf{x}^{(j)}) \phi^{(ij)} d\mathbf{x}^{(j)} d\mathbf{x}^{(i)}\}.$$

The parameter a which appears in this formula is itself given by the formula

$$(8.18) \qquad \exp(-\beta a) = \int_i \exp\{-\beta \sum_{j \neq i} \int_j n(\mathbf{x}^{(j)}) \phi^{(ij)} d\mathbf{x}^{(j)}\} d\mathbf{x}^{(i)}$$

derived by substituting (8. 14) into (8. 10).

The theory of Lennard–Jones & Devonshire corresponds to the first approximation in the procedure outlined above, and can be obtained immediately by substituting (8. 15) into (8. 17) and (8. 18). Nothing is known concerning σ, so this is treated as a constant. To evaluate the expression (8. 18), the molecule at $\mathbf{x}^{(i)}$ is first kept fixed at a distance r from the centre of its cell, and the average potential energy due to the molecules situated at the centres of the neighbouring cells is determined. If c is the number of these 'nearest neighbours', and a the radius of the sphere on which they are situated, one will then have to the required degree of approximation

$$(8.19) \quad \begin{cases} \psi(r) = \sum_j \int n(\mathbf{x}^{(j)}) \phi^{(ij)} d\mathbf{x}^{(j)} \\ \\ \quad = \tfrac{1}{2} c \int_0^\pi \phi\{(r^2 + a^2 - 2ar\cos\theta)^{\frac{1}{2}}\} \sin\theta \, d\theta. \end{cases}$$

This integral was evaluated by Lennard–Jones & Devonshire, using the form of the potential function $\phi(r)$ suggested by (7.3) of chapter III. The remaining integral

$$\int_0^{a/2} \exp\{-\beta\psi(r)\} \cdot 4\pi r^2\, dr$$

required for the evaluation of the right-hand side of (8.18), and its derivative with respect to the specific volume (which is proportional to a^3), were then computed numerically for various densities, at two temperatures in the neighbourhood of the critical point. The free energy and pressure could then also be determined. The results are similar to those of McLellan described in § 5.

The fundamental merit of the work just mentioned lies not so much in the numerical results already obtained—which are, notwithstanding, already in surprisingly good agreement with experiment —but in the possibility which it offers for approaching the true theoretical values as nearly as required. The second approximation to the density function, given by (8.16), is doubtless considerably better than the first, and would yield an appreciably improved formula for the free energy. The exact determination of the parameter σ remains, of course, an unsolved problem; even this difficulty, however, could probably be met by increasing the number of molecules per cell; and this kind of method certainly offers many possibilities for the future.

CHAPTER V

THE STRUCTURE OF FLUIDS IN MOTION

1. General Considerations

In resuming the study of the structure of fluids it is natural to enquire what conclusions can be drawn from the general theory concerning fluids which are in a state of non-uniform motion. This introduces one of the fundamental questions of rheology, on the nature of viscosity and dissipative processes in general.

Just two fundamental dissipative processes are possible in simple fluids: the conversion of mechanical energy into thermal energy through the action of viscosity, and the conduction of heat. Though it is possible to consider these processes separately, little is gained by doing so, for they are of a very similar nature. Both processes are irreversible; but, whereas viscosity has the tendency to smooth out differences in velocity between different parts of the fluid, thermal conduction has the tendency to smooth out differences in temperature. It has already been seen that the entropy of a system necessarily increases under adiabatic conditions if it is not in thermal and mechanical equilibrium. A disturbance of mechanical equilibrium provokes an increase in entropy through the action of viscosity; and a departure from thermal equilibrium induces an increase in entropy through the agency of thermal conduction. It will be shown that the rate of change of entropy in any region of a fluid is

$$R + \lambda \int \left(\frac{1}{T} \frac{\partial T}{\partial \mathbf{x}}\right)^2 d\mathbf{x} + 2\eta \int \frac{1}{T} \left(\overline{\frac{\partial}{\partial \mathbf{x}} \mathbf{u}}\right)^2 d\mathbf{x} + \eta_0 \int \frac{1}{T} \left(\frac{\partial}{\partial \mathbf{x}} \cdot \mathbf{u}\right)^2 d\mathbf{x}$$

where R represents the rate of transfer of entropy through the boundary of the region; λ is the coefficient of thermal conduction, η is the coefficient of viscosity, η_0 is the coefficient of bulk viscosity, $\frac{\partial T}{\partial \mathbf{x}}$ is the temperature gradient, and $\overline{\frac{\partial}{\partial \mathbf{x}}} \mathbf{u}$ is the non-divergent rate of strain tensor. It is obvious from this that there must be a net increase in entropy, unless the temperature gradient and the rate of strain are both everywhere zero.

It is convenient first to note a few relations of macroscopic fluid mechanics. If m is the molecular mass and n the number density of the molecules at a point $\mathbf{x}$, mn is the mass density at that point. The equation of continuity in hydrodynamics is therefore

$$(1.1) \qquad \frac{\partial}{\partial t}(mn) + \frac{\partial}{\partial \mathbf{x}} \cdot (mn\mathbf{u}) = 0,$$

where $\mathbf{u}$ is the macroscopic velocity at the point $\mathbf{x}$. This may be interpreted as a statement that the increase in the number of molecules in any small region is due only to the flux of molecules through the boundary. Further, the equation of motion of hydrodynamics is

$$(1.2) \qquad mn\left(\frac{\partial}{\partial t}\mathbf{u} + \mathbf{u} \cdot \frac{\partial}{\partial \mathbf{x}}\mathbf{u}\right) + \frac{\partial}{\partial \mathbf{x}} \cdot \boldsymbol{p} = \mathbf{F}$$

where $\boldsymbol{p}$ is the pressure tensor, including the so-called *viscous* stresses, and $\mathbf{F}$ the external force acting on unit volume of the fluid. This states that the rate of change of the momentum in a small region moving with the fluid is due partly to external forces, such as gravity, and partly to the pressure acting across the boundary. Finally there is the general equation of conservation of energy; if U_1 is the *internal* energy of the fluid per molecule at the point $\mathbf{x}$, this may be written in the form

$$(1.3) \qquad n\left(\frac{\partial U_1}{\partial t} + \mathbf{u} \cdot \frac{\partial U_1}{\partial \mathbf{x}}\right) + \frac{\partial}{\partial \mathbf{x}} \cdot \mathbf{q} + \boldsymbol{p} : \left(\frac{\partial}{\partial \mathbf{x}}\mathbf{u}\right) = 0$$

where $\mathbf{q}$ is the flux of thermal energy. This states that the increase in internal energy of a small region moving with the fluid is due partly to the flux of energy across the boundary, and partly to the work done by the pressure.

The laws of viscosity may be stated in the form

$$(1.4) \qquad \boldsymbol{p} - p\boldsymbol{\delta} = -2\eta \overline{\frac{\partial}{\partial \mathbf{x}}\mathbf{u}}, \qquad p = p^0 - \eta_0 \frac{\partial}{\partial \mathbf{x}} \cdot \mathbf{u},$$

where $\boldsymbol{\delta}$ is the unit tensor with elements δ_{ij} equal to 1 when $i = j$ and 0 when $i \neq j$ and p^0 is the hydrostatic pressure; the rate of strain tensor $\overline{\frac{\partial}{\partial \mathbf{x}}\mathbf{u}}$ has the elements

$$(1.5) \qquad \begin{cases} \left(\overline{\frac{\partial}{\partial \mathbf{x}}\mathbf{u}}\right)_{11} = \frac{1}{3}\left(2\frac{\partial u_1}{\partial x_1} - \frac{\partial u_2}{\partial x_2} - \frac{\partial u_3}{\partial x_3}\right) \\ \left(\overline{\frac{\partial}{\partial \mathbf{x}}\mathbf{u}}\right)_{12} = \frac{1}{2}\left(\frac{\partial u_2}{\partial x_1} + \frac{\partial u_1}{\partial x_2}\right), \end{cases}$$

etc. Like $\mathbf{p} - p\boldsymbol{\delta}$, $\overline{\frac{\partial}{\partial \mathbf{x}} \mathbf{u}}$ has the property of being non-divergent, i.e., the sum of its diagonal elements is zero. Then the equation (1.4) asserts that the non-divergent parts of the pressure tensor and the rate of strain tensor are proportional to one another, and defines the coefficient of viscosity as one half of the quotient of the two tensors. When the motion is steady and confined to the x_1-direction, this definition reduces to the elementary one expressed by the equation $p_{12} = -\eta \frac{\partial u_1}{\partial x_2}$. The law of thermal conduction assumes the simpler form

$$(1.6) \qquad \mathbf{q} = -\lambda \frac{\partial T}{\partial \mathbf{x}}.$$

This asserts that the thermal flux is proportional to the temperature gradient, and defines the coefficient of thermal conduction as the quotient of the two vectors. One of the results of molecular theory will be to provide a proof of these laws, and expressions for the coefficients of viscosity and thermal conduction in terms of the molecular distribution functions.

The total, or *convective* derivative with respect to time of any quantity $c(t, \mathbf{x})$ depending on time and also on position in the fluid is defined by

$$(1.7) \qquad \frac{dc}{dt} = \frac{\partial c}{\partial t} + \mathbf{u} \cdot \frac{\partial c}{\partial \mathbf{x}}.$$

With this notation, (1.1) and (1.3) can be written

$$(1.8) \qquad \begin{cases} \dfrac{dn}{dt} + n \dfrac{\partial}{\partial \mathbf{x}} \cdot \mathbf{u} = 0, \\ n \dfrac{dU_1}{dt} + \dfrac{\partial}{\partial \mathbf{x}} \cdot \mathbf{q} + \boldsymbol{p} : \left(\dfrac{\partial}{\partial \mathbf{x}} \mathbf{u} \right) = 0. \end{cases}$$

If S_1 is the entropy per molecule, one has, according to the fundamental law of thermodynamics,

$$(1.9) \qquad T \frac{dS_1}{dt} = \frac{dU_1}{dt} + p^0 \frac{d}{dt} \left(\frac{1}{n} \right),$$

since n^{-1} is the volume per molecule. The rate of change of entropy in any region moving with the fluid is therefore

$$(1.10) \quad \begin{cases} \int n \dfrac{dS_1}{dt} d\mathbf{x} = \int \left(\dfrac{n}{T} \dfrac{dU_1}{dt} - \dfrac{p^0}{nT} \dfrac{dn}{dt} \right) d\mathbf{x} \\ \qquad = -\int \dfrac{1}{T} \dfrac{\partial}{\partial \mathbf{x}} \cdot \mathbf{q} \, d\mathbf{x} - \int \dfrac{1}{T} \left(\boldsymbol{p} : \dfrac{\partial}{\partial \mathbf{x}} \mathbf{u} - p^0 \dfrac{\partial}{\partial \mathbf{x}} \cdot \mathbf{u} \right) d\mathbf{x}, \end{cases}$$

with the use of (1.8). The integration in each term extends over the region considered; the first term can be transformed by Gauss' theorem to the sum of $-\int \frac{1}{T^2} \mathbf{q} \cdot \frac{\partial T}{\partial \mathbf{x}} d\mathbf{x}$ and a surface integral over the boundary of the region. Using (1.4) and (1.6), one then obtains

$$(1.11) \quad \begin{cases} \int n \frac{dS_1}{dt} d\mathbf{x} = -\int \frac{1}{T} \mathbf{q} \cdot d\mathbf{S} + \lambda \int \left(\frac{1}{T} \frac{\partial T}{\partial \mathbf{x}}\right)^2 d\mathbf{x} + \\ + 2\eta \int \frac{1}{T} \overline{\left(\frac{\partial}{\partial \mathbf{x}} \mathbf{u}\right)^2} d\mathbf{x} + \eta_0 \int \frac{1}{T} \left(\frac{\partial}{\partial \mathbf{x}} \cdot \mathbf{u}\right)^2 d\mathbf{x}. \end{cases}$$

This is the result which was to be proved, for the first term clearly represents the contribution of the flux of entropy through the boundary.

The result just derived is very useful, for it shows that if one can compute the rate of change of entropy for an arbitrary state of the system, one will immediately have the observed coefficients of viscosity and thermal conduction in this state. Now, it is clear from (1.9) that the rate of change of entropy depends essentially on the rate of change of the internal energy U_1; the hydrostatic pressure p^0 is already known from § 4 of chapter II, and the rate of change of density is given by (1.8). The internal energy per molecule can be inferred from the formula (4.11) of chapter II; it is simply

$$(1.12) \quad U_1 = {}^3/_2 kT + \frac{1}{n} \int n_2 \phi \, d\mathbf{x}^{(2)}.$$

To determine the rate of change of internal energy, it is therefore necessary to study the change in time of the temperature T, and the molecular distribution function n_2.

The above argument shows that viscosity and thermal conduction, like so many equilibrium properties, depend on the structure of the fluid, described in terms of the molecular distribution, but also on the way in which this structure varies with time. This means that under non-uniform conditions it is necessary to examine not only the average configuration of the other molecules relative to a given molecule, but also their average velocities. If $f_2 d\boldsymbol{\xi}^{(1)} d\boldsymbol{\xi}^{(2)} d\mathbf{x}^{(1)} d\mathbf{x}^{(2)}$ is the probability of finding molecules with velocities $\boldsymbol{\xi}^{(1)}$ and $\boldsymbol{\xi}^{(2)}$ in the volume elements $d\mathbf{x}^{(1)}$, $d\mathbf{x}^{(2)}$, the average velocity of the first molecule (at $\mathbf{x}^{(1)}$) when nothing is known concerning the velocity of the second (at $\mathbf{x}^{(2)}$) is given by

$$(1.13) \quad n_2 \mathbf{u}_2^{(1)} = \iint f_2 \boldsymbol{\xi}^{(1)} \, d\boldsymbol{\xi}^{(1)} \, d\boldsymbol{\xi}^{(2)}.$$

In a non-uniform fluid this is generally different from the average velocity **u** determined when it is not known whether there is another molecule at $\mathbf{x}^{(2)}$. The difference, as subsequent developments will show, is closely associated with the process of thermal conduction in fluids; but, before any quantitative results can be obtained, one must examine the behaviour in time of the molecular distributions associated with non-uniform fluids.

2. The Evolution of Molecular Distributions

In this section the fundamental equations satisfied by the molecular distribution functions will be derived. These constitute a simple generalization of Boltzmann's equation in the kinetic theory of gases (cf. CHAPMAN & COWLING [1939] Ch. 3). The method followed here is substantially that adopted by BORN & GREEN [1946]; an alternative derivation of the same results has been given by KIRKWOOD [1946]. The object is to find a relation between the values at two slightly different times of the molecular distribution function f_q, which specifies the probability $f_q \, d\boldsymbol{\xi}^{(1)} \ldots d\boldsymbol{\xi}^{(q)} d\mathbf{x}^{(1)} \ldots d\mathbf{x}^{(q)}$ of finding q molecules with velocities $\boldsymbol{\xi}^{(1)}, \ldots \boldsymbol{\xi}^{(q)}$ respectively in the volume elements $d\mathbf{x}^{(1)}, \ldots d\mathbf{x}^{(q)}$. When this is obtained, a differential equation for the distributions follows at once.

The general principle involved may be stated as follows: if two events A and B at different times are causally related, so that the occurrence of the event A at time t_A implies the occurrence of the event t_B at time B, and *vice versa*, then the separate probabilities p_A and p_B for the occurrence of the two events are equal. This follows naturally from the fact that such pairs of events are in one-to-one correspondence.

If there is a molecule with velocity $\boldsymbol{\xi}$ at the point **x** at time t, then at some slightly previous time $t - \delta t$, there must have existed a molecule with velocity $\boldsymbol{\xi} - \boldsymbol{\eta} \delta t$ at the point $\mathbf{x} - \boldsymbol{\xi} \delta t$ — for *some* value of the acceleration $\boldsymbol{\eta}$. The two events are in one-to-one correspondence; consequently their relative probabilities must be the same. To formulate this mathematically, let $g_1(\boldsymbol{\eta}, \boldsymbol{\xi}, \mathbf{x}, t) d\boldsymbol{\eta} \, d\boldsymbol{\xi} \, d\mathbf{x}$ represent the probability at time t of finding a molecule with acceleration $\boldsymbol{\eta}$, $d\boldsymbol{\eta}$ and velocity $\boldsymbol{\xi}$, $d\boldsymbol{\xi}$ in the volume element $d\mathbf{x}$ at the point **x**.

Then the probability that at time $t - \delta t$ — for some value of the acceleration $\boldsymbol{\eta}$ — there existed a molecule, with velocity $\boldsymbol{\xi} - \boldsymbol{\eta} \delta t$, $d\boldsymbol{\xi}$

in the volume element $d\mathbf{x}$ at the point $\mathbf{x} - \boldsymbol{\xi}\delta t$, is

$$\{\int g_1(\boldsymbol{\eta}, \boldsymbol{\xi} - \boldsymbol{\eta}\, \delta t, \mathbf{x} - \boldsymbol{\xi}\, \delta t, t - \delta t)\, d\boldsymbol{\eta}\}\, d\boldsymbol{\xi}\, d\mathbf{x}.$$

According to the causal principle enunciated above, this must be the same as the probability that, at time t, there exists a molecule, with velocity $\boldsymbol{\xi}$, $d\boldsymbol{\xi}$ in the volume element $d\mathbf{x}$ at the point $\mathbf{x}$. Hence

(2. 1) $\qquad f_1(\boldsymbol{\xi}, \mathbf{x}, t) = \int g_1(\boldsymbol{\eta}, \boldsymbol{\xi} - \boldsymbol{\eta}\, \delta t, \mathbf{x} - \boldsymbol{\xi}\, \delta t, t - \delta t)\, d\boldsymbol{\eta}.$

It follows directly, from the fact that the acceleration of a molecule must have *some* value, that

(2. 2) $\qquad f_1(\boldsymbol{\xi}, \mathbf{x}, t) = \int g_1(\boldsymbol{\eta}, \boldsymbol{\xi}, \mathbf{x}, t)\, d\boldsymbol{\eta}.$

This equation can also be obtained from (2. 1) by setting $\delta t = 0$. Subtracting it from (2. 1), one obtains

$$0 = \int \left(\boldsymbol{\eta} \cdot \frac{\partial g_1}{\partial \boldsymbol{\xi}} + \boldsymbol{\xi} \cdot \frac{\partial g_1}{\partial \mathbf{x}} + \frac{\partial g_1}{\partial t}\right) d\boldsymbol{\eta}\, \delta t,$$

since the square of the small interval δt is negligible. With the help of (2. 2), this result can be written in the form

(2. 3) $\qquad \begin{cases} \dfrac{\partial f_1}{\partial t} + \boldsymbol{\xi} \cdot \dfrac{\partial f_1}{\partial \mathbf{x}} + \dfrac{\partial}{\partial \boldsymbol{\xi}} \cdot (f_1\, \overline{\boldsymbol{\eta}}) = 0, \\ f_1\, \overline{\boldsymbol{\eta}} = \int g_1 \boldsymbol{\eta}\, d\boldsymbol{\eta}, \end{cases}$

in which the quantity $\overline{\boldsymbol{\eta}}$ can obviously be interpreted as the mean acceleration, at time t, of a molecule with velocity $\boldsymbol{\xi}$ at the point $\mathbf{x}$.

This is the fundamental differential equation satisfied by the velocity distribution function f_1; it is an exact form of Boltzmann's equation valid for both gases and liquids. A similar equation can be proved in precisely the same way for the molecular distribution function f_q relating to q molecules; it is

(2. 4) $\qquad \dfrac{\partial f_q}{\partial t} + \sum\limits_{i=1}^{q} \boldsymbol{\xi}^{(i)} \cdot \dfrac{\partial f_q}{\partial \mathbf{x}^{(i)}} + \sum\limits_{i=1}^{q} \dfrac{\partial}{\partial \boldsymbol{\xi}^{(i)}} \cdot (f_q\, \overline{\boldsymbol{\eta}_q^{(i)}}) = 0,$

where

(2. 5) $\qquad \overline{\boldsymbol{\eta}_q^{(i)}} = f_q^{-1} \int \overset{(q)}{\ldots} \int g_q\, \boldsymbol{\eta}^{(i)}\, d\boldsymbol{\eta}^{(1)} \ldots d\boldsymbol{\eta}^{(q)}$

represents the average acceleration of the molecule (i) in a group of q molecules whose positions and velocities are all given.

The mean acceleration of a molecule can be computed from a knowledge of the forces acting on it. Naturally these forces include external forces, such as gravity, but the most important are the attractions and repulsions of the other molecules. The mean force acting on any molecule due to the action of the others can be derived with the help of the distribution function f_2. For, the probability that the volume element $d\mathbf{x}^{(2)}$ is occupied by a molecule with velocity in the range $\boldsymbol{\xi}^{(2)}, d\boldsymbol{\xi}^{(2)}$, when it is already known that the volume element $d\mathbf{x}^{(1)}$ is occupied by a molecule with velocity in the range $\boldsymbol{\xi}^{(1)}, d\boldsymbol{\xi}^{(1)}$, is

$$f_2(\boldsymbol{\xi}^{(1)}, \boldsymbol{\xi}^{(2)}, \mathbf{x}^{(1)}, \mathbf{x}^{(2)}, t)\, d\boldsymbol{\xi}^{(2)}\, d\mathbf{x}^{(2)}/f_1(\boldsymbol{\xi}^{(1)}, \mathbf{x}^{(1)}, t),$$

and the force exerted by a molecule at $\mathbf{x}^{(2)}$ on the molecule at $\mathbf{x}^{(1)}$ is $-\dfrac{\partial \phi^{(12)}}{\partial \mathbf{x}^{(1)}}$, where $\phi^{(12)}$ is the mutual potential energy of two molecules at the distance $|\mathbf{x}^{(1)} - \mathbf{x}^{(2)}|$ from one another. Hence the average force on a molecule with velocity $\boldsymbol{\xi}^{(1)}$ at the point $\mathbf{x}^{(1)}$ is

$$-\frac{1}{f_1} \iint f_2 \frac{\partial \phi^{(12)}}{\partial \mathbf{x}^{(1)}} \, d\mathbf{x}^{(2)} \, d\boldsymbol{\xi}^{(2)}.$$

Assuming that the external force on the same molecule is $\mathbf{F}$, the mean acceleration will therefore be

(2. 6) $$\overline{\boldsymbol{\eta}} = \frac{1}{mf_1}\left(-\iint f_2 \frac{\partial \phi}{\partial \mathbf{x}^{(1)}}\, d\mathbf{x}^{(2)}\, d\boldsymbol{\xi}^{(2)} + f_1 \mathbf{F}\right).$$

If this result is substituted into the equation (2. 3), the result is

(2. 7) $$\frac{\partial f_1}{\partial t} + \boldsymbol{\xi}^{(1)} \cdot \frac{\partial f_1}{\partial \mathbf{x}^{(1)}} + \frac{1}{m} \mathbf{F} \cdot \frac{\partial f_1}{\partial \boldsymbol{\xi}^{(1)}} = \frac{1}{m} \iint \frac{\partial f_2}{\partial \boldsymbol{\xi}^{(1)}} \cdot \frac{\partial \phi}{\partial \mathbf{x}^{(1)}}\, d\mathbf{x}^{(2)}\, d\boldsymbol{\xi}^{(2)}.$$

Unlike (2. 3), this equation contains no reference to the acceleration distribution function g_1; it does, however, involve the velocity distribution function f_2, for which no closed equation has yet been derived.

To obtain the general equation which determines f_q, one requires the value of the mean acceleration of a molecule in a group of other molecules, which appears in (2. 4). This can be obtained by a simple and obvious extension of the argument already used to derive (2. 6). One finds that the mean acceleration of the molecule (i) in a group of q molecules with positions $\mathbf{x}^{(1)}, \mathbf{x}^{(2)}, \ldots, \mathbf{x}^{(q)}$ and velocities $\boldsymbol{\xi}^{(1)}, \boldsymbol{\xi}^{(2)}, \ldots, \boldsymbol{\xi}^{(q)}$ respectively is

(2. 8) $$\overline{\boldsymbol{\eta}_q^{(i)}} = \frac{1}{mf_q}\left(-f_q \frac{\partial \Phi_q}{\partial \mathbf{x}^{(i)}} - \iint f_{q+1} \frac{\partial \phi^{(i,q+1)}}{\partial \mathbf{x}^{(i)}}\, d\mathbf{x}^{(q+1)}\, d\boldsymbol{\xi}^{(q+1)} + f_q \mathbf{F}^{(i)}\right),$$

where

$$\Phi_q = \tfrac{1}{2} \sum_{i,j=1}^{q} \phi^{(ij)}$$

is the mutual potential energy of the group of q molecules. If this formula is substituted into (2. 4), the result is

$$(2.9) \quad \left\{ \begin{array}{l} \dfrac{\partial f_q}{\partial t} + \sum\limits_{i=1}^{q} \left\{ \boldsymbol{\xi}^{(i)} \cdot \dfrac{\partial f_q}{\partial \mathbf{x}^{(i)}} + \dfrac{1}{m} \left(\mathbf{F}^{(i)} - \dfrac{\partial \Phi_q}{\partial \mathbf{x}^{(i)}} \right) \cdot \dfrac{\partial f_q}{\partial \boldsymbol{\xi}^{(i)}} \right\} \\ \qquad = \sum\limits_{i=1}^{q} \dfrac{1}{m} \int\!\!\int \dfrac{\partial f_{q+1}}{\partial \boldsymbol{\xi}^{(i)}} \cdot \dfrac{\partial \phi^{(i,q+1)}}{\partial \mathbf{x}^{(i)}} \, d\mathbf{x}^{(q+1)} \, d\boldsymbol{\xi}^{(q+1)}. \end{array} \right.$$

This is the generalization for a cluster of q molecules in the fluid of the equation (2. 7). It provides a connection between f_2 and f_3, f_3 and $f_4, \ldots, f_{N-1}$ and f_N. Finally, for $q = N$ it reduces to

$$(2.10) \quad \dfrac{\partial f_N}{\partial t} + \sum\limits_{i=1}^{N} \left\{ \boldsymbol{\xi}^{(i)} \cdot \dfrac{\partial f_N}{\partial \mathbf{x}^{(i)}} + \dfrac{1}{m} \left(\mathbf{F}^{(i)} - \dfrac{\partial \Phi}{\partial \mathbf{x}^{(i)}} \right) \cdot \dfrac{\partial f_N}{\partial \boldsymbol{\xi}^{(i)}} \right\} = 0,$$

since the probability f_{N+1} of finding $N + 1$ molecules in any positions whatsoever vanishes.

The equation (2. 10) was the starting point of the theory developed by Kirkwood. The earlier equations (2. 9) can be obtained from it by integrating first over $\mathbf{x}^{(N)}$ and $\boldsymbol{\xi}^{(N)}$, then over $\mathbf{x}^{(N-1)}$ and $\boldsymbol{\xi}^{(N-1)}$, etc. For example, in performing the first integration, one has [cf. (4. 6) of chapter III]

$$\int\!\!\int f_N \, d\mathbf{x}^{(N)} \, d\boldsymbol{\xi}^{(N)} = f_{N-1}$$

so, to derive (2. 9) with $q = N - 1$ from (2. 10), one needs only to show that the integrals

(i) $\quad \int \dfrac{\partial f_N}{\partial \boldsymbol{\xi}^{(N)}} \, d\boldsymbol{\xi}^{(N)}$, (ii) $\quad \int\!\!\int \dfrac{\partial f_N}{\partial \mathbf{x}^{(N)}} \cdot \boldsymbol{\xi}^{(N)} \, d\mathbf{x}^{(N)} \, d\boldsymbol{\xi}^{(N)}$,

vanish. This is easily done by the application of Gauss' theorem (cf. § 4 of chapter I). On transformation to a 'surface' integral, (i) immediately vanishes because the probability of finding a molecule with an infinite velocity is zero; and (ii) becomes $\int\!\!\int f_N \boldsymbol{\xi}^{(N)} \cdot \mathbf{dS}^{(N)} d\boldsymbol{\xi}^{(N)}$, where $\mathbf{dS}^{(N)}$ is an element of the boundary of the region occupied by the fluid. This second integral may therefore be written $\int n_N \overline{\boldsymbol{\xi}_N^{(N)}} \cdot \mathbf{dS}^{(N)}$, where $\overline{\boldsymbol{\xi}_N^{(N)}}$ represents the average velocity of the Nth molecule at a position on the boundary; and, since this mean velocity must be parallel to the boundary at every point, (ii) is also zero.

3. The Theory of Flow

The phenomenon of flow, whether regarded from the macroscopic or from the molecular point of view, is controlled by the requirements of the laws of conservation relating to mass, momentum, and energy. The fact that these laws must assume the same form in macroscopic hydrodynamics and in the molecular theory enables one to study the relations between the motion of the molecules and the fluid as a whole.

From the molecular point of view, the state of motion of the fluid is described by the velocity distribution function (f_1) and its structure by the molecular distribution function (n_2); it will be necessary to study the variation of these distribution functions with time. In the absence of external forces, the velocity distribution function changes according to the equation

$$(3.\ 1) \qquad \frac{\partial f_1}{\partial t} + \boldsymbol{\xi}^{(1)} \cdot \frac{\partial f_1}{\partial \mathbf{x}^{(1)}} = \frac{1}{m} \int\int \frac{\partial f_2}{\partial \boldsymbol{\xi}^{(1)}} \cdot \frac{\partial \phi}{\partial \mathbf{x}^{(1)}}\ d\mathbf{x}^{(2)}\ d\boldsymbol{\xi}^{(2)},$$

which is obtained from (2.7) simply by setting $\mathbf{F} = 0$. When the velocity distribution is changing, the number density (n) in general changes also, at a rate which is obtained by integrating the equation (3.1) over the entire range of velocities. One has $\int f_1 d\boldsymbol{\xi}^{(1)} = n$; also $\int f_1 \boldsymbol{\xi}^{(1)} d\boldsymbol{\xi}^{(1)} = n\mathbf{u}$, where $\mathbf{u}$ is the mean velocity of the molecules at the point $\mathbf{x}^{(1)}$; and, by an application of Gauss' theorem, $\int \frac{\partial f_2}{\partial \boldsymbol{\xi}^{(1)}} d\boldsymbol{\xi}^{(1)} = 0$ because there are no molecules with infinite velocities. Hence

$$(3.\ 2) \qquad \frac{\partial n}{\partial t} + \frac{\partial}{\partial \mathbf{x}^{(1)}} \cdot (n\,\mathbf{u}) = 0.$$

This is the equation of continuity for the fluid, which, in essence, asserts that the total number of molecules remains constant. If it is multiplied by the molecular mass (m), it may also be interpreted in a macroscopic sense, for mn is the mass density, and $\mathbf{u}$ the macroscopic velocity of flow. In this sense, it implies the conservation of mass.

The equation of continuity (3.2) can also be written in the form

$$(3.\ 3) \qquad \frac{dn}{dt} + n\,\frac{\partial}{\partial \mathbf{x}^{(1)}} \cdot \mathbf{u} = 0,$$

if $\frac{d}{dt}$ is understood to mean the 'time derivative following the motion', given by

$$(3.\ 4) \qquad \frac{d}{dt} = \frac{\partial}{\partial t} + \mathbf{u} \cdot \frac{\partial}{\partial \mathbf{x}^{(1)}}.$$

A continuity equation similar to (3.2) can be stated, giving the rate of change with time of the molecular distribution function, which describes the local molecular structure. This equation is obtained, not from (3.1), but from the analogous equation

(3.5)
$$\begin{cases} \dfrac{\partial f_2}{\partial t} + \boldsymbol{\xi}^{(1)} \cdot \dfrac{\partial f_2}{\partial \mathbf{x}^{(1)}} + \boldsymbol{\xi}^{(2)} \cdot \dfrac{\partial f_2}{\partial \mathbf{x}^{(2)}} + \dfrac{1}{m}\left(\dfrac{\partial \phi}{\partial \mathbf{x}^{(1)}} \cdot \dfrac{\partial f_2}{\partial \boldsymbol{\xi}^{(1)}} + \dfrac{\partial \phi}{\partial \mathbf{x}^{(2)}} \cdot \dfrac{\partial f_2}{\partial \boldsymbol{\xi}^{(2)}}\right) \\ = \dfrac{1}{m} \iint \left(\dfrac{\partial f_3}{\partial \boldsymbol{\xi}^{(1)}} \cdot \dfrac{\partial \phi^{(13)}}{\partial \mathbf{x}^{(1)}} + \dfrac{\partial f_3}{\partial \boldsymbol{\xi}^{(2)}} \cdot \dfrac{\partial \phi^{(23)}}{\partial \mathbf{x}^{(2)}}\right) d\mathbf{x}^{(3)} \, d\boldsymbol{\xi}^{(3)} \end{cases}$$

referring to a pair of molecules, which is a specialization of (2.9). If this is integrated over the entire range of each of the velocities $\boldsymbol{\xi}^{(1)}$ and $\boldsymbol{\xi}^{(2)}$, the result is

(3.6)
$$\frac{\partial n_2}{\partial t} + \frac{\partial}{\partial \mathbf{x}^{(1)}} \cdot (n_2 \mathbf{u}_2^{(1)}) + \frac{\partial}{\partial \mathbf{x}^{(2)}} \cdot (n_2 \mathbf{u}_2^{(2)}) = 0,$$

where $\mathbf{u}_2^{(1)}$ and $\mathbf{u}_2^{(2)}$ are defined by

(3.7)
$$\begin{cases} n_2 \mathbf{u}_2^{(1)} = \iint f_2 \, \boldsymbol{\xi}^{(1)} \, d\boldsymbol{\xi}^{(1)} \, d\boldsymbol{\xi}^{(2)}, \\ n_2 \mathbf{u}_2^{(2)} = \iint f_2 \, \boldsymbol{\xi}^{(2)} \, d\boldsymbol{\xi}^{(1)} \, d\boldsymbol{\xi}^{(2)}, \end{cases}$$

and depend on both $\mathbf{x}^{(1)}$ and $\mathbf{x}^{(2)}$; $\mathbf{u}_2^{(1)}$, for example, represents the mean velocity of a molecule at the point $\mathbf{x}^{(1)}$ when a second molecule is located at the point $\mathbf{x}^{(2)}$.

It is also possible to obtain the rate of change of the macroscopic velocity $\mathbf{u}$, directly from the equation (3.1), by multiplying with $\boldsymbol{\xi}^{(1)}$, and then integrating over the entire velocity range. Since, by a second application of Gauss' theorem,

$$\iiint \boldsymbol{\xi}^{(1)} \frac{\partial}{\partial \boldsymbol{\xi}^{(1)}} \cdot \left(f_2 \frac{\partial \phi}{\partial \mathbf{x}^{(1)}}\right) d\mathbf{x}^{(2)} \, d\boldsymbol{\xi}^{(1)} \, d\boldsymbol{\xi}^{(2)} = - \iiint f_2 \frac{\partial \phi}{\partial \mathbf{x}^{(1)}} \, d\mathbf{x}^{(2)} \, d\boldsymbol{\xi}^{(1)} \, d\boldsymbol{\xi}^{(2)}$$
$$= - \int n_2 \frac{\partial \phi}{\partial \mathbf{x}^{(1)}} \, d\mathbf{x}^{(2)},$$

the result is

$$\frac{\partial}{\partial t}(n\mathbf{u}) + \frac{\partial}{\partial \mathbf{x}^{(1)}} \cdot \int f_1 \boldsymbol{\xi}^{(1)} \boldsymbol{\xi}^{(1)} \, d\boldsymbol{\xi}^{(1)} = -\frac{1}{m} \int n_2 \frac{\partial \phi}{\partial \mathbf{x}^{(1)}} \, d\mathbf{x}^{(2)}.$$

To obtain the macroscopic equation of motion, it is necessary to subtract the equation

$$\mathbf{u}\frac{\partial n}{\partial t} - n\mathbf{u} \cdot \frac{\partial}{\partial \mathbf{x}^{(1)}} \mathbf{u} + \frac{\partial}{\partial \mathbf{x}^{(1)}} \cdot (n\,\mathbf{u}\,\mathbf{u}) = 0,$$

obtained simply by multiplying (3.2) with the vector $\mathbf{u}$. Then one has

(3.8) $\quad n \dfrac{d}{dt} \mathbf{u} + \dfrac{\partial}{\partial \mathbf{x}^{(1)}} \cdot \int f_1 (\boldsymbol{\xi}^{(1)} \boldsymbol{\xi}^{(1)} - \mathbf{u}\mathbf{u}) \, d\boldsymbol{\xi}^{(1)} + \dfrac{1}{m} \int n_2 \dfrac{\partial \phi}{\partial \mathbf{x}^{(1)}} \, d\mathbf{x}^{(2)} = 0,$

where $\frac{d}{dt}$ is again the 'derivative following the motion' defined by (3. 4). Now, it has been seen already in § 4 of chapter II that

$$\int n_2 \frac{\partial \phi}{\partial \mathbf{x}^{(1)}} d\mathbf{x}^{(2)} = -\tfrac{1}{2} \frac{\partial}{\partial \mathbf{x}^{(1)}} \cdot \int \bar{n}_2(\mathbf{r}, \mathbf{x}^{(1)}) \phi'(r) r^{-1} \mathbf{rr} \, d\mathbf{r},$$

where $\bar{n}_2(\mathbf{r}, \mathbf{x})$ is the molecular distribution function expressed in terms of the relative position $\mathbf{r} = \mathbf{x}^{(2)} - \mathbf{x}^{(1)}$ and the mean centre $\mathbf{x} = \tfrac{1}{2}(\mathbf{x}^{(1)} + \mathbf{x}^{(2)})$ of the two molecules. Hence the equation (3. 8) can be written

$$(3.\ 9) \qquad mn \frac{d}{dt} \mathbf{u} + \frac{\partial}{\partial \mathbf{x}} \cdot \boldsymbol{p} = 0,$$

where $\boldsymbol{p}$ is the tensor

$$(3.\ 10) \qquad \boldsymbol{p} = m \int f_1 (\boldsymbol{\xi}^{(1)} \boldsymbol{\xi}^{(1)} - \mathbf{u}\mathbf{u}) \, d\boldsymbol{\xi}^{(1)} - \tfrac{1}{2} \int \bar{n}_2 \, \phi'(r) \, r^{-1} \mathbf{rr} \, d\mathbf{r}.$$

The result (3. 9) will be recognized as the equation of motion of ordinary hydrodynamics, where $\boldsymbol{p}$ is identified with the pressure tensor. It has, in fact, been seen already in § 4 of chapter II that the tensor defined by (3. 10) must be the macroscopic pressure tensor. The connection between the motion of the individual molecules and that of the fluid as a whole is now revealed as a purely statistical one, and the hydrodynamical equations as a statistical consequence of the laws of motion of the molecules.

The equation of motion (3. 9) may be regarded in essence as a statement of the law of conservation of momentum. The remaining law — that relating to the conservation of energy — can similarly be expressed in the equation of transport of energy, which will now be investigated. First, by scalar multiplication of (3. 9) with the macroscopic velocity ($\mathbf{u}$), one obtains

$$(3.\ 11) \qquad \tfrac{1}{2} mn \frac{d}{dt} \mathbf{u}^2 + \mathbf{u} \cdot \left(\frac{\partial}{\partial \mathbf{x}^{(1)}} \cdot \boldsymbol{p} \right) = 0,$$

showing that the macroscopic kinetic energy of any small mass of the fluid varies as the rate of working of the pressure forces. The variation of the internal energy of a small mass of the fluid, on the other hand, is partly dependent on the flux of internal energy from the surrounding region, and partly on the internal energy generated by the action of the hydrostatic pressure and the viscous forces. This aspect of the conservation of energy is described by the equation

$$(3.\ 12) \qquad n \frac{dU_1}{dt} + \frac{\partial}{\partial \mathbf{x}^{(1)}} \cdot \mathbf{q} + \boldsymbol{p} : \left(\frac{\partial}{\partial \mathbf{x}^{(1)}} \mathbf{u} \right) = 0$$

wherein U_1 represents the internal energy per molecule, and $\mathbf{q}$ is the thermal flux vector; this is simply a restatement of the equation (1. 3).

The mean total energy of a molecule situated at the point $\mathbf{x}^{(1)}$ is

(3. 13) $$E_1 = U_1 + \tfrac{1}{2} m \mathbf{u}^2;$$

its rate of change is obtained by the addition of the conservation equations (3. 11) and (3. 12), thus:

(3. 14) $$n \frac{dE_1}{dt} + \frac{\partial}{\partial \mathbf{x}^{(1)}} \cdot (\mathbf{q} + \boldsymbol{p} \cdot \mathbf{u}) = 0.$$

This form of the energy conservation equation will now be deduced on the basis of the molecular theory of flow; incidentally an expression for the thermal flux vector ($\mathbf{q}$) will be derived, analogous to the expression (3. 10) for the pressure tensor.

It may be observed at the outset that the explicit formula for the mean total energy of a molecule at the point $\mathbf{x}^{(1)}$ is

(3. 15) $$E_1 = n^{-1} \{ \int f_1 \tfrac{1}{2} m \boldsymbol{\xi}^{(1)2} \, d\boldsymbol{\xi}^{(1)} + \tfrac{1}{2} \int n_2 \phi \, d\mathbf{x}^{(2)} \},$$

the first integral representing the kinetic energy, and the second the potential energy. Hence

$$\frac{\partial}{\partial t} (n E_1) = \tfrac{1}{2} m \int \frac{\partial f_1}{\partial t} \boldsymbol{\xi}^{(1)2} \, d\boldsymbol{\xi}^{(1)} + \tfrac{1}{2} \int \frac{\partial n_2}{\partial t} \phi \, d\mathbf{x}^{(2)}.$$

The first term on the right-hand side can be calculated with the help of (3. 1), and the second with the help of (3. 6). Since

$$\iint \boldsymbol{\xi}^{(1)2} \frac{\partial f_2}{\partial \boldsymbol{\xi}^{(1)}} \, d\boldsymbol{\xi}^{(1)} \, d\boldsymbol{\xi}^{(2)} = -2 \iint f_2 \boldsymbol{\xi}^{(1)} \, d\boldsymbol{\xi}^{(1)} \, d\boldsymbol{\xi}^{(2)} = -2 n_2 \mathbf{u}_2^{(1)},$$

one obtains from (3. 1)

$$-\tfrac{1}{2} m \int \frac{\partial f_1}{\partial t} \boldsymbol{\xi}^{(1)2} \, d\boldsymbol{\xi}^{(1)} = \tfrac{1}{2} m \frac{\partial}{\partial \mathbf{x}^{(1)}} \cdot \int f_1 \boldsymbol{\xi}^{(1)} \boldsymbol{\xi}^{(1)2} \, d\boldsymbol{\xi}^{(1)} + \int n_2 \mathbf{u}_2^{(1)} \cdot \frac{\partial \phi}{\partial \mathbf{x}^{(1)}} \, d\mathbf{x}^{(2)}.$$

Also, from (3. 6),

$$-\int \frac{\partial n_2}{\partial t} \phi \, d\mathbf{x}^{(2)} = \int \left\{ \frac{\partial}{\partial \mathbf{x}^{(1)}} \cdot (n_2 \mathbf{u}_2^{(1)}) + \frac{\partial}{\partial \mathbf{x}^{(2)}} \cdot (n_2 \mathbf{u}_2^{(2)}) \right\} \phi \, d\mathbf{x}^{(2)}$$

$$= \frac{\partial}{\partial \mathbf{x}^{(1)}} \cdot \int n_2 \mathbf{u}_2^{(1)} \phi \, d\mathbf{x}^{(2)} + \int n_2 (\mathbf{u}_2^{(2)} - \mathbf{u}_2^{(1)}) \cdot \frac{\partial \phi}{\partial \mathbf{x}^{(1)}} \, d\mathbf{x}^{(2)}.$$

Hence

$$\frac{\partial}{\partial t} (n E_1) + \frac{\partial}{\partial \mathbf{x}^{(1)}} \cdot \{ \tfrac{1}{2} m \int f_1 \boldsymbol{\xi}^{(1)} \boldsymbol{\xi}^{(1)2} \, d\boldsymbol{\xi}^{(1)} + \tfrac{1}{2} \int n_2 \mathbf{u}_2^{(1)} \phi \, d\mathbf{x}^{(2)} \} +$$
$$+ \tfrac{1}{2} \int n_2 (\mathbf{u}_2^{(1)} + \mathbf{u}_2^{(2)}) \cdot \frac{\partial \phi}{\partial \mathbf{x}^{(1)}} \, d\mathbf{x}^{(2)} = 0.$$

Subtracting the equation

$$E_1 \frac{\partial n}{\partial t} - n\mathbf{u} \cdot \frac{\partial E_1}{\partial \mathbf{x}^{(1)}} + \frac{\partial}{\partial \mathbf{x}^{(1)}} \cdot (n\,\mathbf{u}\,E_1) = 0,$$

which is obtained by multiplying (3. 2) by E_1, one has

$$(3.16) \begin{cases} n\dfrac{dE_1}{dt} + \dfrac{\partial}{\partial \mathbf{x}^{(1)}} \cdot \{\tfrac{1}{2} m \int f_1 (\boldsymbol{\xi}^{(1)} - \mathbf{u}) \boldsymbol{\xi}^{(1)2} \, d\boldsymbol{\xi}^{(1)} + \tfrac{1}{2} \int n_2 (\mathbf{u}_2^{(1)} - \mathbf{u}) \phi \, d\mathbf{x}^{(2)}\} + \\ \qquad\qquad + \tfrac{1}{2} \int n_2 (\mathbf{u}_2^{(1)} + \mathbf{u}_2^{(2)}) \cdot \dfrac{\partial \phi}{\partial \mathbf{x}^{(1)}} \, d\mathbf{x}^{(2)} = 0. \end{cases}$$

In order to make a comparison of this equation with the corresponding macroscopic equation (3. 14), it is necessary to express the last term on the left-hand side as the divergence of a vector. This is done in the following way.

The vector $n_2(\mathbf{u}_2^{(1)} + \mathbf{u}_2^{(2)})$ depends only on the positions $\mathbf{x}^{(1)}$ and $\mathbf{x}^{(2)}$, and can therefore be expressed as a function of the relative position vector $\mathbf{r} = \mathbf{x}^{(2)} - \mathbf{x}^{(1)}$, and the mean centre $\mathbf{x} = \tfrac{1}{2}(\mathbf{x}^{(1)} + \mathbf{x}^{(2)})$. Expressed in this way a bar will be placed over the letters, thus: $\bar{n}_2(\bar{\mathbf{u}}_2^{(1)} + \bar{\mathbf{u}}_2^{(2)})$. Then

$$\{n_2(\mathbf{u}_2^{(1)} + \mathbf{u}_2^{(2)})\}\,(\mathbf{x}^{(1)}, \mathbf{x}^{(2)}) = \{\bar{n}_2(\bar{\mathbf{u}}_2^{(1)} + \bar{\mathbf{u}}_2^{(2)})\}\,(\mathbf{r}, \mathbf{x});$$

also, since $n_2(\mathbf{u}_2^{(1)} + \mathbf{u}_2^{(2)})$ is unchanged when $\mathbf{x}^{(1)}$ and $\mathbf{x}^{(2)}$ are interchanged,

$$(3.17) \qquad \{\bar{n}_2(\bar{\mathbf{u}}_2^{(1)} + \bar{\mathbf{u}}_2^{(2)})\}\,(\mathbf{r}, \mathbf{x}) = \{\bar{n}_2(\bar{\mathbf{u}}_2^{(1)} + \bar{\mathbf{u}}_2^{(2)})\}\,(-\mathbf{r}, \mathbf{x}).$$

Now, since E_1 in (3. 16) is a function of $\mathbf{x}^{(1)}$, it is convenient to have the last integral in the same equation expressed as a function of $\mathbf{x}^{(1)}$ also; this is accomplished by effecting a Taylor's expansion, thus:

$$(3.18) \begin{cases} \{\bar{n}_2(\bar{\mathbf{u}}_2^{(1)} + \bar{\mathbf{u}}_2^{(2)})\}\,(\mathbf{r}, \mathbf{x}) = \{\bar{n}_2(\bar{\mathbf{u}}_2^{(1)} + \bar{\mathbf{u}}_2^{(2)})\}\,(\mathbf{r}, \mathbf{x}^{(1)}) + \\ \qquad + \tfrac{1}{2} \mathbf{r} \cdot \dfrac{\partial}{\partial \mathbf{x}^{(1)}} \{\bar{n}_2(\bar{\mathbf{u}}_2^{(1)} + \bar{\mathbf{u}}_2^{(2)})\}\,(\mathbf{r}, \mathbf{x}^{(1)}) + \ldots, \end{cases}$$

making use of the fact that $\bar{n}_2(\mathbf{u}_2^{(1)} + \mathbf{u}_2^{(2)})$ will vary only slowly with the position of the mean centre, which ensures rapid convergence. When this is substituted in the integral of (3. 16), the first term of the series is seen to give no contribution, because $\dfrac{\partial \phi}{\partial \mathbf{x}^{(1)}} = -\phi'(r)\dfrac{\mathbf{r}}{r}$ is an odd function of $\mathbf{r}$, whilst, according to (3. 17), $\bar{n}_2(\bar{\mathbf{u}}_2^{(1)} + \bar{\mathbf{u}}_2^{(2)})$ is an even function. Hence

$$(3.19) \quad \begin{cases} \int n_2 (\mathbf{u}_2^{(1)} + \mathbf{u}_2^{(2)}) \cdot \dfrac{\partial \phi}{\partial \mathbf{x}^{(1)}} \, d\mathbf{x}^{(2)} \\ \qquad = -\tfrac{1}{2} \dfrac{\partial}{\partial \mathbf{x}^{(1)}} \cdot \int \mathbf{r} \, \bar{n}_2 (\overline{\mathbf{u}}_2^{(1)} + \overline{\mathbf{u}}_2^{(2)}) \cdot \mathbf{r} \, r^{-1} \, \phi'(r) \, d\mathbf{r}, \end{cases}$$

with neglect only of third and higher order spacial derivatives. At this stage comparing (3. 16) with (3. 14), one sees that

$$(3.20) \quad \begin{cases} \mathbf{q} + \mathbf{p} \cdot \mathbf{u} = \tfrac{1}{2} m \int f_1 (\boldsymbol{\xi}^{(1)} - \mathbf{u}) \, \boldsymbol{\xi}^{(1)2} \, d\boldsymbol{\xi}^{(1)} + \tfrac{1}{2} \int n_2 (\mathbf{u}_2^{(1)} - \mathbf{u}) \, \phi \, d\mathbf{x}^{(2)} - \\ \qquad - \tfrac{1}{4} \int \mathbf{r} \, \bar{n}_2 (\overline{\mathbf{u}}_2^{(1)} + \overline{\mathbf{u}}_2^{(2)}) \cdot \mathbf{r} \, r^{-1} \, \phi'(r) \, d\mathbf{r}. \end{cases}$$

The pressure tensor $\mathbf{p}$ is given by the formula (3. 10); hence the thermal flux vector $\mathbf{q}$ is

$$(3.21) \quad \begin{cases} \mathbf{q} = \tfrac{1}{2} m \int f_1 (\boldsymbol{\xi}^{(1)} - \mathbf{u}) \, (\boldsymbol{\xi}^{(1)2} - 2 \boldsymbol{\xi}^{(1)} \cdot \mathbf{u}) \, d\boldsymbol{\xi}^{(1)} + \\ \qquad + \tfrac{1}{2} \int n_2 (\mathbf{u}_2^{(1)} - \mathbf{u}) \, \phi \, d\mathbf{x}^{(2)} - \tfrac{1}{4} \int \mathbf{r} \, \bar{n}_2 (\overline{\mathbf{u}}_2^{(1)} + \overline{\mathbf{u}}_2^{(2)} - 2\mathbf{u}) \cdot \mathbf{r} \, r^{-1} \phi'(r) \, d\mathbf{r}. \end{cases}$$

In this way the macroscopic quantity $\mathbf{q}$ has been expressed entirely in terms of the molecular distribution functions and intermolecular forces. The formula (3. 21) is strictly analogous to (3. 10). Like the pressure tensor, the thermal flux vector is compounded of two parts, one of which arises from the thermal motion of the molecules, and the second from the intermolecular forces. Indeed, the first term of (3. 21) is rather obviously the flux of the kinetic energy transported by the molecules on account of their motion; the origin of the second term is more occult, since it represents the transfer of energy arising from the distant action of the molecules on one another. If one wishes to have a mechanical picture of this mode of transport of energy, one may imagine the molecules tied together by elastic strings, the tension in which varies in a rather eccentric way so as to simulate the attractive forces. Then, as the molecules move, the elastic energy of the strings will vary and by this means energy can be transported from one part of the assembly to another without actually being carried by the molecules themselves. In rare gases, where most of the molecules are outside the sphere of influence of one another, the energy carried by this method is very small. In liquids, on the other hand, it is certainly the most important mode of energy transfer.

4. Viscosity and Thermal Conduction

It is now possible to examine in some detail the connection between the molecular structure of a fluid, and the phenomena of viscosity and thermal conduction.

4 1 Viscosity

Viscosity, like most other mechanical processes in fluids, is the outcome of two distinct causes: the motion of the molecules, and the action of the intermolecular forces. The former is operative mainly in gases, where the molecules are free to traverse relatively long distances (of the order of 10^{-5} cm at ordinary pressures) in between collisions with one another. Supposing that one has a fluid in which two adjacent layers are moving at different velocities, the average velocities of the molecules in these two layers will differ by the same amount. This means that the molecules arriving in the faster moving layer from the other will have, on the average, a smaller velocity than those already present, and conversely, the molecules arriving in the more slowly moving layer from the other will have, on the average, a greater velocity than those already present. This tendency to remove inequalities in the macroscopic velocity distribution is the salient characteristic of viscosity; in the absence of externally imposed stresses to maintain the flow, it would ultimately bring the entire fluid to a state of uniform motion or rest.

In liquids, however, the molecular mechanism which tends to dissipate the motion of the fluid is essentially different from that already described as operative in gases. Being surrounded by other molecules, the individual molecule is not able to travel freely from one layer to another, and actually follows a very erratic path which is not well adapted to the transport of momentum. There is available, however, in liquids, another, very much more effective method of exchange of momentum, which depends on the continuous action of the intermolecular forces. Perhaps the easiest method of visualizing the nature of the process involved is again to imagine the molecules connected by elastic strings which simulate the action of the van der Waals' forces; it will then be seen that, if two adjacent layers in the fluid move with different velocities, each will tend to drag the other in such a manner as to dissipate the state of relative motion, in the absence of sustaining external stresses. This method of exchange of momentum is most effective when the molecules are close together, which explains why it is rather unimportant in gases, yet the dominant process in liquids.

As a result of the drag exerted on one another by molecules in adjacent layers of the fluid moving with different velocities, the molecular structure of the fluid is deformed. If one considers the mean configuration of the molecules surrounding a given molecule, one will find, instead of a radial distribution such as obtains in equilibrium, a somewhat ellipsoidal distribution with principal axes determined by the local velocity gradient. Moreover, the greater the velocity gradient, the greater will be the deviation from the radial distribution. A study of the deformation of the molecular structure will enable one to establish the quantitative relations which determine the coefficient of viscosity.

It has been seen that the molecular distribution function $\bar{n}_2$, which, for fluids in equilibrium depends only on the relative position $\mathbf{r} = \mathbf{x}^{(2)} - \mathbf{x}^{(1)}$, for fluids in motion is affected by the existence of a velocity gradient. It may therefore be regarded as a function of the velocity gradient, and, if the latter is not too large, can be expanded in the form of a Taylor's series, of which the $(i + 1)$th term involves the ith power of the velocity gradient, thus:

$$(4.1) \qquad \bar{n}_2 = \bar{n}_2^0 + \bar{n}_2' + \bar{n}_2'' + \ldots,$$

where $\bar{n}_2^0$ is independent of the velocity gradient, $\bar{n}_2'$ is linear in the velocity gradient, and so on. When the fluid is at rest, so that the velocity gradient vanishes, $\bar{n}_2 = \bar{n}_2^0$; hence $\bar{n}_2^0$ must be the radial distribution function for the fluid at rest. Since the term $\bar{n}_2'$ is linear in the velocity gradient, it can be written in the form

$$(4.2) \quad \begin{cases} \bar{n}_2' = \nu\, \mathbf{r} \cdot \left(\dfrac{\partial}{\partial \mathbf{x}} \mathbf{u}\right) \cdot \mathbf{r}/r^2 + (\nu_0 - \tfrac{1}{3}\nu)\dfrac{\partial}{\partial \mathbf{x}} \cdot \mathbf{u} \\ \phantom{\bar{n}_2'} = \nu\, \mathbf{r} \cdot \overline{\left(\dfrac{\partial}{\partial \mathbf{x}} \mathbf{u}\right)} \cdot \mathbf{r}/r^2 + \nu_0 \dfrac{\partial}{\partial \mathbf{x}} \cdot \mathbf{u}, \end{cases}$$

where the coefficients ν and ν_0 of the Taylor's expansion may of course depend on r. It will be noticed that in the above formula the coefficient of $\dfrac{\partial}{\partial \mathbf{x}} \cdot \mathbf{u}$ has been written in the form $\nu_0 - \tfrac{1}{3}\nu$ so as to provide a more compact expression in terms of the non-divergent rate of strain tensor $\overline{\dfrac{\partial}{\partial \mathbf{x}} \mathbf{u}}$.

Not only the molecular distribution, but also the velocity distribution will be affected by the existence of a velocity gradient. It follows that f_1 must likewise be a function of the velocity gradient, as well as of the molecular velocity $\mathbf{v} = \mathbf{v}^{(1)} = \mathbf{\xi}^{(1)} - \mathbf{u}$, measured relative to the local macroscopic velocity $\mathbf{u}$ [1]). If follows that one may expand f_1, in exactly the same way as $\bar{n}_2$, in a Taylor's series of ascending powers of the velocity gradient, thus:

(4. 3) $$f_1 = f_1^0 + f_1' + f_1'' + \ldots.$$

The first term, which is independent of the velocity gradient, is necessarily the equilibrium velocity distribution function; and since $\int f_1 d\boldsymbol{\xi}^{(1)}$ and $\int f_1^0 d\boldsymbol{\xi}^{(1)}$ are both equal to the local density n, one must have

(4. 4) $$\int f_1' d\boldsymbol{\xi}^{(1)} = 0.$$

This constraint on f_1' requires that it should be of the form

(4. 5) $$\begin{cases} f_1' = \omega \left\{ \mathbf{v} \cdot \left(\frac{\partial}{\partial \mathbf{x}} \mathbf{u} \right) \cdot \mathbf{v}/v^2 - \tfrac{1}{3} \frac{\partial}{\partial \mathbf{x}} \cdot \mathbf{u} \right\} \\ = \omega \, \mathbf{v} \cdot \overline{\left(\frac{\partial}{\partial \mathbf{x}} \mathbf{u} \right)} \cdot \mathbf{v}/v^2 \end{cases}$$

without any additional term such as that which occurs in (4. 2): for it will be found on substitution that (4. 5) satisfies (4. 4) identically, and any additional term would spoil this property. The coefficient ω in (4. 5) may of course depend on the magnitude of the velocity v.

In the foregoing, it has been assumed that f_1 can depend only on the relative molecular velocity $\mathbf{v}$ and the macroscopic velocity gradient $\frac{\partial}{\partial \mathbf{x}} \mathbf{u}$; if, however, there were a gradient of density or temperature in the fluid, f_1 would be affected also by these, and f_1' would involve terms linear in the density gradient or temperature gradient, as well as the velocity gradient. For the present it is therefore assumed that only the velocity gradient exists. It is interesting to note at this stage, however, that even if a gradient of density or temperature existed, the expression for $\bar{n}_2'$ would be unaffected. For, any additional terms would necessarily be proportional to $\mathbf{r} \cdot \frac{\partial n}{\partial \mathbf{x}}$ or $\mathbf{r} \cdot \frac{\partial T}{\partial \mathbf{x}}$, where $\mathbf{r} = \mathbf{x}^{(2)} - \mathbf{x}^{(1)}$; and these could not form part of an expression for $\bar{n}_2'$ because they change sign when $\mathbf{x}^{(1)}$ and $\mathbf{x}^{(2)}$ are interchanged, whilst $\bar{n}_2'$ itself must remain unaltered.

Now that the general nature of the distribution functions has been determined for a fluid in a state of motion, one can proceed to evaluate

---

[1]) The fact that f_1 can depend only on *differences* of velocity, such as the difference between the molecular and macroscopic velocity, is a consequence of the obvious fact that the velocity distribution remains intrinsically unchanged when the fluid as a whole moves with uniform velocity relative to the observer.

the pressure tensor as given by the equation (3. 10). According to the formula (4. 15) of chapter I,

$$\int f_1' \mathbf{v}\mathbf{v}\, d\mathbf{v} = \int \frac{\omega}{v^2} \mathbf{v} \cdot \overline{\left(\frac{\partial}{\partial \mathbf{x}}\mathbf{u}\right)} \cdot \mathbf{v}\,\mathbf{v}\mathbf{v}\, d\mathbf{v}$$

$$= \frac{2}{15} \overline{\left(\frac{\partial}{\partial \mathbf{x}}\mathbf{u}\right)} \int \omega v^2\, d\mathbf{v},$$

$$\int \bar{n}_2' \mathbf{r}\mathbf{r}\, r^{-1}\, \phi'(r)\, d\mathbf{r} = \int \left\{\frac{v}{r^3}\mathbf{r}\cdot\overline{\left(\frac{\partial}{\partial \mathbf{x}}\mathbf{u}\right)}\cdot \mathbf{r} + \frac{v_0}{r}\frac{\partial}{\partial \mathbf{x}}\cdot\mathbf{u}\right\} \mathbf{r}\mathbf{r}\,\phi'(r)\, d\mathbf{r}$$

$$= \frac{2}{15}\overline{\left(\frac{\partial}{\partial \mathbf{x}}\mathbf{u}\right)}\int v r\,\phi'(r)\, d\mathbf{r} + \left(\frac{\partial}{\partial \mathbf{x}}\cdot\mathbf{u}\right)\boldsymbol{\delta}\int v_0 r\,\phi'(r)\, d\mathbf{r},$$

where $\boldsymbol{\delta}$ is the unit tensor. Hence

(4. 6) $$\mathbf{p} = \left(p^0 - \eta_0 \frac{\partial}{\partial \mathbf{x}}\cdot\mathbf{u}\right)\boldsymbol{\delta} - 2\eta\,\overline{\frac{\partial}{\partial \mathbf{x}}\mathbf{u}},$$

where

(4. 7) $$\eta = \tfrac{1}{30}\int v r\,\phi'(r)\,d\mathbf{r} - \tfrac{1}{15}\int \omega v^2\,d\mathbf{v},$$

(4. 8) $$\eta_0 = \tfrac{1}{2}\int v_0 r\,\phi'(r)\,d\mathbf{r}$$

and p^0 is the equilibrium value of the hydrostatic pressure.

The equation (4. 6) is the symbolic statement of the macroscopic law of viscosity, according to which the non-divergent part of the pressure tensor is proportional to the rate of strain tensor. The constant of proportionality η is identified with the coefficient of viscosity, and the constant η_0, which determines the deviation of the hydrostatic pressure from its equilibrium value, with the coefficient of bulk viscosity. These quantities are given by (4. 7) and (4. 8) in terms of the molecular distribution functions.

The functions v and v_0 of (4. 2) measure the deformation of the molecular distribution from its equilibrium configuration. As such, they must be proportional to the radial distribution function of equilibrium, and though there is of course another factor involved, this will not vary very rapidly with r. Accordingly one may write

(4. 9) $$v = a\,\bar{n}_2^0, \qquad v_0 = a_0\,\bar{n}_2^0,$$

and assume that neither a nor a_0 is a very rapidly varying function of r. Now $\bar{n}_2^0$ is very nearly zero for all values of r less than a molecular diameter, and $\phi'(r)$ is nearly zero for values of r somewhat greater than a molecular diameter. Thus the main contribution to the space integrals in (4. 7) and (4. 8) will *usually* come from values of r slightly greater than a molecular diameter. The integrals themselves will therefore vary as $\bar{n}_2^0(r_0)$, where r_0 is some value of r in this region. For the liquid, the second integral of (4. 7), which represents the contribution of the thermal motion to the viscosity, can be neglected; it follows that, for liquids, both η and η_1 are effectively proportional to $\bar{n}_2^0(r_0)$. If now one approximates $\bar{n}_2^0(r_0)$ by $n^2 \{\exp-\beta\phi(r_0)\}$, one obtains for the temperature dependence of the viscosity

$$(4.\ 10) \qquad \eta \sim \exp\{-\beta\phi(r_0)\}$$

— a relation which is the basis of ANDRADE's law of viscosity [1934]. This temperature dependence is exhibited by the viscosity of many liquids over wide ranges of temperature. To investigate the *limitations* of Andrade's formula, it is necessary to examine the integrals of (4. 7) and (4. 8) more closely. It is clear that if, in any fluid, the first co-ordination shell of the radial distribution should overlap, even slightly, with the repulsive part of the intermolecular field of force, these integrals will no longer vary simply with $\bar{n}_2^0(r_0)$, but will have a value which is the difference between a positive and a negative contribution, and not easy to represent analytically. This situation will obtain especially in the region of high pressures, where the electron shells of the molecules are forced closely together. In this region it is found experimentally that there are indeed considerable deviations from Andrade's formula.

It is possible to trace the relation of Andrade's theory of viscosity to some earlier considerations of van der Waals, in a way which has been described by Burgers. The present discussion will, however, be limited to such considerations as are useful in obtaining information concerning the parameters a and a_0, defined in (4. 9). It is clear that if these parameters were known, the coefficients of viscosity would be fully determined.

From (4. 2) it can be seen that ν and ν_0 both have the dimensions $time/(volume)^2$. The parameters a and a_0 defined in (4. 9) have therefore both the dimensions of time. Also, the only quantities on which they can depend are $\beta^{-1} = kT$, which has the dimensions of energy, the molecular mass m, the density n, and the distance r. It follows that both a and a_0 must have the form

$$n^{-\frac{1}{2}} (m\beta)^{\frac{1}{2}} \times \text{function of } (rn^{\frac{1}{3}}).$$

If one admits the argument that a and a_0 can depend at most weakly on r, the function of $(rn^{\frac{1}{3}})$ can be replaced by a constant, which is actually of unit order of magnitude.

Determined in this way, a is of the order of magnitude of the period of vibration of the molecules in the crystal, at the melting point; and Andrade suggested that, if similar vibrations occurred in the liquid, the coefficient of viscosity might also be proportional to the period of the vibrations there. It is now known that vibrations of the kind experienced by the molecules of a crystal are possible in the liquid only very near the melting point, and therefore no definite period can usually be assigned. From a qualitative point of view, however, it is quite useful to extend Andrade's idea by associating a with the 'relaxation time' of a liquid. This concept is most easily understood in relation to the theory of elasticity, and is discussed in the appropriate section (§ 1 of chapter VII).

4. 2 Thermal Conduction

The theory of thermal conduction is in many respects parallel to that of viscosity which has just been considered. Again one has two fundamental processes to examine. In gases, the essential process is the distribution of energy by the motion of the molecules. If one has two adjacent layers, one of which has a higher temperature than the other, the mean kinetic energy of the molecules in this layer is greater than in the other. In consequence, the molecules arriving in the warmer layer from the colder have, on the average, a smaller kinetic energy than those already present, and conversely. Thus the effect of the motion of the molecules is a tendency to reduce differences in temperature — a tendency which is recognized macroscopically as essential to thermal conduction. In the absence of externally applied temperature differences, the redistribution of energy effected by the motion of the molecules would eventually reduce the gas to a state of uniform temperature.

The mechanism just described does not suffice, however, to account for energy transport in liquids. It has been seen already in § 3 that the thermal flux includes a term which depends on the intermolecular forces, and which, though small in gases, in liquids can be expected to account unaided for almost the entire energy transfer. It remains to examine how the existence of a temperature gradient can influence the organization of the molecules in such a way as to secure a transfer of energy by means of the distant action of the molecules on one another.

In a fluid whose temperature is uniform, the mean velocity of the centre of gravity of a pair of molecules is independent of the distance between them; it is, in fact, the same as the velocity of motion of the fluid (**u**) at the position of their centre of gravity. The same is no longer true if there is a temperature gradient, for the following reason. The force acting on both molecules is the resultant of the force due to the other, and that due to the remainder in the vicinity. The forces exerted by each molecule on the other are equal and opposite, and can therefore have no effect on the motion of the centre of gravity of the two molecules. The average forces exerted by the remainder, however, depend on the local temperature, and will be slightly different in magnitude for the two molecules if they are aligned in the direction of the temperature gradient; their centre of gravity will therefore tend to be displaced. Specifically, the force exerted by one of the molecules on the other is opposed by the average force due to the remainder, to an extent which, in liquids, increases with the local temperature. Hence, if two molecules in the attractive part of one another's field are aligned in the direction of the temperature gradient, their centre of gravity will tend to drift in the direction of the temperature gradient. To compensate, if they are aligned in a direction perpendicular to the temperature gradient, their mass centre will tend to drift in a direction opposite to the temperature gradient, relative to the other molecules. The mean total momentum of a pair of molecules therefore depends to an appreciable extent on their configuration relative to one another and the direction of the temperature gradient. Generally speaking, if they are in one another's attractive field and aligned parallel with the temperature gradient, their resultant momentum will be in the direction of the temperature gradient, whilst if their alignment is perpendicular to the temperature gradient, their resultant momentum will be in the opposite direction to the temperature gradient.

To formulate these considerations quantitatively, one may observe that the mean resultant of the velocities of two molecules, whose relative displacement is **r** and mean centre is at the point $\mathbf{x} = \frac{1}{2}(\mathbf{x}^{(1)} + \mathbf{x}^{(2)})$ is $(\mathbf{\bar{u}}_2^{(1)} + \mathbf{\bar{u}}_2^{(2)})(\mathbf{r}, \mathbf{x})$; and the probability of finding them in this configuration is proportional to $\bar{n}_2(\mathbf{r}, \mathbf{x})$. Then, according to the foregoing, the vector $\bar{n}_2(\mathbf{\bar{u}}_2^{(1)} + \mathbf{\bar{u}}_2^{(2)})$ will depend on the temperature gradient, and can be expanded in the form of a Taylor's series, thus:

(4. 11) $\quad \bar{n}_2(\overline{\mathbf{u}}_2^{(1)} + \overline{\mathbf{u}}_2^{(2)}) = \{\bar{n}_2(\overline{\mathbf{u}}_2^{(1)} + \overline{\mathbf{u}}_2^{(2)})\}^0 + \{\bar{n}_2(\overline{\mathbf{u}}_2^{(1)} + \overline{\mathbf{u}}_2^{(2)})\}' + \ldots,$

where the first term is independent of the temperature gradient, the second term is linear in the temperature gradient, and so on. In thermal equilibrium, where there is no temperature gradient, $\bar{n}_2(\overline{\mathbf{u}}_2^{(1)} + \overline{\mathbf{u}}_2^{(2)})$ reduces to $\bar{n}_2 \cdot 2\mathbf{u}$, where $\mathbf{u}$ — even when there is a velocity gradient, — is the local macroscopic velocity at the mean centre $\mathbf{x}$ of the two molecules. Hence

$$\{\bar{n}_2(\overline{\mathbf{u}}_2^{(1)} + \overline{\mathbf{u}}_2^{(2)})\}^0 = 2\bar{n}_2 \mathbf{u}.$$

Since $2\mathbf{u}$ is always the mean resultant velocity of a pair of molecules whose mean centre is at the point $\mathbf{x}$, one must have

(4. 12) $\quad \begin{cases} \int \bar{n}_2(\overline{\mathbf{u}}_2^{(1)} + \overline{\mathbf{u}}_2^{(2)} - 2\mathbf{u})\, d\mathbf{r} = 0, \\ \int \{\bar{n}_2(\overline{\mathbf{u}}_2^{(1)} + \overline{\mathbf{u}}_2^{(2)})\}'\, d\mathbf{r} = 0. \end{cases}$

This condition requires that the term $\{\bar{n}_2(\overline{\mathbf{u}}_2^{(1)} + \overline{\mathbf{u}}_2^{(2)})\}'$ linear in the temperature gradient should be of the form

(4. 13) $\quad \{\bar{n}_2(\overline{\mathbf{u}}_2^{(1)} + \overline{\mathbf{u}}_2^{(2)})\}' = \tau\left(\mathbf{r} \cdot \frac{\partial T}{\partial \mathbf{x}}\, \mathbf{r}/r^2 - \frac{1}{3}\frac{\partial T}{\partial \mathbf{x}}\right),$

where τ may depend on r.

Even if the fluid were in non-uniform motion, the expression for $\bar{n}_2(\overline{\mathbf{u}}_2^{(1)} + \overline{\mathbf{u}}_2^{(2)})$ could not include any term linear in the velocity gradient; for, such a term would necessarily be proportional to either $\mathbf{r} \cdot \left(\overline{\frac{\partial}{\partial \mathbf{x}}\mathbf{u}}\right)$, $\mathbf{r} \cdot \left(\overline{\frac{\partial}{\partial \mathbf{x}}\mathbf{u}}\right) \cdot \mathbf{rr}$, or $\frac{\partial}{\partial \mathbf{x}} \cdot \mathbf{ur}$, and all these expressions change sign when $\mathbf{x}^{(1)}$ and $\mathbf{x}^{(2)}$ are interchanged, whereas $\bar{n}_2(\overline{\mathbf{u}}_2^{(1)} + \overline{\mathbf{u}}_2^{(2)})$ is unchanged. The same is not true, however, of the expression

$$\bar{n}_2(\overline{\mathbf{u}}_2^{(2)} - \overline{\mathbf{u}}_2^{(1)})$$

which measures the average relative velocity of two molecules whose relative displacement is $\mathbf{r}$ and mean centre is at $\mathbf{x}$. This expression will vanish in equilibrium, but not when there is a velocity gradient in the fluid. In general one will have

(4. 14) $\quad \begin{cases} \{\bar{n}_2(\overline{\mathbf{u}}_2^{(2)} - \overline{\mathbf{u}}_2^{(1)})\}' = \Big\{v_2 \mathbf{r} \cdot \left(\overline{\frac{\partial}{\partial \mathbf{x}}\mathbf{u}}\right) \cdot \mathbf{r}/r^2 + v_0 \frac{\partial}{\partial \mathbf{x}} \cdot \mathbf{u}\Big\}\mathbf{r} + \\ \qquad\qquad\qquad\qquad + v_1 \mathbf{r} \cdot \left(\overline{\frac{\partial}{\partial \mathbf{x}}\mathbf{u}}\right)/r. \end{cases}$

Here no term linear in the temperature gradient could occur, because such a term would have to be similar to those appearing in (4. 13), which are unaltered when $\mathbf{x}^{(1)}$ and $\mathbf{x}^{(2)}$ are interchanged; whereas $\bar{n}_2(\bar{\mathbf{u}}_2^{(2)} - \bar{\mathbf{u}}_2^{(1)})$ changes sign when $\mathbf{x}^{(1)}$ and $\mathbf{x}^{(2)}$ are interchanged.

The formulae already derived enable one to show that the term $\frac{1}{2}\int n_2(\mathbf{u}_2^{(1)} - \mathbf{u})\phi d\mathbf{x}^{(2)}$ in the expression (3. 21) for the thermal flux makes no effective contribution. For if one substitutes

$$(n_2 \mathbf{u}_2^{(1)})(\mathbf{x}^{(1)}, \mathbf{x}^{(2)}) = (\bar{n}_2 \bar{\mathbf{u}}_2^{(1)})(\mathbf{r}, \mathbf{x})$$
$$= (\bar{n}_2 \bar{\mathbf{u}}_2^{(1)})(\mathbf{r}, \mathbf{x}^{(1)}) + \tfrac{1}{2}\mathbf{r} \cdot \frac{\partial}{\partial \mathbf{x}^{(1)}}(\bar{n}_2 \mathbf{u}_2^{(1)})(\mathbf{r}, \mathbf{x}^{(1)}),$$

and

$$n_2(\mathbf{x}^{(1)}, \mathbf{x}^{(2)}) = \bar{n}_2(\mathbf{r}, \mathbf{x}^{(1)}) + \tfrac{1}{2}\mathbf{r} \cdot \frac{\partial}{\partial \mathbf{x}^{(1)}} \bar{n}_2(\mathbf{r}, \mathbf{x}^{(1)}),$$

one finds, with the help of (4. 13) and (4. 14) that every term of the integral $\int n_2(\mathbf{u}_2^{(1)} - \mathbf{u})\phi d\mathbf{x}^{(2)}$ vanishes by reason of symmetry. On the other hand, it follows from (4. 13) that

$$(4.\ 15) \quad -\frac{1}{4}\int \mathbf{r}\,\bar{n}_2(\bar{\mathbf{u}}_2^{(1)} + \bar{\mathbf{u}}_2^{(2)} - 2\mathbf{u}) \cdot \mathbf{r}\, r^{-1}\phi'(r)\,d\mathbf{r} = -\frac{1}{18}\frac{\partial T}{\partial \mathbf{x}}\int \tau r\,\phi'(r)\,d\mathbf{r}.$$

This, then, is the contribution of the action of the intermolecular forces to the thermal flux.

To take account of the effect of the thermal motion of the molecules on the transport of energy, one requires a formula for the velocity distribution function when there is a temperature gradient in the fluid. It has already been noticed that (4. 5) is correct only if the temperature is uniform, and generally a term proportional to the temperature gradient must be supplied, thus:

$$(4.\ 16) \qquad f_1' = \omega \mathbf{v} \cdot \left(\overline{\frac{\partial}{\partial \mathbf{x}}\mathbf{u}}\right) \cdot \mathbf{v}/v^2 + \chi \mathbf{v} \cdot \frac{\partial T}{\partial \mathbf{x}}\Big/ v.$$

Now this formula gives

$$\tfrac{1}{2} m \int f_1 \mathbf{v}\, v^2\, dv = \tfrac{1}{6} m \frac{\partial T}{\partial \mathbf{x}} \int \chi v^3\, dv;$$

hence, combining this result with (4. 15), the equation (3. 21) reduces to

$$(4.\ 17) \qquad \mathbf{q} = -\lambda \frac{\partial T}{\partial \mathbf{x}},$$

where

$$(4.\ 18) \qquad \lambda = \frac{1}{18}\int \tau r\, \phi'(r)\,d\mathbf{r} - \frac{1}{6} m \int \chi v^3\, d\mathbf{v}.$$

The final result (4.17) is a symbolic statement of the law of thermal conduction, according to which the thermal flux is proportional to the temperature gradient. The constant of proportionality λ is identified macroscopically with the coefficient of thermal conduction, and it is expressed by (4.18) in terms of the molecular distributions and the intermolecular force.

It should be remarked that the theory of thermal conduction outlined above assumes implicitly that the pressure distribution is uniform, as it certainly is under the experimental conditions of measurement. When the pressure (p) is uniform, there is a direct relation between the density and temperature gradient, namely

$$\frac{\partial n}{\partial \mathbf{x}} = -\left(\frac{\partial p}{\partial T}\bigg/\frac{\partial p}{\partial n}\right)\frac{\partial T}{\partial \mathbf{x}},$$

where the quotient $\left(\frac{\partial p}{\partial T}\big/\frac{\partial p}{\partial n}\right)$ is, in fact, proportional to the volumetric coefficient of expansion of the fluid. On account of this relation, it is superfluous to include in the formulae (4.13) and (4.16) terms proportional to the density gradient. Under other, more general conditions, where the pressure varied from point to point in the fluid, there would no longer be a definite relation between the density gradient and temperature gradient, and it would be necessary to represent the effects of inhomogenity of density and temperature by separate terms in the equations (4.13) and (4.16). The general theory, applicable to fluids in which there are irregular variations of temperature, density and macroscopic velocity, is considered briefly in the following section. There it will also be possible to formulate the equations which determine the distribution functions f_1 and f_2, and therefore, in principle, the coefficients ν, ω, τ and χ introduced in this section.

5. Theory of Non-Uniform Fluids

In the preceding analysis of the phenomena of viscosity and thermal conduction, it was found necessary to consider special situations in which there was some inhomogeneity of the fluid, represented by spatial variations in the local temperature and macroscopic velocity of flow. It is natural to proceed to consider the fluid in its most general macroscopic inhomogeneous state, where there may exist simultaneously irregularities in the distributions of density, temperature and macroscopic velocity. Such a state might, for example, be obtained physically by first stirring and heating the fluid in an irregular manner, and then allowing it gradually to approach, without actually reaching, a state of equilibrium. It is, of course, possible to conceive of states of the fluid even further removed from equilibrium than those which might be obtained in this way, but they represent abnormal conditions which could only be produced by specialized methods which directly affect the molecular distributions. For example, with the help of a powerful radioactive source of α-particles, one might obtain helium gas in a state in which the velocity distribution was far removed from the normal Maxwellian type. However, such states cannot be obtained by normal macroscopic methods, and they are therefore excluded from consideration here. A 'normal' state is sufficiently specified by a statement of the density, temperature and macroscopic velocity distribution throughout the fluid; it is the *most probable* state with such a specification, just as the equilibrium state is the most probable one with a specified *mean* density and temperature.

Instead of specifying the values of n, T and $\mathbf{u}$ at every point in the fluid, one may specify their values at any point $\mathbf{x}$, together with the values of their spatial derivatives $\frac{\partial n}{\partial \mathbf{x}}$, $\frac{\partial T}{\partial \mathbf{x}}$, $\frac{\partial \mathbf{u}}{\partial \mathbf{x}}$, $\frac{\partial^2 n}{\partial \mathbf{x} \partial \mathbf{x}}$, etc. at the same point; for these, in conjunction with the appropriate Taylor's series, suffice to determine the variation of n, T and $\mathbf{u}$ throughout at least a fairly large volume of the fluid. For the sake of brevity, let

(5..1) $\qquad \lambda_1 = u_1,\ \lambda_2 = u_2,\ \lambda_3 = u_3,\ \lambda_4 = n,\ \lambda_5 = T;$

then, in any non-uniform fluid, the various distribution functions will depend on the values of the λ_k, and their derivatives, at any point $\mathbf{x}$, thus:

(5. 2) $\qquad f_q(\boldsymbol{\xi}^{(i)}, \mathbf{x}^{(i)}, t) = \bar{f}_q\left(\boldsymbol{\xi}^{(i)} - \mathbf{u},\ \mathbf{x}^{(i)},\ \lambda_k,\ \frac{\partial \lambda_k}{\partial \mathbf{x}},\ \frac{\partial^2 \lambda_k}{\partial \mathbf{x}\, \partial \mathbf{x}},\ \ldots\right).$

A number of remarks need to be made concerning this last formula:

(i) It will be noticed that the velocity dependence of $\bar{f}_q$ has been represented through the variables $\mathbf{v}^{(i)} = \boldsymbol{\xi}^{(i)} - \mathbf{u}$. When this is done, the macroscopic velocity $\mathbf{u}$ will not appear elsewhere in the expression, except in the spatial derivatives $\frac{\partial \lambda_k}{\partial \mathbf{x}}$, etc. The reason is that the distribution function f_q must be intrinsically unchanged if the whole fluid moves with a uniform velocity relative to the observer. Analytically, one must have

(5. 3) $\qquad \dfrac{\partial \bar{f}_q}{\partial \mathbf{u}} + \sum\limits_{i=1}^{q} \dfrac{\partial \bar{f}_q}{\partial \boldsymbol{\xi}^{(i)}} = 0.$

Hence $\overline{f}_q$ must be a function of the combinations $\boldsymbol{\xi}^{(i)} - \mathbf{u}$ only. A special instance of this appeared in the formulae (4.5) and (4.16).

(ii) The point $\mathbf{x}$ at which the λ_k and their derivatives are supposed to be measured can be chosen arbitrarily. It is convenient to make the choice in such a way that it coincides with the centre of gravity, or — what is the same thing for molecules all of the same kind — the mean centre of the q molecules considered. Then

(5.4) $$q\mathbf{x} = \mathbf{x}^{(1)} + \mathbf{x}^{(2)} + \ldots + \mathbf{x}^{(q)}.$$

As a result of this convention, $\overline{f}_q$ will depend explicitly only on the relative coordinates $\mathbf{x}^{(i)} - \mathbf{x}^{(j)}$, thus:

(5.5) $$\sum_{i=1}^{q} \frac{\partial \overline{f}_q}{\partial \mathbf{x}^{(i)}} = 0.$$

For the dependence of $\overline{f}_q$ on the mean centre arises solely on account of the variation of temperature, density, and velocity in the fluid, which is fully described as soon as it is admitted that the values of the λ_k and their derivatives depend on the position of the mean centre.

(iii) Similarly, the function $\overline{f}_q$ does not depend explicitly on the time, because, in a normal state, the variation with time is essentially due to the variation of the parameters λ_k and their derivatives; since these are explicitly represented in the function $\overline{f}_q$, full account has already been taken of the time dependence of this distribution function.

(iv) As a matter of notation, the distribution functions, when regarded as functions of $\boldsymbol{\xi}^{(1)} \ldots \boldsymbol{\xi}^{(q)}$, $\mathbf{x}^{(1)} \ldots \mathbf{x}^{(q)}$ and t, are represented by ordinary letters, thus: f_q, n_q, $\mathbf{u}_q^{(i)}$, etc.; when, however, they are regarded as functions of the relative coordinates and velocities, and the parameters λ_k and their spatial derivatives, a bar is placed over these letters, thus: $\overline{f}_q$, $\overline{n}_q$, $\overline{\mathbf{u}}_q^{(i)}$ etc.

Now, the gradients $\frac{\partial \lambda_k}{\partial \mathbf{x}}$ and higher spatial derivatives of the λ_k are normally quite small, and it is therefore possible to expand the distribution functions in Taylor's series, of which the $(i+1)$th term involves i spatial derivatives, thus:

(5.6) $$\overline{f}_q = \overline{f}_q^0 + \overline{f}_q' + \overline{f}_q'' + \ldots,$$

where $\overline{f}_q^0$ is independent of the derivatives of the λ_k, $\overline{f}_q'$ is linear in their first derivatives, $\overline{f}_q''$ is quadratic in their first derivatives and linear in their second derivatives, and so on. When there is no spatial gradient of the density, temperature, or macroscopic velocity, $\bar{f}_q$ reduces to $\bar{f}_q^0$; the latter is therefore always the equilibrium distribution function, given by

$$(5.7) \quad \bar{f}_q^0 = \bar{n}_q^0 (m\beta/2\pi)^{3q/2} \exp\{-\beta \, (\sum_{i=1}^{q} \tfrac{1}{2} \, mv^{(i)2})\},$$

where $\mathbf{v}^{(i)} = \boldsymbol{\xi}^{(i)} - \mathbf{u}$. The second term $\bar{f}_q'$ in (5.6) is more complicated; it has already been seen in (4.2) and (4.14) that

$$(5.8) \quad \int \bar{f}_2' \, d\boldsymbol{\xi}^{(1)} \, d\boldsymbol{\xi}^{(2)} = \bar{n}_2' = \nu \, r^{-2} \, \mathbf{r} \cdot \left(\overline{\frac{\partial}{\partial \mathbf{x}} \mathbf{u}}\right) \cdot \mathbf{r} + \nu_0 \frac{\partial}{\partial \mathbf{x}} \cdot \mathbf{u},$$

and

$$(5.9) \quad \begin{cases} \int \bar{f}_2'(\mathbf{v}^{(2)} - \mathbf{v}^{(1)}) \, d\boldsymbol{\xi}^{(1)} \, d\boldsymbol{\xi}^{(2)} = \{\bar{n}_2(\bar{\mathbf{u}}_2^{(2)} - \bar{\mathbf{u}}_2^{(1)})\}' \\ = \left\{v_2 \, r^{-2} \, \mathbf{r} \cdot \left(\overline{\frac{\partial}{\partial \mathbf{x}} \mathbf{u}}\right) \cdot \mathbf{r} + v_0 \frac{\partial}{\partial \mathbf{x}} \cdot \mathbf{u}\right\} \frac{\mathbf{r}}{r} + v_1 \, r^{-1} \, \mathbf{r} \cdot \left(\overline{\frac{\partial}{\partial \mathbf{x}} \mathbf{u}}\right). \end{cases}$$

Also,

$$(5.10) \quad \begin{cases} \int \bar{f}_2'(\mathbf{v}^{(1)} + \mathbf{v}^{(2)}) \, d\boldsymbol{\xi}^{(1)} \, d\boldsymbol{\xi}^{(2)} = \{\bar{n}_2(\bar{\mathbf{u}}_2^{(1)} + \bar{\mathbf{u}}_2^{(2)} - 2\mathbf{u})\}' \\ = \tau_0 \left(r^{-2} \, \mathbf{r} \cdot \frac{\partial T}{\partial \mathbf{x}} \, \mathbf{r} - \frac{1}{3} \frac{\partial T}{\partial \mathbf{x}}\right) + \tau_0' \left(r^{-2} \, \mathbf{r} \cdot \frac{\partial n}{\partial \mathbf{x}} \, \mathbf{r} - \frac{1}{3} \frac{\partial n}{\partial \mathbf{x}}\right), \end{cases}$$

where τ_0 and τ_0' are related to the coefficient τ appearing in (4.13) by

$$(5.11) \quad \tau = \tau_0 + \left(\frac{\partial p}{\partial T}\bigg/\frac{\partial p}{\partial n}\right) \tau_0'.$$

Similarly

$$(5.12) \quad \bar{f}_1' = \omega \, \mathbf{v}^{(1)} \cdot \left(\overline{\frac{\partial}{\partial \mathbf{x}} \mathbf{u}}\right) \cdot \mathbf{v}^{(1)}/v^{(1)2} + \chi_0 \, \mathbf{v}^{(1)} \cdot \frac{\partial T}{\partial \mathbf{x}} \bigg/ v^{(1)} + \chi_0' \, \mathbf{v}^{(1)} \cdot \frac{\partial n}{\partial \mathbf{x}} \bigg/ v^{(1)},$$

where χ_0 and χ_0' are related to the coefficient χ in (4.16) by

$$(5.13) \quad \chi = \chi_0 + \left(\frac{\partial p}{\partial T}\bigg/\frac{\partial p}{\partial n}\right) \chi_0'.$$

To obtain the equations which determine the distribution functions $\bar{f}_q'$ from those already derived for f_q (eqs. 2.9), one uses the formulae

$$(5.14) \quad \begin{cases} \dfrac{\partial f_q}{\partial t} = \sum_k \left\{ \dfrac{\partial \bar{f}_q}{\partial \lambda_k} \dfrac{\partial \lambda_k}{\partial t} + \dfrac{\partial \bar{f}_q}{(\partial \lambda_k/\partial \mathbf{x})} \dfrac{\partial^2 \lambda_k}{\partial t \, \partial \mathbf{x}} + \cdots \right\}, \\ \dfrac{\partial f_q}{\partial \mathbf{x}^{(i)}} = \sum_k \dfrac{1}{q} \left\{ \dfrac{\partial \bar{f}_q}{\partial \lambda_k} \dfrac{\partial \lambda_k}{\partial \mathbf{x}} + \dfrac{\partial \bar{f}_q}{(\partial \lambda_k/\partial \mathbf{x})} \dfrac{\partial^2 \lambda_k}{\partial \mathbf{x} \, \partial \mathbf{x}} + \cdots \right\} + \dfrac{\partial \bar{f}_q}{\partial \mathbf{x}^{(i)}}. \end{cases}$$

The first of these relations involves the time derivatives of the λ_k, which must be eliminated by means of the equations

$$\frac{\partial n}{\partial t} = - \mathbf{u} \cdot \frac{\partial n}{\partial \mathbf{x}} - n \frac{\partial}{\partial \mathbf{x}} \cdot \mathbf{u},$$

$$\frac{\partial \mathbf{u}}{\partial t} = - \mathbf{u} \cdot \frac{\partial}{\partial \mathbf{x}} \mathbf{u} - \frac{1}{mn} \frac{\partial}{\partial \mathbf{x}} \cdot \boldsymbol{p},$$

$$\frac{\partial U_1}{\partial T} \frac{\partial T}{\partial t} + \frac{\partial U_1}{\partial n} \frac{\partial n}{\partial t} = \frac{\partial U_1}{\partial t} = - \mathbf{u} \cdot \frac{\partial U_1}{\partial \mathbf{x}} - \frac{1}{n} \left\{ \boldsymbol{p} : \left(\frac{\partial}{\partial \mathbf{x}} \mathbf{u} \right) + \frac{\partial}{\partial \mathbf{x}} \cdot \mathbf{q} \right\},$$

obtained from (3. 2), (3. 9) and (3. 12) respectively. If one substitutes these formulae in (5. 14), and selects from the resulting equations terms linear in the gradients of the λ_k, one obtains

15) $\left\{ \begin{aligned} & \left\{ \frac{\partial f_q}{\partial t} + \sum_{i=1}^{q} \left(\boldsymbol{\xi}^{(i)} \cdot \frac{\partial f_q}{\partial \mathbf{x}^{(i)}} - \frac{1}{m} \frac{\partial \Phi_q}{\partial \mathbf{x}^{(i)}} \cdot \frac{\partial f_q}{\partial \boldsymbol{\xi}^{(i)}} \right) \right\}' = \sum_{i=1}^{q} \left(\mathbf{v}^{(i)} \cdot \frac{\partial \bar{f}'_q}{\partial \mathbf{x}^{(i)}} - \frac{1}{m} \frac{\partial \Phi_q}{\partial \mathbf{x}^{(i)}} \cdot \frac{\partial \bar{f}'_q}{\partial \mathbf{v}^{(i)}} \right) + \\ & + \bar{f}^0_q \left[\left(\frac{n}{\bar{n}^0_q} \frac{\partial \bar{n}^0_q}{\partial n} \right) \left\{ \frac{1}{q} \left(\sum_{i=1}^{q} \mathbf{v}^{(i)} \right) \cdot \left(\frac{1}{n} \frac{\partial n}{\partial \mathbf{x}} \right) - \frac{\partial}{\partial \mathbf{x}} \cdot \mathbf{u} \right\} + \\ & + \left(\frac{T}{\bar{n}^0_q} \frac{\partial \bar{n}^0_q}{\partial T} + \beta \sum_{i=1}^{q} \tfrac{1}{2} m v^{(i)2} - \frac{3q}{2} \right) \left\{ \frac{1}{q} \left(\sum_{i=1}^{q} \mathbf{v}^{(i)} \right) \cdot \left(\frac{1}{T} \frac{\partial T}{\partial \mathbf{x}} \right) + \gamma \frac{\partial}{\partial \mathbf{x}} \cdot \mathbf{u} \right\} + \\ & + \beta \left(\sum_{i=1}^{q} \mathbf{v}^{(i)} \right) \cdot \left\{ \frac{m}{q} \left(\frac{\partial}{\partial \mathbf{x}} \mathbf{u} \right) \cdot \left(\sum_{i=1}^{q} \mathbf{v}^{(i)} \right) - \frac{1}{n} \frac{\partial p^0}{\partial \mathbf{x}} \right\} \right], \end{aligned} \right.$

where

(5. 16) $$\gamma = \left(n \frac{\partial U_1}{\partial n} - \frac{p^0}{n} \right) \Big/ \left(T \frac{\partial U_1}{\partial T} \right).$$

Thus the terms linear in the gradients of the λ_k on the left-hand side of the equation (2. 9) have been evaluated. For $q = 1$, these reduce to

(5. 17) $\left\{ \begin{aligned} & \left\{ \frac{\partial f_1}{\partial t} + \boldsymbol{\xi}^{(1)} \cdot \frac{\partial f_1}{\partial \mathbf{x}} \right\}' = \bar{f}^0_1 \left[\left(1 - \beta \frac{\partial p^0}{\partial n} \right) \mathbf{v}^{(1)} \cdot \left(\frac{1}{n} \frac{\partial n}{\partial \mathbf{x}} \right) + \right. \\ & + \left(\tfrac{1}{2} \beta m v^{(1)2} - \tfrac{3}{2} - \frac{\beta T}{n} \frac{\partial p^0}{\partial T} \right) \mathbf{v}^{(1)} \cdot \left(\frac{1}{T} \frac{\partial T}{\partial \mathbf{x}} \right) + \\ & + \left. (\tfrac{1}{3} \beta m v^{(1)2} - \tfrac{3}{2}) (\gamma + \tfrac{2}{3}) \frac{\partial}{\partial \mathbf{x}} \cdot \mathbf{u} + \beta m \mathbf{v}^{(1)} \cdot \overline{\left(\frac{\partial}{\partial \mathbf{x}} \mathbf{u} \right)} \cdot \mathbf{v}^{(1)} \right]. \end{aligned} \right.$

It remains to evaluate the expression

(5. 18) $$\left\{ \sum_{i=1}^{q} \frac{1}{m} \iint \frac{\partial f_{q+1}}{\partial \boldsymbol{\xi}^{(i)}} \cdot \frac{\partial \Phi^{(i\,q+1)}}{\partial \mathbf{x}^{(i)}} \, d\mathbf{x}^{(q+1)} \, d\boldsymbol{\xi}^{(q+1)} \right\}'$$

derived from the right-hand side of (2. 9), in a similar way. One has to remember that, whereas the values of the λ_k appearing in (5. 15) are those obtaining at the mean centre $\mathbf{x}$ of the q positions $\mathbf{x}^{(1)}, \mathbf{x}^{(2)} \ldots \mathbf{x}^{(q)}$, the corresponding values λ_k^* contained in the function $\bar{f}_{p+1}$ are those measured at the mean centre

$$\mathbf{x} + \sum_{i=1}^{q} (\mathbf{x}^{(q+1)} - \mathbf{x}^{(i)})/q(q+1)$$

of the $q + 1$ positions $\mathbf{x}^{(1)}, \mathbf{x}^{(2)} \ldots \mathbf{x}^{(q+1)}$. In order that the result obtained from (5.18) should be comparable with (5.15), one must therefore substitute in the former expression

(5.19) $$f'_{q+1} = \bar{f}_{q+1} + \left\{ \sum_k \frac{\partial \bar{f}^0_{q+1}}{\partial \lambda_k} \frac{\partial \lambda_k}{\partial \mathbf{x}} \right\} \cdot \left\{ \sum_{i=1}^{q} \frac{(\mathbf{x}^{(q+1)} - \mathbf{x}^{(i)})}{q(q+1)} \right\}$$

In particular, for $q = 1$, one substitutes

(5.20) $$f'_2 = \bar{f}'_2 + \tfrac{1}{2} \sum_k \frac{\partial f^0_2}{\partial \lambda_k} \frac{\partial \lambda_k}{\partial \mathbf{x}} \cdot \mathbf{r},$$

obtaining

(5.21) $$\left\{ \frac{1}{m} \iint \frac{\partial f_2}{\partial \boldsymbol{\xi}^{(2)}} \cdot \frac{\partial \phi}{\partial \mathbf{x}^{(1)}} d\mathbf{x}^{(2)} d\boldsymbol{\xi}^{(2)} \right\}' = \frac{1}{m} \iint \frac{\partial \bar{f}'_2}{\partial \mathbf{v}^{(1)}} \cdot \frac{\partial \phi}{\partial \mathbf{x}^{(1)}} d\mathbf{r} \, d\mathbf{v}^{(2)} + \\ + \frac{1}{2m} \sum_k \frac{\partial \lambda_k}{\partial \mathbf{x}} \cdot \frac{\partial}{\partial \lambda_k} \int \mathbf{r} \frac{\partial \bar{f}^0_2}{\partial \mathbf{v}^{(1)}} \cdot \frac{\partial \phi}{\partial \mathbf{x}^{(1)}} d\mathbf{r} \, d\mathbf{v}^{(2)}.$$

Since

$$\frac{1}{2m} \int \mathbf{r} \frac{\partial \bar{f}^0_2}{\partial \mathbf{v}^{(1)}} \cdot \frac{\partial \phi}{\partial \mathbf{x}^{(1)}} d\mathbf{r} \, d\mathbf{v}^{(2)} = \frac{\beta \bar{f}^0_1}{2n} \mathbf{v}^{(1)} \cdot \int \mathbf{r} \, \bar{n}^0_2 \, \phi'(r) \, r^{-1} \mathbf{r} \, d\mathbf{r}$$
$$= \bar{f}^0_1 \mathbf{v}^{(1)} (1 - \beta \, p^0/n),$$

it follows from a comparison of (5.17) and (5.21) that

(5.22) $$\left\{ \frac{\beta p^0 \bar{f}^0_1}{n} \left\{ (\tfrac{1}{2} \beta m v^{(1)2} - \tfrac{5}{2}) \mathbf{v}^{(1)} \cdot \left(\frac{1}{T} \frac{\partial T}{\partial \mathbf{x}} \right) + \beta m \mathbf{v}^{(1)} \cdot \left(\frac{\partial}{\partial \mathbf{x}} \mathbf{u} \right) \cdot \mathbf{v}^{(1)} \right\} + \right. \\ \left. + (\gamma + \tfrac{2}{3} \beta p^0/n) \bar{f}^0_1 (\tfrac{1}{2} \beta m v^{(1)2} - \tfrac{3}{2}) \frac{\partial}{\partial \mathbf{x}} \cdot \mathbf{u} = \frac{1}{m} \iint \frac{\partial \bar{f}'_2}{\partial \mathbf{v}^{(1)}} \cdot \frac{\partial \phi}{\partial \mathbf{x}^{(1)}} d\mathbf{r} \, d\mathbf{v}^{(2)} \right.$$

This result is the basis of the kinetic theory of non-uniform fluids, which will be developed in chapter VIII. A more general equation, to determine, $\bar{f}'_q$ for values of q greater than 1, is obtained by equating (5.15) with (5.18); no solution of this general equation has, however, so far been obtained.

6. The Friction Constant

Another interesting development of the fundamental equation (2. 7) was initiated by KIRKWOOD [1946]. His method differs from that so far followed in this chapter in that it depends on the comparison of states of the fluid at appreciably different instants of time. However, it will be seen in the next section that it provides a new approach to the determination of the coefficient of viscosity.

If one could follow the motion of the individual molecules in a liquid, in the same way as one can follow the motion of a Brownian particle, one would observe continuous but erratic trajectories, with a large degree of correlation between positions and velocities at closely following instants, but little or no correlation between configurations separated by sufficiently long intervals of time. Owing to the interaction with the other molecules, the velocity of a molecule, though known exactly at one instant, will become more and more unpredictable in the course of time; its most probable value will ultimately approximate to that (**u**) of the fluid as a whole. On the average, therefore, there is a tendency for a molecule whose velocity is specified at some initial time to be brought to rest relative to the mass motion of the fluid; the effect is the same as if the molecule were retarded by any force of a frictional nature. The force in question is, of course, none other than the resultant force due to the other molecules. It is true that the mean value of this force vanishes in a fluid in equilibrium; however, the mean *rate of change* of the resultant force does not vanish, so a molecule which is not at rest relative to the fluid will experience a deceleration during a finite interval of time, even when the fluid is in thermal and mechanical equilibrium.

Kirkwood has attempted to determine the mean deceleration experienced in a finite time interval by a molecule whose initial position and velocity in the fluid are known, by the use of a method suggested by the theory of the motion of a Brownian particle. In doing so, he found it necessary to appropriate some assumptions which are made in the theory of the Brownian motion — and which are sufficiently well justified in that connection.

Suppose that, at some initial time (t_0), the positions and velocities of the N molecules in the fluid are $\mathbf{x}_0^{(i)}$ and $\boldsymbol{\xi}_0^{(i)}$ ($i = 1, 2, \ldots N$). As a result of the motion and mutual interactions of the molecules, their positions and velocities at any later time (t) must have values ($\mathbf{x}^{(i)}$ and $\boldsymbol{\xi}^{(i)}$) which depend on the initial values ($\mathbf{x}_0^{(i)}$ and $\boldsymbol{\xi}_0^{(i)}$), and also on the time interval $\tau = t - t_0$ which has elapsed since the initial time. This dependence of $\mathbf{x}^{(i)}$ and $\boldsymbol{\xi}^{(i)}$ on $\mathbf{x}_0^{(i)}$, $\boldsymbol{\xi}_0^{(i)}$ and τ may be expressed by writing

(6. 1) $\qquad \boldsymbol{\xi}^{(i)} = \boldsymbol{\xi}^{(i)}(\boldsymbol{\xi}_0, \mathbf{x}_0, \tau), \qquad \mathbf{x}^{(i)} = \mathbf{x}^{(i)}(\boldsymbol{\xi}_0, \mathbf{x}_0, \tau),$

where $\boldsymbol{\xi}^{(i)}(\ )$ and $\mathbf{x}^{(i)}(\ )$ are certain functions of the variables indicated. The precise nature of these functions will not be required at present, but could be determined in principle by integrating the equations of motion of the system of N molecules. Instead of regarding the final positions and velocities $\mathbf{x}^{(i)}$ and $\boldsymbol{\xi}^{(i)}$ as functions of $\mathbf{x}_0^{(i)}$, $\boldsymbol{\xi}_0^{(i)}$ and τ, one may equally well regard the initial positions and velocities $\mathbf{x}_0^{(i)}$ and $\boldsymbol{\xi}_0^{(i)}$ as functions of $\mathbf{x}^{(i)}$, $\boldsymbol{\xi}^{(i)}$ and τ, thus:

(6. 2) $\qquad \boldsymbol{\xi}_0^{(i)} = \boldsymbol{\xi}_0^{(i)}(\boldsymbol{\xi}, \mathbf{x}, \tau), \qquad \mathbf{x}_0^{(i)} = \mathbf{x}_0^{(i)}(\boldsymbol{\xi}, \mathbf{x}, \tau).$

Now, if $\Phi(\mathbf{x})$ is the total potential energy of the system of molecules, expressed as a function of their positions $\mathbf{x}^{(i)}$, the instantaneous acceleration of the molecule at $\mathbf{x}^{(j)}$ is

$$-\frac{1}{m} \frac{\partial \Phi}{\partial \mathbf{x}^{(j)}};$$

hence, if the positions $\mathbf{x}^{(i)}$ are supposed to vary with the time (t) in the way indicated by (6. 1), the velocities at time t are given by

$$\boldsymbol{\xi}^{(j)} = \boldsymbol{\xi}_0^{(j)} - \int_{t_0}^{t} \frac{1}{m} \frac{\partial \Phi}{\partial \mathbf{x}^{(j)}} \, dt.$$

In particular, the change of velocity of the molecule situated at the point $\mathbf{x}^{(1)}$ at time t, during the interval τ, is

(6. 3) $\qquad \boldsymbol{\rho} = \boldsymbol{\xi}^{(1)} - \boldsymbol{\xi}_0^{(1)} = - \int_{t_0}^{t} \frac{1}{m} \frac{\partial \Phi}{\partial \mathbf{x}^{(1)}} \, dt.$

To determine the *average* change of velocity $[\overline{\boldsymbol{\rho}}(\boldsymbol{\xi}_0^{(1)}, \mathbf{x}_0^{(1)})]$ in the interval τ, of a molecule with velocity $\boldsymbol{\xi}_0^{(1)}$ and position $\mathbf{x}_0^{(1)}$ at time t_0, it is necessary to multiply $\boldsymbol{\rho}$ by the relative distribution function

$$f_N(\boldsymbol{\xi}_0, \mathbf{x}_0, t_0)/(N-1)! \, f_1(\boldsymbol{\xi}_0, \mathbf{x}_0, t_0),$$

and integrate over the initial positions and velocities of all molecules except the first:

(6. 4) $\qquad \overline{\boldsymbol{\rho}}(\boldsymbol{\xi}_0^{(1)}, \mathbf{x}_0^{(1)}) = \frac{N}{N!} \int \overset{(2N-2)}{\cdots} \int \frac{f_N(\boldsymbol{\xi}_0, \mathbf{x}_0, t_0)}{f_1(\boldsymbol{\xi}_0, \mathbf{x}_0, t_0)} \boldsymbol{\rho} \, d\boldsymbol{\xi}_0^{(2)} \, d\mathbf{x}_0^{(2)} \ldots d\boldsymbol{\xi}_0^{(N)} \, d\mathbf{x}_0^{(N)}.$

To evaluate the last expression, it is convenient to make use of some properties of the δ-function. This is defined in such a way that $\delta(\mathbf{x})$ is effectively zero when $\mathbf{x} \neq 0$, but so large near $\mathbf{x} = 0$ that

$$\int \delta(\mathbf{x}) \, d\mathbf{x} = 1.$$

Then an integral of the type

$$\int F(\mathbf{x}) \, \delta(\mathbf{x} - \mathbf{x}') \, d\mathbf{x},$$

where F is any function, will depend only on the value of F at the pole $\mathbf{x} = \mathbf{x}'$ of the δ-function, and

(6. 5) $\quad \int F(\mathbf{x}) \, \delta(\mathbf{x} - \mathbf{x}') \, d\mathbf{x} = F(\mathbf{x}') \int \delta(\mathbf{x} - \mathbf{x}') \, d\mathbf{x} = F(\mathbf{x}').$

Using the property (6. 5), it follows from (6. 4) that

(6. 6) $\quad \overline{\rho}(\boldsymbol{\xi}_t^{(1)}, \mathbf{x}_t^{(1)}) = \frac{N}{N!} \int \overset{(2N)}{\cdots\cdots} \int \frac{f_N(\boldsymbol{\xi}_0, \mathbf{x}_0, t_0)}{f_1(\boldsymbol{\xi}_0, \mathbf{x}_0, t_0)} \rho \, \delta(\boldsymbol{\xi}_0^{(1)} - \boldsymbol{\xi}_t^{(1)}) \, \delta(\mathbf{x}_0^{(1)} - \mathbf{x}_t^{(1)}) \times$
$\qquad \times d\boldsymbol{\xi}_0^{(1)} \, d\mathbf{x}_0^{(1)} \ldots d\boldsymbol{\xi}_0^{(N)} \, d\mathbf{x}_0^{(N)}.$

The next step is to change the variables of integration from $\mathbf{x}_0^{(i)}$ and $\boldsymbol{\xi}_0^{(i)}$ to $\mathbf{x}^{(i)}$ and $\boldsymbol{\xi}^{(i)}$. In doing so, one may use the fact that, since a set of molecules with positions $\mathbf{x}_0^{(i)}$ and velocities $\boldsymbol{\xi}_0^{(i)}$ at time t_0 must necessarily have the positions $\mathbf{x}^{(i)}$ and velocities $\boldsymbol{\xi}^{(i)}$, given by the equations (6. 1), at time t, the probabilities for the occurrence of these causally connected configurations must be the same:

(6. 7) $\quad \begin{cases} f_N(\boldsymbol{\xi}_0, \mathbf{x}_0, t_0) \, d\boldsymbol{\xi}_0^{(1)} \, d\mathbf{x}_0^{(1)} \ldots d\boldsymbol{\xi}_0^{(N)} \, d\mathbf{x}_0^{(N)} = \\ = f_N(\boldsymbol{\xi}, \mathbf{x}, t) \, d\boldsymbol{\xi}^{(1)} d\mathbf{x}^{(1)} \ldots d\boldsymbol{\xi}^{(N)} \, d\mathbf{x}^{(N)}. \end{cases}$

The principle used here is the same as that introduced in § 2. It then follows from (6. 6) that

(6. 8) $\quad \overline{\rho}(\boldsymbol{\xi}_t^{(1)}, \mathbf{x}_t^{(1)}) = \frac{N}{N!} \int \overset{(2N)}{\cdots\cdots} \int \frac{f_N(\boldsymbol{\xi}, \mathbf{x}, t)}{f_1(\boldsymbol{\xi}_0, \mathbf{x}_0, t_0)} \rho \, \delta(\boldsymbol{\xi}_0^{(1)} - \boldsymbol{\xi}_t^{(1)}) \, \delta(\mathbf{x}_0^{(1)} - \mathbf{x}_t^{(1)}) \times$
$\qquad \times d\boldsymbol{\xi}^{(1)} \, d\mathbf{x}^{(1)} \ldots d\boldsymbol{\xi}^{(N)} \, d\mathbf{x}^{(N)}.$

Clearly $\boldsymbol{\xi}_0^{(1)}$ and $\mathbf{x}_0^{(1)}$ now have to be regarded as functions of $\mathbf{x}^{(1)}$, $\boldsymbol{\xi}^{(1)}$ and τ, as indicated by (6. 2).

The formula (6. 8) is true for all values of t. One now introduces the supposition that $\tau = t - t_0$ is *small* by macroscopic standards. Then quantities of order τ can be neglected in comparison with those of order $\sqrt{\tau}$, etc. The *mean* change in the velocity of a molecule (ρ) in the interval τ is proportional to τ. On the other hand, the *actual* change of velocity of an individual molecule (ρ), measured by the mean square deviation from the mean, is normally much larger: the theory of Brownian motion (cf. § 3 of chapter VII, where these assumptions are analysed in detail) indicates that it is usually of order $\sqrt{\tau}$. Thus, although the difference τ between t_0 and t in (6. 8) can be neglected in comparison with $\sqrt{\tau}$, and the change $\mathbf{x}^{(1)} - \mathbf{x}_0^{(1)}$ in position of the molecule in the interval τ, which is proportional to τ, can also be neglected, only the *square* of the change in velocity $\boldsymbol{\rho} = \boldsymbol{\xi}^{(1)} - \boldsymbol{\xi}_0^{(1)}$ is negligible. So t_0 may be replaced by t in the denominator $f_1(\boldsymbol{\xi}_0, \mathbf{x}_0, t_0)$ of (6. 8), and $\mathbf{x}_0^{(1)}$ may be replaced by $\mathbf{x}^{(1)}$ in both the denominator $f_1(\boldsymbol{\xi}_0, \mathbf{x}_0, t)$ and the δ-function $\delta(\mathbf{x}_0^{(1)} - \mathbf{x}_t^{(1)})$; but it is necessary to expand

$$\delta(\boldsymbol{\xi}_0^{(1)} - \boldsymbol{\xi}_t^{(1)})/f_1(\boldsymbol{\xi}_0, \mathbf{x}, t)$$

in powers of $\boldsymbol{\rho} = \boldsymbol{\xi}^{(1)} - \boldsymbol{\xi}_0^{(1)}$ and retain the term proportional to $\boldsymbol{\rho}$. The part of the Taylor's expansion required is, in fact,

(6. 9) $\quad \dfrac{\delta(\boldsymbol{\xi}_0^{(1)} - \boldsymbol{\xi}_t^{(1)})}{f_1(\boldsymbol{\xi}_0, \mathbf{x}, t)} = \dfrac{\delta(\boldsymbol{\xi}^{(1)} - \boldsymbol{\xi}_t^{(1)})}{f_1(\boldsymbol{\xi}, \mathbf{x}, t)} + \dfrac{\boldsymbol{\rho}}{f_1} \cdot \left\{ \dfrac{\delta(\boldsymbol{\xi}^{(1)} - \boldsymbol{\xi}_t^{(1)})}{f_1(\boldsymbol{\xi}, \mathbf{x}, t)} \dfrac{\partial f_1}{\partial \boldsymbol{\xi}^{(1)}} - \dfrac{\partial}{\partial \boldsymbol{\xi}^{(1)}} \delta(\boldsymbol{\xi}^{(1)} - \boldsymbol{\xi}_t^{(1)}) \right\}.$

Substituting this in (6. 8), and performing an integration by parts to free the δ-function of differentiation, one obtains

(6. 10) $\quad \overline{\boldsymbol{\rho}}(\boldsymbol{\xi}_t^{(1)}, \mathbf{x}_t^{(1)}) = \dfrac{N}{N!} \int^{(2N)} \int \left\{ \dfrac{f_N}{f_1} \boldsymbol{\rho} + \left(\dfrac{1}{f_1} \dfrac{\partial f_1}{\partial \boldsymbol{\xi}^{(1)}} + \dfrac{\partial}{\partial \boldsymbol{\xi}^{(1)}} \right) \cdot \left(\dfrac{f_N}{f_1} \boldsymbol{\rho} \boldsymbol{\rho} \right) \right\} \delta(\boldsymbol{\xi}^{(1)} - \boldsymbol{\xi}_t^{(1)})$
$\hspace{3em} \times \, \delta(\mathbf{x}^{(1)} - \mathbf{x}_t^{(1)}) \, d\boldsymbol{\xi}^{(1)} \, d\mathbf{x}^{(1)} \dots d\boldsymbol{\xi}^{(N)} \, d\mathbf{x}^{(N)}$

The δ-functions can now be removed by a second application of the formula (6. 5); and, replacing the variables $\boldsymbol{\xi}_t^{(1)}$ and $\mathbf{x}_t^{(1)}$ by $\boldsymbol{\xi}^{(1)}$ and $\mathbf{x}^{(1)}$, one has finally

(6. 11) $\quad \overline{\boldsymbol{\rho}}(\boldsymbol{\xi}^{(1)}, \mathbf{x}^{(1)}) = \dfrac{N}{N!} \int^{(2N-2)} \int \left\{ \dfrac{f_N}{f_1} \boldsymbol{\rho} + \left(\dfrac{1}{f_1} \dfrac{\partial f_1}{\partial \boldsymbol{\xi}^{(1)}} + \dfrac{\partial}{\partial \boldsymbol{\xi}^{(1)}} \right) \cdot \left(\dfrac{f_N}{f_1} \boldsymbol{\rho} \boldsymbol{\rho} \right) \right\} \times$
$\hspace{3em} \times \, d\boldsymbol{\xi}^{(2)} \, d\mathbf{x}^{(2)} \dots d\boldsymbol{\xi}^{(N)} \, d\mathbf{x}^{(N)}.$

This may be compared with the corresponding formula obtained by replacing $\boldsymbol{\xi}_0^{(i)}$, $\mathbf{x}_0^{(i)}$ and t_0 by $\boldsymbol{\xi}^{(i)}$, $\mathbf{x}^{(i)}$ and t in (6. 4):

(6. 12) $\quad \overline{\boldsymbol{\rho}}(\boldsymbol{\xi}^{(1)}, \mathbf{x}^{(1)}) = \dfrac{N}{N!} \int^{(2N-2)} \int \dfrac{f_N}{f_1} \boldsymbol{\rho}' \, d\boldsymbol{\xi}^{(2)} \, d\mathbf{x}^{(2)} \dots d\boldsymbol{\xi}^{(N)} \, d\mathbf{x}^{(N)},$

where $\boldsymbol{\rho}'$ is the change in the velocity of the molecule in an interval τ *following* the time t. This is given by

(6. 13) $\quad \boldsymbol{\rho}' = - \int\limits_{t}^{t+\tau} \dfrac{1}{m} \dfrac{\partial \Phi}{\partial \mathbf{x}^{(1)}} \, dt.$

Taking half the sum of (6. 11) and (6. 12), one obtains the required result

(6. 14) $\quad \bar{\rho}(\boldsymbol{\xi}^{(1)}, \mathbf{x}^{(1)}) = \tau \left\{ \bar{\boldsymbol{\eta}}_1 + \frac{kT}{m} \left(\frac{1}{f_1} \frac{\partial f_1}{\partial \boldsymbol{\xi}^{(1)}} + \frac{\partial}{\partial \boldsymbol{\xi}^{(1)}} \right) \cdot \boldsymbol{\zeta}_1 \right\}$

where $\bar{\boldsymbol{\eta}}_1$ and $\boldsymbol{\zeta}_1$ are given by

(6. 15) $\quad \bar{\boldsymbol{\eta}}_1 = \frac{N}{N!} \int \overset{(2N-2)}{\cdots\cdots} \int \frac{f_N}{f_1} \frac{(\boldsymbol{\rho}+\boldsymbol{\rho}')}{2\tau} d\boldsymbol{\xi}^{(2)} d\mathbf{x}^{(2)} \ldots d\boldsymbol{\xi}^{(N)} d\mathbf{x}^{(N)},$

and [1])

(6. 16) $\quad \boldsymbol{\zeta}_1 = \frac{m\beta}{2\tau} \frac{N}{N!} \int \overset{(2N-2)}{\cdots\cdots} \int \frac{f_N}{f_1} \boldsymbol{\rho}\boldsymbol{\rho}\, d\boldsymbol{\xi}^{(2)} d\mathbf{x}^{(2)} \ldots d\boldsymbol{\xi}^{(N)} d\mathbf{x}^{(N)}.$

According to (6. 3) and (6. 13),

$$\frac{(\boldsymbol{\rho}+\boldsymbol{\rho}')}{2\tau} = -\frac{1}{2\tau} \int_{t-\tau}^{t+\tau} \frac{1}{m} \frac{\partial \Phi}{\partial \mathbf{x}^{(1)}} \, dt$$

is the mean acceleration in the interval between the times $t - \tau$ and $t + \tau$; and, as it changes sign when the signs of the relative coordinates $\mathbf{x}^{(i)} - \mathbf{x}^{(j)}$ and molecular velocities $\boldsymbol{\xi}^{(i)}$ are all reversed, whereas, *in equilibrium*, the distribution function f_N/f_1 is unchanged, the average value $\bar{\boldsymbol{\eta}}_1$, as defined in (6. 15), must vanish in equilibrium. If it is assumed, in addition, that in equilibrium $\boldsymbol{\zeta}_1$ reduces to some multiple ζ_1^0 of the unit tensor $\boldsymbol{\delta}$, (6. 14) reduces to

(6. 17) $\quad \begin{cases} \dfrac{\bar{\boldsymbol{\rho}}}{\tau} = \dfrac{kT}{m} \left(\dfrac{1}{f_1} \dfrac{\partial f_1}{\partial \boldsymbol{\xi}^{(1)}} + \dfrac{\partial}{\partial \boldsymbol{\xi}^{(1)}} \right) \zeta_1^0 \\ \quad = -\zeta_1^0 \boldsymbol{\xi}^{(1)} + \dfrac{kT}{m} \dfrac{\partial \zeta_1^0}{\partial \boldsymbol{\xi}^{(1)}} \end{cases}$

in equilibrium, the scalar quantity ζ_1^0 being given by

(6. 18) $\quad \zeta_1^0 = \frac{m\beta}{6\tau} \frac{N}{N!} \int \overset{(2N-2)}{\cdots\cdots} \int \frac{f_N}{f_1} \boldsymbol{\rho}^2 \, d\boldsymbol{\xi}^{(2)} d\mathbf{x}^{(2)} \ldots d\boldsymbol{\xi}^{(N)} d\mathbf{x}^{(N)}.$

The tensor $\boldsymbol{\zeta}_1$ is known as the friction tensor, and ζ_1^0 is the friction constant. The only restriction which has been imposed on τ is that it should be small by macroscopic standards. When $\tau \to 0$, $\boldsymbol{\rho} \sim \tau$, so ζ_1^0 vanishes. Kirkwood, however, guided by the theory of Brownian motion, supposes that there is a wide range of values of τ, in between the very small values where ζ_1^0 is small and the large values for which (6. 17) no longer holds, where ζ_1^0 is almost constant in value. He also

---
[1]) It is clear from the definition that $\boldsymbol{\zeta}_1$ must be a tensor.

considers that this 'plateau' value will be almost independent of the velocity of the molecule. Then (6.17) suggests an almost perfect analogue of Langevin's equation in the theory of Brownian motion, namely

(6. 19) $$\frac{d}{dt}\boldsymbol{\xi} = -\zeta_1^0 \boldsymbol{\xi} + \mathbf{g}(t)$$

This states simply that the acceleration of a molecule must be the resultant of the average value $-\zeta_1^0 \boldsymbol{\xi}$ indicated by (6. 17), and a second term $\mathbf{g}(t)$ which fluctuates rapidly in such a way that its average value vanishes in equilibrium for intervals of time comparable with τ.

The great merit of Kirkwood's work lies in the expression (6. 18) which it provides for the friction constant in terms of the molecular forces. The plateau value of this expression is not, however, very easy to determine. In a subsequent paper [1950], KIRKWOOD, BUFF & GREEN quote an estimate

(6. 20) $$\zeta_1^0 = \left[\frac{4\pi a_0}{3\,mn}\int_0^\infty n_2(r)\,\frac{d}{dr}\{r^2\,\phi'(r)\}\,dr\right]^{\frac{1}{2}}$$

which they made in order to verify their formula for the coefficient of viscosity. The indications are that this provides at least a reliable order of magnitude for the friction constant.

7. Applications of the Theory

Next the application of Kirkwood's theory of the friction constant to non-uniform fluids will be considered. For this purpose it is necessary to return to the main result (6. 14) of the previous section, which can also be written in the form

(7. 1) $$\overline{\boldsymbol{\eta}}_1 = \frac{\overline{\boldsymbol{\rho}}}{\tau} - \frac{kT}{m}\left(\frac{1}{f_1}\frac{\partial f_1}{\partial \boldsymbol{\xi}^{(1)}} + \frac{\partial}{\partial \boldsymbol{\xi}^{(1)}}\right)\cdot \boldsymbol{\zeta}_1.$$

It will be supposed that there are no external forces; then, since $\overline{\boldsymbol{\eta}}_1$ vanishes in equilibrium, one may subtract from the right-hand side of this equation the part which does not depend on the gradients of density (n), temperature (T) and macroscopic velocity $(\mathbf{u})$, namely

$$\left(\frac{\overline{\boldsymbol{\rho}}}{\tau}\right)^0 + \left(\boldsymbol{\xi}^{(1)} - \mathbf{u} - \frac{kT}{m}\frac{\partial}{\partial \boldsymbol{\xi}^{(1)}}\right)\cdot \boldsymbol{\zeta}_1.$$

The result is

(7. 2) $$\overline{\boldsymbol{\eta}}_1 = \mathbf{g}_1 - \left(\boldsymbol{\xi}^{(1)} - \mathbf{u} + \frac{kT}{mf_1}\frac{\partial f_1}{\partial \boldsymbol{\xi}^{(1)}}\right)\cdot \boldsymbol{\zeta}_1,$$

if one represents by g_1 the part of $\overline{\rho}/\tau$ which is due to departures from equilibrium. Deviations of the friction tensor from its equilibrium value are neglected.

The formula (7. 2), for the mean acceleration of a molecule whose position and velocity at time t are $\mathbf{x} = \mathbf{x}^{(1)}$ and $\boldsymbol{\xi} = \boldsymbol{\xi}^{(1)}$ respectively, can be substituted into (2. 3); the result is

$$(7.3) \quad \frac{\partial f_1}{\partial t} + \boldsymbol{\xi} \cdot \frac{\partial f_1}{\partial \mathbf{x}} + \frac{\partial}{\partial \boldsymbol{\xi}} \cdot \left[f_1 \mathbf{g}_1 - \left\{ (\boldsymbol{\xi} - \mathbf{u}) f_1 + \frac{kT}{m} \frac{\partial f_1}{\partial \boldsymbol{\xi}} \right\} \cdot \boldsymbol{\zeta}_1 \right] = 0.$$

This may be compared with a very similar equation [(3. 8) of chapter VII] which appears in the theory of the Brownian motion. It is naturally to be expected that, by making assumptions equivalent to those on which the theory of the Brownian motion is founded, closely similar results should emerge. In the usual theory of Brownian motion, however, no account is taken of the nature of the interaction between the molecules and the Brownian particle, with the result that the friction constant has to be taken from experience. Here it is possible to trace the molecular mechanism which determines the diffusion of a particular molecule through the fluid. Whether the assumptions introduced in the theoretical study of Brownian particles are really applicable in the ultra-microscopic domain can only be decided by the ultimate success of Kirkwood's theory.

An equation, completely analogous to (7. 3), can be derived for the distribution function f_q relating to q molecules with assigned positions and velocities. If one repeats the calculation of the previous section to determine the mean acceleration of a molecule, situated in a group of q molecules whose positions and velocities are $\mathbf{x}^{(i)}$ and $\boldsymbol{\xi}^{(i)}$ ($i = 1, 2, \ldots q$) at time t, one obtains

$$(7.4) \quad \overline{\boldsymbol{\eta}_q^{(j)}} = \frac{\overline{\boldsymbol{\rho}_q^{(j)}}}{\tau} - \frac{kT}{m} \sum_{i=1}^{q} \left(\frac{1}{f_q} \frac{\partial f_q}{\partial \boldsymbol{\xi}^{(i)}} + \frac{\partial}{\partial \boldsymbol{\xi}^{(i)}} \right) \cdot \boldsymbol{\zeta}_q^{(ij)}$$

where $\overline{\boldsymbol{\rho}_q^{(j)}}$ is the mean value of $\boldsymbol{\rho}^{(j)}$, the change of velocity in time τ of the molecule at $\mathbf{x}^{(j)}$; and the generalized friction tensor $\boldsymbol{\zeta}_q^{(ij)}$ is defined by

$$(7.5) \quad (N-q)!\, \boldsymbol{\zeta}_q^{(ij)} = \frac{m\beta}{2\tau} \int^{2(N-q)} \int \frac{f_N}{f_q} \boldsymbol{\rho}^{(i)} \boldsymbol{\rho}^{(j)}\, d\boldsymbol{\xi}^{(q+1)}\, d\mathbf{x}^{(q+1)} \ldots d\boldsymbol{\xi}^{(N)}\, d\mathbf{x}^{(N)}.$$

In equilibrium, the mean force acting on the molecule at $\mathbf{x}^{(j)}$ is

$$(7.6) \quad m\,(\overline{\boldsymbol{\eta}_q^{(j)}})^0 = -\frac{\partial \Phi_q}{\partial \mathbf{x}^{(j)}} - \int \frac{n_{q+1}}{n_q} \frac{\partial \Phi^{(j,q+1)}}{\partial \mathbf{x}^{(j)}}\, d\mathbf{x}^{(q+1)},$$

and (7.4) reduces to

$$(\overline{\eta_q^{(j)}})^0 = \left(\overline{\frac{\rho_q^{(j)}}{\tau}}\right)^0 + \sum_{i=1}^{q} \left(\xi^{(i)} - \mathbf{u} - \frac{kT}{m}\frac{\partial}{\partial \xi^{(i)}}\right) \cdot \zeta_q^{(ij)};$$

hence, in general,

(7.7) $$\overline{\eta_q^{(j)}} = (\overline{\eta_q^{(j)}})^0 + \mathbf{g}_q^{(j)} - \sum_{i=1}^{q} \left(\xi^{(i)} - \mathbf{u} + \frac{kT}{mf_q}\frac{\partial f_q}{\partial \xi^{(i)}}\right) \cdot \zeta_q^{(ij)},$$

where $\mathbf{g}_q^{(j)}$ denotes the departure of $\overline{\rho_q^{(j)}}/\tau$ from its equilibrium value. When the formula (7.6) is substituted into (2.4), one obtains, as the generalization of (7.3),

(7.8) $$\begin{cases} \frac{\partial f_q}{\partial t} + \sum_{i=1}^{q} \xi^{(i)} \cdot \frac{\partial f_q}{\partial \mathbf{x}^{(i)}} + \sum_{j=1}^{q} \frac{\partial}{\partial \xi^{(j)}} \cdot \left[f_q(\overline{\eta_q^{(j)}})^0 + f_q \mathbf{g}_q^{(j)} - \right. \\ \left. - \sum_{i=1}^{q} \left\{(\xi^{(i)} - \mathbf{u}) f_q + \frac{kT}{m}\frac{\partial f_q}{\partial \xi^{(i)}}\right\} \cdot \zeta_q^{(ij)}\right] = 0. \end{cases}$$

By setting $q = 2$ in (7.8), one obtains an equation which, in principle, determines f_2, and hence n_2, when the friction tensor $\zeta_q^{(ij)}$ is regarded as known. Since the deviation of n_2 from its equilibrium value determines the coefficient of viscosity in liquids, one may hope to derive an explicit formula for this coefficient by the solution of a differential equation. Some progress towards this end has been made by KIRKWOOD, BUFF & GREEN [1949]; these authors, however, found it necessary to make the following approximations:

(i) The term $\mathbf{g}_q^{(j)}$ in (7.7) and (7.8) was neglected. Thus it was assumed that the mean change of velocity $\overline{\rho_q^{(j)}}$ in the time interval τ is not affected by the existence of a velocity gradient. This seems a reasonable assumption when τ has the moderately large values required by the theory, though its precise effect is difficult to determine. It should be observed, also, that $\mathbf{g}_q^{(j)}$ may *not* be neglected when there is a gradient of temperature or density in the fluid. Further, $\mathbf{g}_q^{(j)}$ cannot be zero when the fluid is suffering dilatation; this would invalidate any formula for the coefficient of bulk viscosity derived by the neglect of the term in question.

(ii) The friction tensor $\zeta_q^{(ij)}$ was approximated by the diagonal tensor $\zeta_1^0 \delta_{ij} \boldsymbol{\delta}$, which vanishes when $i \neq j$, and has the equilibrium value $\zeta_1^0 \boldsymbol{\delta}$ appropriate to a single molecule when $i = j$. The error involved in this approximation is probably small, and might in principle be eliminated by a careful study of the generalized friction tensors.

With these approximations, (7. 8) reduces to

$$
(7.9) \quad \left\{ \frac{\partial f_q}{\partial t} + \sum_{i=1}^{q} \left[\boldsymbol{\xi}^{(i)} \cdot \frac{\partial f_q}{\partial \mathbf{x}^{(i)}} + \frac{\partial}{\partial \boldsymbol{\xi}^{(i)}} \cdot \left\{ f_q (\overline{\boldsymbol{\eta}_q^{(i)}})^0 - \zeta_1^0 (\boldsymbol{\xi}^{(i)} - \mathbf{u}) f_q - \zeta_1^0 \frac{kT}{m} \frac{\partial f_q}{\partial \boldsymbol{\xi}^{(i)}} \right\} \right] \right\} = 0.
$$

Taking $q = 1$, one may multiply (7. 9) by the tensor $m(\boldsymbol{\xi} - \mathbf{u})(\boldsymbol{\xi} - \mathbf{u})$ and integrate over all values of $\boldsymbol{\xi}$. The expression

$$
\frac{\partial}{\partial \mathbf{x}} \cdot \int f_1 (\boldsymbol{\xi} - \mathbf{u})(\boldsymbol{\xi} - \mathbf{u})(\boldsymbol{\xi} - \mathbf{u}) \, d\boldsymbol{\xi}
$$

can involve only gradients of density and temperature, and therefore vanishes under the conditions now contemplated. Thus one obtains

$$
(7.10) \quad \frac{\partial \mathbf{k}}{\partial t} + \frac{\partial}{\partial \mathbf{x}} \cdot (\mathbf{u}\,\mathbf{k}) + 2\,\mathbf{k} \cdot \left(\overline{\frac{\partial}{\partial \mathbf{x}} \mathbf{u}} \right) + 2\,\zeta_1^0 (\mathbf{k} - n\,kT\,\boldsymbol{\delta}) = 0,
$$

where

$$
(7.11) \quad \mathbf{k} = m \int f_1 (\boldsymbol{\xi} - \mathbf{u})(\boldsymbol{\xi} - \mathbf{u}) \, d\boldsymbol{\xi}
$$

is the part of the pressure tensor due to the thermal motion. Separating from (7. 10) the part linear in the velocity gradient, one has

$$
(7.12) \quad \mathbf{k}' = - (nkT/\zeta_1^0) \left(\overline{\frac{\partial}{\partial \mathbf{x}} \mathbf{u}} \right).
$$

In a rare gas, the coefficient of viscosity should therefore be $\tfrac{1}{2} nkT/\zeta_1^0$

It should also be possible to determine the deformation (n_2') of the molecular distribution function from (7. 9) with $q = 2$, without further approximation, and thus to evaluate the coefficient of viscosity in liquids. No exact solution of (7. 9) has yet been obtained, however, and in the meantime it does not seem possible to improve on an argument derived from macroscopic hydrodynamics. It is there shown that the force exerted on a sphere of radius a_0, moving with velocity $\boldsymbol{\xi}$ in a fluid of viscosity η', is $-6\pi\eta' a_0 \boldsymbol{\xi}$. If one assumes that this formula would remain correct for a sphere of molecular dimensions, one may infer from (6. 19) the relation

$$
m\zeta_1^0 = 6\pi\eta' a_0
$$

between the friction constant ζ_1^0 and η'. The viscosity derived in this way, however, evidently would not include the effect of the thermal motion; hence, in general, one would expect

(7. 13) $$\eta = m\zeta_1^0/6\pi a_0 + \tfrac{1}{2} nkT/\zeta_1^0$$

where a_c is the molecular radius. This is, essentially, the result obtained by Kirkwood and his collaborators by more devious considerations. If one employs their estimated formula (6. 20) for the friction constant, one obtains, in liquid argon at 89° K., the value 4.84×10^{-10} gm/sec for $m\zeta_1^0$. Then, taking $a_0 = 1.9$ Å, which is one half of the actual distance between nearest neighbours in the fluid, one finds $\eta = 1 \cdot 4 \times 10^{-3}$ poise, against the experimental value of $2 \cdot 39 \times 10^{-3}$ poise. The discrepancy, which is not large in view of the approximations employed, may be due either to the failure of the assumption on which (7. 13) was derived, or to an underestimate of the value of ζ_1^0.

It has not so far been possible to correlate Kirkwood's expression for the coefficient of viscosity with Andrade's older formula. The kinetic theory of fluids (§ 6 of chapter VIII) leads to the formula

$$\eta = \frac{5}{64}\left(\frac{m}{\pi\beta}\right)^{\frac{1}{2}} \frac{1}{a_0^2}$$

for the coefficient of viscosity of a gas with hard spherical molecules of radius a_c, so comparison with (7. 12) would lead to the identification

$$\zeta_1^0 = \frac{32}{5}\left(\frac{\pi}{\beta m}\right)^{\frac{1}{2}} na_0^2$$

for gases. On the other hand, the average number of collisions per unit time of a molecule with other molecules in a gas is $16\pi^{\frac{1}{2}}(\beta m)^{-\frac{1}{2}}na_0^2$, so ζ_1^0 is a measure of the collision frequency. One has in this connection, therefore, a quantity in some sense analogous to the vibrational frequency which (cf. § 4. 1) appears in Andrade's formula. How the analogy is preserved in condensed systems is not known.

8. Motion of Molecular Clusters

The equations of continuity, motion and thermal flow, represented by (3. 3), (3. 9) and (3. 12) respectively, have so far been regarded as formulating the macroscopic laws of conservation of matter, momentum and energy. When, however, the same equations are regarded simply as specifying the rate of change with time of the variables n, $\mathbf{u}$ and T, they obviously have another interpretation. The variable $n(t, \mathbf{x})$ was originally defined as the probability per unit volume at time t of finding a molecule at the point $\mathbf{x}$. Similarly, $\mathbf{u}(t, \mathbf{x})$ and

$\frac{1}{2}kT(t, \mathbf{x})$ were defined as the mean velocity and mean kinetic energy respectively at time t of a molecule at the point $\mathbf{x}$. From this point of view, the conservation equations evidently relate to the mean motion of a single molecule in the fluid. This interpretation of the macroscopic laws suggests the possibility of describing in a similar way the mean motion of *clusters* of molecules in the fluid.

One might, for example, hope to obtain in this way information concerning the rate of change in time of the molecular distribution functions n_q relating to a group of two or more molecules in the fluid, and also the way in which their mean velocities and energies are determined by their relative positions and the motion of the fluid as a whole. The equations (3. 2) and (3. 6) are clearly particular instances of the generalized conservation law

(8. 1) $$\frac{\partial n_q}{\partial t} + \sum_{i=1}^{q} \frac{\partial}{\partial \mathbf{x}^{(i)}} \cdot (n_q \mathbf{u}_q^{(i)}) = 0,$$

in which $\mathbf{u}_q^{(i)}$ is the mean velocity of the molecule at $\mathbf{x}^{(i)}$, in a group of q molecules with specified positions $\mathbf{x}^{(1)}, \ldots \mathbf{x}^{(q)}$, given by

(8. 2) $$n_q \mathbf{u}_q^{(i)} = \int \overset{(q)}{\cdots} \int f_q \boldsymbol{\xi}^{(i)} d\boldsymbol{\xi}^{(1)} \ldots d\boldsymbol{\xi}^{(q)}.$$

The equation (8. 1) is, in fact, easily obtained by integrating (2. 4) or (2. 9) over all values of the velocities $\boldsymbol{\xi}^{(1)}, \ldots \boldsymbol{\xi}^{(q)}$. If $\left(\frac{d}{dt}\right)_q$ represents the operator $\frac{\partial}{\partial t} + \sum_{i=1}^{q} \mathbf{u}_q^{(i)} \cdot \frac{\partial}{\partial \mathbf{x}^{(i)}}$, (8. 1) can be written in the form

(8. 3) $$\left(\frac{d}{dt}\right)_q n_q + n_q \sum_{i=1}^{q} \frac{\partial}{\partial \mathbf{x}^{(i)}} \cdot \mathbf{u}_q^{(i)} = 0,$$

analogous to (3. 3).

A generalized equation of motion analogous to (3. 8) can be obtained in a similar way, by first multiplying (2. 9) by the velocity $\boldsymbol{\xi}^{(i)}$, and then integrating over all velocities; the result is

(8. 4) $$\begin{cases} n_q \left(\frac{d}{dt}\right)_q \mathbf{u}_q^{(i)} + \sum_{j=1}^{q} \frac{\partial}{\partial \mathbf{x}^{(j)}} \cdot \int \overset{(q)}{\cdots} \int f_q (\boldsymbol{\xi}^{(j)} \boldsymbol{\xi}^{(i)} - \mathbf{u}_q^{(j)} \mathbf{u}_q^{(i)}) d\boldsymbol{\xi}^{(1)} \ldots d\boldsymbol{\xi}^{(q)} \\ + \left\{ \frac{n_q}{m} \sum_{j=1}^{q} \frac{\partial \phi^{(ij)}}{\partial \mathbf{x}^{(i)}} + \int \frac{n_{q+1}}{m} \frac{\partial \phi^{(i, q+1)}}{\partial \mathbf{x}^{(i)}} d\mathbf{x}^{(q+1)} \right\} = 0. \end{cases}$$

By a suitable definition of $\mathbf{p}_q^{(ij)}$, this equation can be written in the form

(8. 5) $$mn_q \left(\frac{d}{dt}\right)_q \mathbf{u}_q^{(i)} + \sum_{j=1}^{q} \frac{\partial}{\partial \mathbf{x}^{(j)}} \cdot \mathbf{p}_q^{(ji)} = 0,$$

which is analogous to (3. 9). The mean motion of the molecule at $\mathbf{x}^{(i)}$ may thus be regarded as due to the resultant of a number of forces, including the gradient of the pressure tensor $\mathbf{p}_q^{(ii)}$ as modified by the presence of the other molecules, and the gradients of the tensors $\mathbf{p}_q^{(ji)}$ ($j \neq i$) which represent the direct action of these molecules.

Finally, a generalized equation of energy transfer, analogous to (3. 16), can be obtained by (i) multiplying (2. 9) by $\frac{1}{2}m\boldsymbol{\xi}^{(i)2}$ and integrating over all velocities, (ii) forming the time derivative of the mean potential energy, weighted with the factor $\frac{1}{2}n_q$:

$$\frac{1}{2} n_q \sum_{j=1}^{q} \phi^{(ij)} + \frac{1}{2} \int n_{q+1} \, \phi^{(i,\,q+1)} \, d\mathbf{x}^{(q+1)}$$

of the molecule at $\mathbf{x}^{(i)}$, with the help of (8. 1), and (iii) adding the results. One obtains in this way

$$(8.\,6) \quad \begin{cases} n_q \left(\dfrac{d}{dt}\right)_q E_q^{(i)} + \sum_{j=1}^{q} \dfrac{\partial}{\partial \mathbf{x}^{(j)}} \cdot \left\{ \int \overset{(q)}{\ldots} \int f_q (\boldsymbol{\xi}^{(j)} - \mathbf{u}_q^{(j)}) \, \tfrac{1}{2} m \, \boldsymbol{\xi}^{(i)2} \, d\boldsymbol{\xi}^{(1)} \ldots d\boldsymbol{\xi}^{(q)} + \right. \\ \left. + \tfrac{1}{2} \int n_{q+1} (\mathbf{u}_{q+1}^{(j)} - \dot{\mathbf{u}}_q^{(j)}) \, \phi^{(i,\,q+1)} \, d\mathbf{x}^{(q+1)} \right\} + \tfrac{1}{2} n_q \sum_{j=1}^{q} (\mathbf{u}_q^{(j)} + \mathbf{u}_q^{(i)}) \cdot \dfrac{\partial \phi^{(ij)}}{\partial \mathbf{x}^{(i)}} + \\ + \tfrac{1}{2} \int n_{q+1} (\mathbf{u}_{q+1}^{(j)} + \mathbf{u}_{q+1}^{(q+1)}) \cdot \dfrac{\partial \phi^{(i,\,q+1)}}{\partial \mathbf{x}^{(i)}} \, d\mathbf{x}^{(q+1)} = 0. \end{cases}$$

This, again, can be written in the form

$$(8.\,7) \qquad n_q \left(\frac{d}{dt}\right)_q E_q^{(i)} + \sum_{j=1}^{q} \frac{\partial}{\partial \mathbf{x}^{(j)}} \cdot (\mathbf{q}_q^{(ji)} + \mathbf{p}_q^{(ji)} \cdot \mathbf{u}_q^{(j)}) = 0$$

analogous to (3. 14), provided $\mathbf{q}_q^{(ji)}$, the generalized thermal flux vector, is defined in the appropriate way.

The three generalized conservation equations (8. 3), (8. 5) and (8. 7) describe the motion of a group of q molecules with the instantaneous positions $\mathbf{x}^{(1)}, \ldots \mathbf{x}^{(q)}$, among the wider assembly of molecules which constitutes the fluid. Of course, all the quantities n_q, $\mathbf{u}_q^{(i)}$, $E_q^{(i)}$, $\mathbf{p}_q^{(ij)}$ and $\mathbf{q}_q^{(ij)}$ depend on the relative coordinates of the molecules: the motion of each molecule is greatly influenced by the positions of the other molecules in the cluster.

CHAPTER VI

COMPLEX FLUIDS AND FLUID MIXTURES

1. Complex Structures

It was pointed out in § 7 of chapter III that in most fluids the intermolecular forces have a rather complicated character, and, except for the inert substances and other rather special types of fluids, it is an idealization to represent the mutual potential energy of two molecules as a function of the distance between their mean centres. The advantage, which has been exploited in this book, of devoting attention mainly to fluids with effectively spherical molecules, is that one can thereby isolate for study those features of molecular flow which do not depend on the structure of the molecules themselves. From the theoretical point of view, the investigation of the properties of fluids whose molecules have a simple structure is logically prior to the study of the complications which arise when the molecules have a more complex structure. However, the experimental study of 'complex' fluids has disclosed many features of interest which will ultimately require theoretical explanation. It cannot be said that much of a rigorous, quantitative nature has yet been achieved in this direction, and progress will probably be limited to somewhat qualitative or formal developments for some time. The reason is that the structure of even a single molecule containing more than a few atoms is not easy to investigate in its quantitative aspects; although the principles involved — which are mainly of a quantum-mechanical character — are fully understood, their detailed application is beset by mathematical difficulties. The question of the interaction between two or more molecules presents the same type of difficulty, where forces of a chemical nature are involved; in fact, the only problems which are strictly amenable to treatment by classical methods are those in which it is legitimate to formulate a potential energy function and derive the equations of motion on the basis of Newtonian mechanics.

Some of the general results established in the foregoing chapters can be applied to fluids, irrespective of their complexity. This is true, for example, of Boltzmann's law in the general form derived in § 5. 1 of chapter II. The probability of finding a molecule in a rare gas in a state of energy E is $\exp \beta(F_1 - E)$, where $\beta = 1/kT$ and F_1 is the free energy per molecule. The energy levels of a molecule are not, however, in general continuous; there usually exists a set of discrete energy states, and transitions between two states of different energy are necessarily discontinuous. This has an important effect on such macroscopic properties as the specific heat of a gas. If the molecules have z atoms, and therefore $3z$ degrees of freedom, one would expect the internal energy per molecule to be $3zkT/2$ on the basis of the classical theory. Often, however, many of the degrees of freedom are, partly or entirely, inhibited by the existence of the 'gaps' in the energy spectrum, and the observed specific heat falls considerably below the classical prediction. This could be explained quantitatively only by the determination of the quantum-mechanical states of the system and their associated energies. At low temperatures, most molecules may probably be regarded as rigid structures; at very high temperatures, they approximate more closely to the classical model.

It would obviously be difficult to take full account of such intricacies of molecular structure in formulating a theory of condensed fluids; but it has to be recognized that, without doing so, discrepancies between the theoretical and experimental values of certain macroscopic quantities could not be eliminated. Foremost among the macroscopic quantities which are sensitive to molecular structure are the internal energy and specific heat already mentioned, and the coefficient of thermal conduction. The hydrostatic pressure and the coefficient of viscosity, on the other hand, are not so gravely affected.

On this understanding, one may proceed to describe the configuration of any molecule by a number of coordinates $x_1, x_2, \ldots x_s$, of which x_1, x_2, x_3 are as usual the coordinates of the centre of gravity of the molecule, and the remainder specify its internal configuration: the orientation of the molecule in space, and the distance between any two atoms, if this is a genuine variable. The potential energy of the ith molecule in isolation is then a function $\psi^{(i)} = \psi(\mathbf{x}^{(i)})$ of its coordinates $x_a^{(i)}$; and the total kinetic energy of the molecule is a quadratic function $\frac{1}{2} \sum_{a,b} g_{ab}^{(i)} \xi_a^{(i)} \xi_b^{(i)}$ of the derivatives $\xi_a^{(i)}$ of the coordinates $x_a^{(i)}$, with respect to time. Similarly the mutual potential energy of two molecules, whose internal coordinates are $x_a^{(i)}$ and $x_a^{(j)}$, may be represented as a function $\phi^{(ij)} = \phi(\mathbf{x}^{(i)}, \mathbf{x}^{(j)})$ of the external and internal coordinates of both. The total energy of a system of N molecules is then

$$(1.1) \qquad E = \tfrac{1}{2} \sum_{i,a,b} g_{ab}^{(i)} \xi_a^{(i)} \xi_b^{(i)} + \sum_i \psi^{(i)} + \tfrac{1}{2} \sum_{i,j} \phi^{(ij)},$$

and the Lagrangian equations of motion of the system are

$$(1.2) \qquad \sum_b g_{ab}^{(i)} \frac{d\xi_b^{(i)}}{dt} + \sum_j \frac{\partial \phi^{(ij)}}{\partial x_a^{(i)}} = F_a^{(i)},$$

where

$$(1.3) \qquad F_a^{(i)} = - \left\{ \frac{\partial \psi^{(i)}}{\partial x_a^{(i)}} + \tfrac{1}{2} \sum_{b,c} \left(2 \frac{\partial g_{ab}^{(i)}}{\partial x_c^{(i)}} - \frac{\partial g_{bc}^{(i)}}{\partial x_a^{(i)}} \right) \xi_b^{(i)} \xi_c^{(i)} \right\}$$

is the generalized force for the ith molecule, in the absence of the others. If $\gamma_{ab}^{(i)}$ is the matrix 'reciprocal' to $g_{ab}^{(i)}$, obtained by solving the algebraic equations

$$(1.4) \qquad \sum_c \gamma_{ac}^{(i)} g_{cb}^{(i)} = \begin{cases} 1, & a = b \\ 0, & a \neq b, \end{cases}$$

the generalized acceleration is obtained from (1.2) in the form

$$(1.5) \qquad \eta_a^{(i)} = \frac{d\xi_a^{(i)}}{dt} = \sum_b \gamma_{ab}^{(i)} \left(F_b^{(i)} - \sum_j \frac{\partial \phi^{(ij)}}{\partial x_b^{(i)}} \right).$$

To obtain a formal generalization of the theory of the preceding chapter, it is necessary only to define a set of generalized molecular and velocity distribution functions, in the following way. Let the 'volume element' of coordinate space, containing points whose coordinates lie in the intervals x_1 to $x_1 + dx_1$, x_2 to $x_2 + dx_2$, ... x_s to $x_s + dx_s$ respectively, be represented by $d\mathbf{x}_s$. Also, let the probability of finding a molecule with coordinates within this generalized volume element be $n d\mathbf{x}_s$. Similarly, let the probability of finding q molecules with coordinates within the volume elements $d\mathbf{x}_s^{(1)} \ldots d\mathbf{x}_s^{(q)}$ respectively be $n_q d\mathbf{x}_s^{(1)} \ldots d\mathbf{x}_s^{(q)}$. Then n_q will depend on the time t, as well as the internal and external coordinates of each of the q molecules. Further, let the element of generalized velocity space in which the components of the generalized velocity vector have values in the intervals ξ_1 to $\xi_1 + d\xi_1$, ... ξ_s to $\xi_s + d\xi_s$ respectively be represented by $d\boldsymbol{\xi}_s$. Also, let the probability at time t of finding q molecules in the elements $d\boldsymbol{\xi}_s^{(1)} d\mathbf{x}_s^{(1)}, \ldots d\boldsymbol{\xi}_s^{(q)} d\mathbf{x}_s^{(q)}$ of phase space be $f_q d\boldsymbol{\xi}_s^{(1)} \ldots d\boldsymbol{\xi}_s^{(q)} d\mathbf{x}_s^{(1)} \ldots d\mathbf{x}_s^{(q)}$.

Then the generalized velocity distribution functions f_q will satisfy

(1.6) $$\int \overset{(q)}{\cdots} \int f_q \, d\boldsymbol{\xi}_s^{(1)} \ldots d\boldsymbol{\xi}_s^{(q)} = n_q$$

and

(1.7) $$\iint f_{q+1} \, d\boldsymbol{\xi}_s^{(q+1)} \, d\mathbf{x}_s^{(q+1)} = (N-q) f_q.$$

To obtain a generalized continuity equation, it is necessary to consider the evolution of the molecular distribution in time. If a molecule is in the element $d\mathbf{x}_s^{(i)}$ of coordinate space at time t, it must have had the coordinates $x_a^{(i)} - \xi_a^{(i)} \delta t$ at time $t - \delta t$, if $\xi_a^{(i)}$ was the generalized velocity corresponding to the coordinate $x_a^{(i)}$. Hence

(1.8) $$\int \overset{(q)}{\cdots} \int f_q (\boldsymbol{\xi}^{(i)}, \mathbf{x}^{(i)} - \boldsymbol{\xi}^{(i)} \delta t, t - \delta t) \, d\boldsymbol{\xi}_s^{(1)} \ldots d\boldsymbol{\xi}_s^{(q)} = n_q.$$

Subtracting this result from (1.6), one obtains

(1.9) $$\frac{\partial n_q}{\partial t} + \sum_{i,a} \frac{\partial (n_q u_{qa}^{(i)})}{\partial x_a^{(i)}} = 0,$$

where $u_{qa}^{(i)}$ is the mean generalized velocity, given by

(1.10) $$n_q u_{qa}^{(i)} = \int \overset{(q)}{\cdots} \int f_q \xi_a^{(i)} \, d\boldsymbol{\xi}_s^{(1)} \ldots d\boldsymbol{\xi}_s^{(q)}.$$

By a closely similar argument, one may obtain

(1.11) $$\frac{\partial f_q}{\partial t} + \sum_{i,a} \left\{ \xi_a^{(i)} \frac{\partial f_q}{\partial x_a^{(i)}} + \frac{\partial}{\partial \xi_a^{(i)}} (f_q \overline{\eta_{qa}^{(i)}}) \right\} = 0,$$

where $\overline{\eta_{qa}^{(i)}}$ is the mean value of the generalized acceleration $\eta_a^{(i)}$ for the ith molecule in a group of q molecules whose coordinates and velocities are specified. According to (1.5) this may be expressed by the formula

(1.12) $$\overline{\eta_{qa}^{(i)}} = \sum_b \gamma_{ab}^{(i)} \left(F_b^{(i)} - \sum_j \frac{\partial \phi^{(ij)}}{\partial x_b^{(i)}} - \iint \frac{f_{q+1}}{f_q} \frac{\partial \phi^{(i,q+1)}}{\partial x_b^{(i)}} \, d\boldsymbol{\xi}_s^{(q+1)} \, d\mathbf{x}_s^{(q+1)} \right).$$

The first term represents the external force and self-force, the second is the force due to the other molecules in the group of q molecules whose positions and velocities are known, and the third term is the average force due to the remaining molecules, whose positions and velocities are not known.

Under equilibrium conditions, the velocity distribution function f_N is given by the usual formula

$$f_N = \exp \beta (F - E)$$

with the total energy E expressed by (1.1). From this general result, it is in principle possible to obtain the equilibrium values of all other distribution functions by integration. If there are no external forces, so that $\psi(\mathbf{x})$ depends only on the internal coordinates, one obtains

(1. 13) $\qquad f_1 = \exp \beta \{F_g - \tfrac{1}{2} \sum_{a,b} g_{ab} \xi_a \xi_b - \psi(\mathbf{x})\}$

where

(1. 14) $\quad \begin{cases} \exp(-\beta F_g) = n^{-1} \int \exp(-\tfrac{1}{2} \beta \sum_{a,b} g_{ab} \xi_a \xi_b) \, d\mathbf{\xi}_s \cdot \\ \qquad \cdot \int^{(s-3)} \int \exp\{-\beta \psi(\mathbf{x})\} \, dx_4 \ldots dx_s \end{cases}$

in equilibrium. The mean total kinetic energy of a molecule is therefore $skT/2$.

The equations (1. 11) and (1. 12) are obvious generalizations of (2. 4) and (2. 6) respectively of the last chapter, and many of the results of chapter V admit of an easy generalization for molecules of the more complex structure now being considered. For example, if the distribution function n_q is integrated over all the internal coordinates $x_4^{(i)} \ldots x_s^{(i)}$ of the q molecules, a distribution function

(1. 15) $\qquad \overline{n}_q = \int^{q(s-3)} \int n_q \, dx_4^{(1)} \ldots dx_s^{(1)} \ldots dx_4^{(q)} \ldots dx_s^{(q)}$

is obtained which depends only on the positions of the mass centres, and satisfies the continuity equation

(1. 16) $\qquad \dfrac{\partial \overline{n}_q}{\partial t} + \sum_{i=1}^{q} \dfrac{\partial}{\partial \mathbf{x}^{(i)}} \cdot (\overline{n}_q \, \overline{\mathbf{u}}_q^{(i)}) = 0,$

identical in form with (8. 1) of the last chapter.

It is, however, impossible in general to disregard the internal structure in considering the effect of their mutual interaction, and on this question only a few qualitative conclusions can be drawn. In condensed fluids, deviations from spherical symmetry in the shape of the molecules must obviously favour certain configurations in the relative distribution of the molecules, in which the total potential energy has nearly its minimum value. The molecules will then tend to interlock with one another, and cybotactic groups of molecules with a rather regular, lattice-like arrangement will be more probable than in simple fluids under the same conditions. This will lead to larger coefficients of viscosity and thermal conduction than would be expected on the basis of the theory of simple fluids. Similar conclusions apply where the interaction between two molecules is asymmetrical, on account of localized forces of a chemical nature, or the polarization of the molecules. Associated and polar fluids both show unusually large coefficients of viscosity and thermal conduction. Much remains to be done, however, towards providing a quantitative explanation of the experimental values.

2. Fluid Mixtures

A further complication of the theory of fluids described in previous chapters arises when two or more kinds of molecules are present, so that the fluid under consideration is strictly a mixture of the simple fluids studied previously. From the theoretical point of view, the new features which arise in this connection may be attributed to two circumstances: the masses of molecules of different species are, in general, different from one another; and the force which they exert on one another is, in general, different from that exerted between two molecules of the same species, at the same distance. From the experimental point of view, it is known that no simple connection exists in general between the macroscopic properties of the mixture and those of its components. For example, it is true that the law of partial pressures enables one to predict the pressure exerted by a mixture of gases at low density; but there is no corresponding law for condensed gases or liquids. The specific heat of a liquid mixture also bears no simple relation, in general, to the values obtained for the separate components; and the same is true of the coefficients of viscosity and thermal conduction. To account for the macroscopic properties of fluid mixtures, it will obviously be necessary to re-examine the whole of the molecular theory.

For simplicity, attention will be restricted to mixtures of two simple fluids; no essentially new features arise in the easy generalization to mixtures of several components. The masses associated with individuals of the two species of molecules will be represented by m_a, m_b respectively. It is necessary to assume that no interaction of a chemical nature takes place between the components of the mixture considered. It is also supposed that deviations from spherical symmetry of the molecules are unimportant. Then the mutual potential energy of two molecules of different species at distance r from one another can be represented by a function $\psi(r)$ of the same type as the functions $\phi_a(r)$ and $\phi_b(r)$ which specify the interaction between two similar molecules of each species. The total potential energy of a fluid containing M and N molecules respectively of the two species, situated at the points $\mathbf{x}_a^{(i)}$ ($i = 1 \ldots M$) and $\mathbf{x}_b^{(i)}$ ($i = 1 \ldots N$), will be

(2. 1) $$\tfrac{1}{2}\sum_{i=1}^{M}\sum_{j=1}^{M}\phi_a^{(ij)} + \sum_{i=1}^{M}\sum_{j=1}^{N}\psi^{(ij)} + \tfrac{1}{2}\sum_{i=1}^{N}\sum_{j=1}^{N}\phi_b^{(ij)},$$

where $\phi_a^{(ij)}$, $\phi_b^{(ij)}$ represent the values of $\phi_a(r)$, $\phi_b(r)$ when r has the values $r_a^{(ij)} = |\mathbf{x}_a^{(j)} - \mathbf{x}_a^{(i)}|$, $r_b^{(ij)} = |\mathbf{x}_b^{(j)} - \mathbf{x}_b^{(i)}|$ respectively, and $\psi^{(ij)}$ represents the value of $\psi(r)$ when r has the value $|\mathbf{x}_b^{(j)} - \mathbf{x}_a^{(i)}|$.

The macroscopic variables in such a fluid mixture are the mass densities ($m_a n_a$, $m_b n_b$) of the two components, the macroscopic velocity ($\mathbf{u}$), and the temperature ($T = 1/k\beta$). The variables n_a and n_b, obtained by dividing the mass densities $m_a n_a$ and $m_b n_b$ by the molecular masses m_a and m_b respectively, may obviously be interpreted as the number densities of the two molecular species. These may be defined alternatively by saying that the probability of finding a molecule of the first kind in the volume element $d\mathbf{x}_a$ is $n_a d\mathbf{x}_a$, and the probability of finding a molecule of the second species in the volume element $d\mathbf{x}_b$ is $n_b d\mathbf{x}_b$. One can, however, no longer say simply that the macroscopic velocity $\mathbf{u}$ of the fluid is the same as the mean molecular velocity; it may happen that the local mean velocities $\mathbf{u}_a$ and $\mathbf{u}_b$ of molecules of the two species are not identical. The macroscopic velocity is actually the momentum per unit volume, divided by the mass density, so

(2. 2) $$(m_a n_a + m_b n_b)\mathbf{u} = m_a n_a \mathbf{u}_a + m_b n_b \mathbf{u}_b.$$

It is possible also that the mean kinetic energies of molecules of the two species may differ from one another at the same point in the fluid. Thus the local temperature T has to be defined in molecular terms by the equation

(2. 3) $$(n_a + n_b)T = n_a T_a + n_b T_b$$

where $3kT_a/2$ and $3kT_b/2$ are the mean kinetic energies of molecules of the first and second kinds respectively. It is to be expected that, in thermal and mechanical equilibrium, $\mathbf{u}_a$ and $\mathbf{u}_b$ will be the same, and T_a and T_b will be the same. This may, in fact be inferred immediately from the fact that the probability of finding molecules of the first and second species with velocities $\boldsymbol{\xi}_a$ and $\boldsymbol{\xi}_b$ respectively must be proportional to $\exp -\tfrac{1}{2}\beta m_a(\boldsymbol{\xi}_a - \mathbf{u})^2$ and $\exp -\tfrac{1}{2}\beta m_b(\boldsymbol{\xi}_b - \mathbf{u})^2$ in equilibrium, according to Boltzmann's law. This is not so in general, however.

In the theory of simple fluids, two distribution functions were found to be of particular interest and importance: the velocity distribution function f_1 and the molecular distribution function n_2. In the theory of mixtures, these are increased to five: the velocity distribution functions f_a and f_b for each of the two molecular species, and the molecular distribution functions n_{aa}, n_{ab} and n_{bb}; these are specified by the following definitions. The probability of finding a molecule of the first species, with velocity in the range $\pmb{\xi}$, $d\pmb{\xi}$, in the volume element $d\mathbf{x}$ at time t is $f_a(\pmb{\xi}, \mathbf{x}, t)\, d\pmb{\xi} d\mathbf{x}$. The probability of finding two molecules of the first species in the volume elements $d\mathbf{x}^{(1)}$, $d\mathbf{x}^{(2)}$ respectively at time t is $n_{aa}^{(12)} d\mathbf{x}^{(1)} d\mathbf{x}^{(2)}$, or $n_{aa}(\mathbf{x}^{(1)}, \mathbf{x}^{(2)}, t)\, d\mathbf{x}^{(1)} d\mathbf{x}^{(2)}$. The probability of finding a molecule of the first species in the volume element $d\mathbf{x}_a$, and a molecule of the second species in the volume element $d\mathbf{x}_b$, at time t is $n_{ab} d\mathbf{x}_a d\mathbf{x}_b$ or $n_{ab}(\mathbf{x}_a, \mathbf{x}_b, t)\, d\mathbf{x}_a d\mathbf{x}_b$. The corresponding quantity n_{ba} is defined by

(2. 4) $$n_{ba}(\mathbf{x}_b, \mathbf{x}_a, t) = n_{ab}(\mathbf{x}_a, \mathbf{x}_b, t).$$

The distribution functions f_b and n_{bb}, referring to molecules of the second species only, are of course defined like f_a and n_{aa}. The properties of these, and the distribution functions of higher order, for fluid mixtures, have been investigated by YANG [1949b].

It should be remarked that, although the distribution function f_a refers, apparently, only to molecules of the first species, its value will in general depend on n_b, as well as n_a, $\mathbf{u}$ and T; only in equilibrium are the velocities of the molecules of one species independent of the presence of the others. Similarly, the value of n_{aa} depends on n_b, as well as n_a and T; this is true even in equilibrium, for compressed gases and liquids. As a result, even the equilibrium properties of fluid mixtures cannot in general be predicted from the corresponding properties of the simple fluids.

It will be recalled that the pressure tensor inside any fluid is defined macroscopically as the tensor whose divergence gives the mean rate of change of momentum per unit volume, due to causes originating within the fluid itself. It must therefore be computed by determining the rate of change of momentum within a closed surface moving with the fluid; this is due partly to the transfer of momentum by molecules crossing the surface, and partly to the action of the molecules outside the surface on those within. Just as for a simple fluid, therefore, the pressure will consist of two parts, associated with the thermal motion and the intermolecular forces respectively. The first part is simply the resultant momentum flux

$$\int \{m_a f_a(\boldsymbol{\xi}) + m_b f_b(\boldsymbol{\xi})\} (\boldsymbol{\xi} - \mathbf{u})(\boldsymbol{\xi} - \mathbf{u}) \, d\boldsymbol{\xi}$$

which is obviously compounded of the momentum flux of each of the two molecular species separately. The second part of the pressure tensor, due to the intermolecular forces, can be derived by a simple extension of the method of § 4 of chapter II, as follows. The mean force acting on a molecule of the first species, situated at the point $\mathbf{x}^{(1)}$, is

$$-\frac{1}{n_a} \int \left(n_{aa}^{(12)} \frac{\partial \phi_a^{(12)}}{\partial \mathbf{x}^{(1)}} + n_{ab}^{(12)} \frac{\partial \psi^{(12)}}{\partial \mathbf{x}^{(1)}} \right) d\mathbf{x}^{(2)}$$

and this contributes a term

$$-\tfrac{1}{2} \int \{n_{aa}^{(12)} \phi_a'(r) + n_{ab}^{(12)} \psi'(r)\} r^{-1} \mathbf{rr} \, d\mathbf{r}$$

to the pressure tensor; a similar term

$$-\tfrac{1}{2} \int \{n_{bb}^{(12)} \phi_b'(r) + n_{ba}^{(12)} \psi'(r)\} r^{-1} \mathbf{rr} \, d\mathbf{r}$$

arises from the action on molecules of the second species. The resultant pressure tensor is therefore

$$(2.5) \quad \begin{cases} \boldsymbol{P} = \boldsymbol{P}_a + \boldsymbol{P}_b, \\ \boldsymbol{P}_a = \int m_a f_a(\boldsymbol{\xi})(\boldsymbol{\xi}-\mathbf{u})(\boldsymbol{\xi}-\mathbf{u}) \, d\boldsymbol{\xi} - \\ \qquad -\tfrac{1}{2} \int \{n_{aa}^{(12)} \phi_a'(r) + n_{ab}^{(12)} \psi'(r)\} r^{-1} \mathbf{rr} \, d\mathbf{r}, \\ \boldsymbol{P}_b = \int m_b f_b(\boldsymbol{\xi})(\boldsymbol{\xi}-\mathbf{u})(\boldsymbol{\xi}-\mathbf{u}) \, d\boldsymbol{\xi} - \\ \qquad -\tfrac{1}{2} \int \{n_{bb}^{(12)} \phi_b'(r) + n_{ba}^{(12)} \psi'(r)\} r^{-1} \mathbf{rr} \, d\mathbf{r}. \end{cases}$$

From this it follows immediately that the hydrostatic pressure is

$$(2.6) \quad \begin{cases} p = p_a + p_b, \\ p_a = n_a k T_a - \tfrac{1}{6} \int \{n_{aa}(r) \phi_a'(r) + n_{ab}(r) \psi'(r)\} r \, d\mathbf{r}, \\ p_b = n_b k T_b - \tfrac{1}{6} \int \{n_{bb}(r) \phi_b'(r) + n_{ba}(r) \psi'(r)\} r \, d\mathbf{r}. \end{cases}$$

A similar expression can be obtained for the internal energy per unit volume. The mean energy of a molecule of the first kind, situated at the point $\mathbf{x}^{(1)}$, is

$$U_a = n_a^{-1} [\int f_a(\boldsymbol{\xi}) \tfrac{1}{2} m_a (\boldsymbol{\xi}-\mathbf{u})^2 \, d\boldsymbol{\xi} + \tfrac{1}{2} \int \{n_{aa}^{(12)} \phi_a(r) + n_{ab}^{(12)} \psi(r)\} \, d\mathbf{r}]$$
$$= \tfrac{3}{2} k T_a + \tfrac{1}{2} n_a^{-1} \int \{n_{aa}^{(12)} \phi_a(r) + n_{ab}^{(12)} \psi(r)\} \, d\mathbf{r},$$

and a similar expression for the mean energy of a molecule of the second kind is obtained by interchanging the suffixes a and b. The internal energy per unit volume is then

(2.7) $$(n_a + n_b)\, U = n_a\, U_a + n_b\, U_b.$$

The formula (2. 5) enables one to draw some qualitative conclusions concerning the coefficient of viscosity in fluid mixtures. The coefficient of ordinary viscosity (η) and the coefficient of bulk viscosity (η_0) are still defined by the equation

$$\mathbf{P} = \left(p^0 - \eta_0\, \frac{\partial}{\partial \mathbf{x}} \cdot \mathbf{u}\right) \mathbf{\delta} - 2\,\eta \left(\overline{\frac{\partial}{\partial \mathbf{x}}\, \mathbf{u}}\right),$$

where $\mathbf{\delta}$ is the unit tensor, and $\overline{\frac{\partial}{\partial \mathbf{x}}\, \mathbf{u}}$ the non-divergent rate of strain tensor with components given by (1. 5) of chapter V. If the deviations f_a', f_b', n_{aa}', n_{ab}', n_{bb}' of f_a, f_b, n_{aa}, n_{ab} and n_{bb} from their equilibrium values were known, one would be able to calculate η_0 and η directly from the formula (2. 5). Even without determining these deviations, one may conclude from considerations of symmetry that, when n_a, n_b and T have the same values throughout the fluid, f_a' must be of the form

(2.8) $$f_a' = \omega_a\, \mathbf{v} \cdot \left(\overline{\frac{\partial}{\partial \mathbf{x}}\, \mathbf{u}}\right) \cdot \mathbf{v}, \qquad \mathbf{v} = \mathbf{\xi} - \mathbf{u},$$

where ω_a can depend only on the magnitude v of the velocity relative to the mean motion, apart from n_a, n_b and T. Similarly n_{aa}' and $n_{ab}' + n_{ba}'$ must be given by

(2.9) $$\begin{cases} n_{aa}' = \nu_a\, \mathbf{r} \cdot \left(\overline{\frac{\partial}{\partial \mathbf{x}}\, \mathbf{u}}\right) \cdot \mathbf{r}/r^2 + \nu_{a0}\, \frac{\partial}{\partial \mathbf{x}} \cdot \mathbf{u}, \\ \tfrac{1}{2}(n_{ab}' + n_{ba}') = \nu_c\, \mathbf{r} \cdot \left(\overline{\frac{\partial}{\partial \mathbf{x}}\, \mathbf{u}}\right) \cdot \mathbf{r}/r^2 + \nu_{c0}\, \frac{\partial}{\partial \mathbf{x}} \cdot \mathbf{u}, \end{cases}$$

where ν_a, ν_{a0}, ν_c and ν_{c0} depend only on r, apart from n_a, n_b and T. Also, it can be seen from (2. 4) that $n_{ab}' - n_{ba}'$ can depend only on the gradients of n_a, n_b and T, which vanish when these quantities have the same values throughout the fluid. By analogy with (4. 7) and (4. 8) of chapter V, the coefficients of viscosity are then obviously given by

(2.10) $$\begin{cases} \eta = \tfrac{1}{30} \int \{\nu_a\, \phi_a'(r) + 2\,\nu_c\, \psi'(r) + \nu_b\, \phi_b'(r)\}\, r d\mathbf{r} - \\ \quad - \tfrac{1}{15} \int (\omega_a + \omega_b)\, v^2\, d\mathbf{v} \end{cases}$$

and

(2.11) $$\eta_0 = \tfrac{1}{2} \int \{\nu_{a0}\, \phi_a'(r) + 2\,\nu_{c0}\, \psi'(r) + \nu_{b0}\, \phi_b'(r)\}\, r\, d\mathbf{r}.$$

It can be seen from these formulae that the viscosity of fluid mixtures depends on all three of the intermolecular potential functions $\phi_a(r)$, $\psi(r)$ and $\phi_b(r)$. This could have been anticipated on the basis of the intuitive concept of viscosity developed in the previous chapter. As a result of the intermolecular forces, molecules in parts of the fluid moving with different mean velocities tend to 'drag' one another in such a way as to dissipate the macroscopic motion. The effectiveness of this process is obviously independent of the precise nature of the molecules, so long as they have the same type of interaction. Precisely the same arguments can be advanced to elucidate the nature of thermal conduction in fluid mixtures. No essentially new process is involved, and it is only when one proceeds to consider the phenomena of diffusion that features peculiar to fluid mixtures are encountered.

The equations of conservation of mass, momentum and energy, represented by (3. 3), (3. 9), and (3. 12) of chapter V for simple fluids, are easily reestablished for fluid mixtures. The number densities n_a and n_b satisfy the separate conservation equations

(2. 12)
$$\begin{cases} \frac{\partial n_a}{\partial t} + \frac{\partial}{\partial \mathbf{x}} \cdot (n_a \mathbf{u}_a) = 0, \\ \frac{\partial n_b}{\partial t} + \frac{\partial}{\partial \mathbf{x}} \cdot (n_b \mathbf{u}_b) = 0. \end{cases}$$

By virtue of (2. 2), these combine to give

$$\frac{\partial \varrho}{\partial t} + \frac{\partial}{\partial \mathbf{x}} \cdot (\varrho \mathbf{u}) = 0,$$

where ϱ is the mass density $m_a n_a + m_b n_b$. This equation, which expresses the conservation of mass, can also be written in the form

(2. 13) $$\frac{d\varrho}{dt} + \varrho \frac{\partial}{\partial \mathbf{x}} \cdot \mathbf{u} = 0,$$

where $\frac{d}{dt}$ represents as usual the convective derivative.

The equation of motion is, as usual,

$$\varrho \frac{d}{dt} \mathbf{u} + \frac{\partial}{\partial \mathbf{x}} \cdot \boldsymbol{p} = 0,$$

with $\boldsymbol{p}$ given by (2. 5). The equation of energy transfer, on the other hand, has the somewhat more complicated form

(2. 14) $$\varrho \frac{d}{dt} \{(n_a + n_b) U/\varrho\} + \frac{\partial}{\partial \mathbf{x}} \cdot \mathbf{q} + \boldsymbol{p} : \left(\frac{\partial}{\partial \mathbf{x}} \mathbf{u}\right) = 0$$

if U is defined by (2. 7). This then, is the form assumed by the equation of conservation of energy in fluid mixtures.

3. Theory of Diffusion

In a simple fluid, it has been seen that two essentially irreversible processes are possible: viscosity and thermal conduction. In a fluid mixture, these processes may still occur, and there exists in addition the possibility of a third irreversible process: the diffusion of one component relative to the other.

The phenomenon of diffusion is complicated wherever there are gradients of temperature or pressure; in considering ordinary diffusion it is convenient, therefore, to imagine that both the temperature and the pressure are uniform, and that only the relative concentrations of the two fluid components vary from point to point in the fluid mixture. In fact, when the pressure and temperature are fixed, there exists a definite relation between the number densities n_a and n_b of the two species of molecules; the only additional variable is therefore the ratio

(3. 1)
$$\begin{cases} c_a = n_a/(n_a + n_b), \text{ or} \\ c_b = n_b/(n_a + n_b) = 1 - c_a \end{cases}$$

of the number of molecules of one species to the total number, per unit volume. Supposing then that c_a varies from point to point in the fluid, it can be seen that the fluid cannot be in equilibrium. Owing to the thermal motion, molecules of each type will make their way from point to point in the fluid, and more will pass from regions where they are in excess to regions in which they are in defect, than in the opposite direction. Thus, even if the distribution of molecular velocities is Maxwellian [with $f_a = n_a \exp\{-\tfrac{1}{2}\beta m_a (\boldsymbol{\xi} - \mathbf{u})^2\}$ and $f_b = n_b \exp\{-\tfrac{1}{2}\beta m_b (\boldsymbol{\xi} - \mathbf{u})^2\}$] at some initial time, it will not remain so; there will be a flow of molecules of the first species, relative to the mass motion, in a direction opposite to the concentration gradient, and a corresponding flow of molecules of the second species in the opposite direction. In such circumstances, the mean velocities $\mathbf{u}_a$ and $\mathbf{u}_b$ of the two species will differ from the mass velocity $\mathbf{u}$, and the difference will depend on the concentration gradient. Assuming that the concentration gradient is small, so that terms quadratic or of higher powers in $\partial c_a/\partial \mathbf{x}$ are negligible, one will have

(3. 2) $$\mathbf{u}_a - \mathbf{u}_b = - D_0 \frac{(n_a+n_b)^2}{n_a n_b} \frac{\partial c_a}{\partial \mathbf{x}} = D_0 \frac{n_a}{n_b} \frac{\partial}{\partial \mathbf{x}} \left(\frac{n_b}{n_a}\right)$$

where the coefficient D_0 is known as the coefficient of diffusion. It may depend on the concentration ratio n_a/n_b, as well as density and temperature, and has the dimensions *area/time*. Since

(3. 3) $$\mathbf{u}_a - \mathbf{u}_b = \int \{n_a^{-1} f_a(\boldsymbol{\xi}) - n_b^{-1} f_b(\boldsymbol{\xi})\} \boldsymbol{\xi} \, d\boldsymbol{\xi},$$

the coefficient of diffusion will be determinate if the deviations of the velocity distribution functions f_a and f_b from their equilibrium values are known. Unlike the coefficients of viscosity and thermal conduction, therefore, D_0 does not depend directly on the molecular structure of the fluid; and the intermolecular forces are important only in so far as they affect the thermal motion of the molecules. That is why diffusion is a comparatively slow process in liquids, where viscosity and thermal conduction are much more effective than in gases. In general the rate of diffusion will be lowered as the temperature decreases; also, the greater the density, the more the diffusion of the molecules is impeded.

Ordinary diffusion is perhaps the most easily comprehended of the transport processes. The phenomenon of *thermal diffusion*, on the other hand, was not investigated experimentally till 1916, five years after its existence had been predicted, on the basis of molecular theory, by ENSKOG [1911]. In liquids, where it is called the Soret effect, it is less effective than in gases. However, it has become very important in recent years as a means of separating mixtures of chemically similar substances or isotopes. The characteristic feature of thermal diffusion is the tendency of a fluid to separate into its components, under the influence of strong temperature gradients. It occurs even under conditions of uniform pressure, and is best differentiated from ordinary diffusion by considering a mixture in which the concentration ratio n_a/n_b is everywhere the same. Then there is a diffusive motion of one component relative to the other, described by the formula

(3. 4) $$\mathbf{u}_a - \mathbf{u}_b = D_T \frac{(n_a+n_b)^2}{n_a n_b} \frac{1}{T} \frac{\partial T}{\partial \mathbf{x}},$$

where D_T is called the coefficient of thermal diffusion. The factor T^{-1} is introduced to give D_T the same dimensions as D_0.

Unlike the coefficient of ordinary diffusion, D_T is not necessarily positive, and may even change sign as the mean temperature is varied. It can be completely understood only by considering in detail the mechanics of molecular motion. It generally happens that, in a fluid mixture, one molecular species is more mobile than the other; owing to a difference in molecular mass, or in the molecular forces, the thermal motion of one species is impeded less than that of the other. Molecules of the former species are then able to move from colder to warmer regions of the fluid more rapidly than their cumbersome competitors, even though the average speed of molecules of the two species is initially the same in the same locality. Reaching the warmer region first, the more mobile molecules gain in speed and tend ultimately to accumulate where the temperature is highest. Molecules of the other species, on the other hand, are left in the cooler parts of the fluid and tend to accumulate there. Ultimately — when the concentration gradient and the temperature gradient are in the ratio TD_0/D_T — a balance is reached between the processes of ordinary and thermal diffusion, and no further change is possible, apart from the continual flow of thermal energy under the influence of the temperature gradient.

There is one other natural mode of diffusion, which can operate even when there is no gradient of concentration or temperature. This may be referred to as pressure diffusion, and is possible wherever there is a gradient of the hydrostatic pressure. It is this form of diffusion which is responsible for the tendency of a mixture of fluids with different molecular weights to separate into its components under the influence of gravity. The gravitational force itself accelerates all molecules to the same degree, and would not alone suffice to effect any separation. The pressure gradient, however, which normally balances the gravitational force, accelerates the lighter molecules more than the heavier ones, with the result that the former tend to migrate to higher levels, leaving the heavier molecules down below.

A similar effect is noticeable wherever an external force accelerates different molecules to a different extent. For example, in an ionized fluid, the ions must diffuse relative to the uncharged molecules in the presence of an electric field.

The exact theory of diffusion in rare gases has been well known since the work of Chapman and Enskog. A general theory of diffusion in fluids has been developed by YANG [1949a], who has traced the connection with the old free path theories by relating ordinary diffusion to the self-diffusion of the molecules of different species in the fluid mixture. The method bears some resemblance to Kirkwood's method (cf. § 6 of chapter V) for investigating the transport processes by a development of the analogy between the motion of individual molecules and the motion of a Brownian particle.

It is necessary to specify the probability that a molecule of the first species, located at the point $\mathbf{x}_0$ in the fluid at some initial time t_0, will be found in the volume element $d\mathbf{x}$ at the point $\mathbf{x} = \mathbf{x}_0 + \mathbf{r}$, at time $t = t_0 + \tau$. This probability will be denoted by $\psi_a(\mathbf{x}_0, \mathbf{r}, \tau)d\mathbf{x}$, and, as the molecule must be located somewhere at the end of the interval τ, ψ_a must satisfy

(3. 5) $$\int \psi_a(\mathbf{x}_0, \mathbf{r}, \tau)\, d\mathbf{r} = 1.$$

The total probability $n_a(\mathbf{x}, t)d\mathbf{x}$ of finding a molecule of this species in the volume element $d\mathbf{x}$ at time t may then be obtained by multiplying $\psi_a(\mathbf{x}_0, \mathbf{r}, \tau)d\mathbf{x}$ by the probability $n_a(\mathbf{x}_0, t_0)d\mathbf{x}_0$ that the molecule is in the volume element $d\mathbf{x}_0$ at the point $\mathbf{x}_0$ at time t_0, and then summing over all volume elements $d\mathbf{x}_0$. Hence

(3. 6) $$\begin{cases} n_a(\mathbf{x}, t) = \int n_a(\mathbf{x}_0, t_0)\, \psi_a(\mathbf{x}_0, \mathbf{r}, \tau)\, d\mathbf{x}_0 \\ \qquad\quad = \int n_a(\mathbf{x} - \mathbf{r}, t - \tau)\, \psi_a(\mathbf{x} - \mathbf{r}, \mathbf{r}, \tau)\, d\mathbf{r}. \end{cases}$$

The interval τ must be chosen sufficiently long for the motion of the molecule at time t to be completely independent of its initial motion at time t_0. In accordance with the theory of the Brownian motion, it may then be assumed that the mean displacement of a molecule

$$\bar{\mathbf{r}}_a = \int \mathbf{r}\, \psi_a(\mathbf{x}, \mathbf{r}, \tau)\, d\mathbf{r}$$

and the mean square deviation of this displacement

$$\overline{\mathbf{r}_a \mathbf{r}_a} - \bar{\mathbf{r}}_a \bar{\mathbf{r}}_a = \int (\mathbf{r} - \bar{\mathbf{r}}_a)(\mathbf{r} - \bar{\mathbf{r}}_a)\, \psi_a(\mathbf{x}, \mathbf{r}, \tau)\, d\mathbf{r}$$

from the mean, in the interval τ, are both proportional to τ. Accordingly one may write

(3. 7) $$\bar{\mathbf{r}}_a = \beta_a \tau; \quad \overline{\mathbf{r}_a \mathbf{r}_a} - \bar{\mathbf{r}}_a \bar{\mathbf{r}}_a = 2\alpha_a \tau,$$

where α_a is a tensor quantity. If, in the integral of (3. 6), one effects the expansion

(3. 8) $$\begin{cases} n_a(\mathbf{x}-\mathbf{r}, t-\tau)\, \psi_a(\mathbf{x}-\mathbf{r}, \mathbf{r}, \tau) = \left\{ n_a(\mathbf{x}, t) - \tau\, \frac{\partial n_a(\mathbf{x}, t)}{\partial t} \right\} \psi_a(\mathbf{x}, \mathbf{r}, \tau) - \\ \qquad\qquad - \mathbf{r} \cdot \frac{\partial}{\partial \mathbf{x}} \left(1 - \tfrac{1}{2}\, \mathbf{r} \cdot \frac{\partial}{\partial \mathbf{x}} \right) \{ n_a(\mathbf{x}, t)\, \psi_a(\mathbf{x}, \mathbf{r}, \tau) \}, \end{cases}$$

neglecting terms quadratic in τ or cubic in **r**, one obtains, with the help of (3. 5) and (3. 7),

$$\text{(3. 9)} \qquad \frac{\partial n_a}{\partial t} + \frac{\partial}{\partial \mathbf{x}} \cdot (n_a \boldsymbol{\beta}_a) + \frac{\partial^2}{\partial \mathbf{x} \partial \mathbf{x}} : (n_a \boldsymbol{\alpha}_a) = 0.$$

By comparison of (3. 9) with (2. 12), one sees that

$$\text{(3. 10)} \qquad \mathbf{u}_a = \boldsymbol{\beta}_a + \frac{1}{n_a} \frac{\partial}{\partial \mathbf{x}} \cdot (n_a \boldsymbol{\alpha}_a).$$

Further, by subtracting a similar equation with the suffix b in place of a, one has

$$\text{(3. 11)} \quad \mathbf{u}_a - \mathbf{u}_b = \boldsymbol{\beta}_a - \boldsymbol{\beta}_b + \frac{\partial}{\partial \mathbf{x}} \cdot (\boldsymbol{\alpha}_a - \boldsymbol{\alpha}_b) + \frac{\boldsymbol{\alpha}_a}{n_a} \cdot \frac{\partial n_a}{\partial \mathbf{x}} - \frac{\boldsymbol{\alpha}_b}{n_b} \cdot \frac{\partial n_b}{\partial \mathbf{x}}.$$

To determine the coefficients of diffusion it is necessary to select from the right-hand side of (3. 11) terms linear in the gradients of the concentration, temperature and pressure. For this purpose, it is sufficient to know the equilibrium values of $\boldsymbol{\alpha}_a$ and $\boldsymbol{\alpha}_b$, expressed as functions of n_a, n_b and T; but the deviations of $\boldsymbol{\beta}_a$ and $\boldsymbol{\beta}_b$ from equilibrium are required as well as their equilibrium values. In equilibrium, $\boldsymbol{\alpha}_a$ and $\boldsymbol{\alpha}_b$ are simple multiples, say $\tfrac{1}{3} D_a$ and $\tfrac{1}{3} D_b$, of the unit tensor $\boldsymbol{\delta}$, and $\boldsymbol{\beta}_a$ and $\boldsymbol{\beta}_b$ each have the value **u**:

$$\text{(3. 12)} \qquad \boldsymbol{\alpha}_a^0 = \tfrac{1}{3} D_a \boldsymbol{\delta}, \quad \boldsymbol{\alpha}_b^0 = \tfrac{1}{3} D_b \boldsymbol{\delta}, \quad \boldsymbol{\beta}_a^0 = \boldsymbol{\beta}_b^0 = \mathbf{u}.$$

The deviations of $\boldsymbol{\beta}_a$ and $\boldsymbol{\beta}_b$ from equilibrium cannot be determined exactly without knowledge of the corresponding departures of f_a and f_b from equilibrium; however, there are grounds for supposing that they will be rather small. For, the motion of a given molecule during the interval τ depends mainly on the momentum transferred to it by molecules of both species; and, as the average momentum per unit mass of these molecules is **u**, it is reasonable to suppose that the mean velocity of a molecule over a 'fairly long' period τ will be independent of the initial velocity distribution, and approximate to **u**. On this assumption, (3. 11) will reduce to

$$\text{(3. 13)} \qquad \mathbf{u}_a - \mathbf{u}_b = \tfrac{1}{3} \frac{\partial}{\partial \mathbf{x}} (D_a - D_b) + \tfrac{1}{3} \left(\frac{D_a}{n_a} \frac{\partial n_a}{\partial \mathbf{x}} - \frac{D_b}{n_b} \frac{\partial n_b}{\partial \mathbf{x}} \right).$$

It should be remarked that D_a and D_b are called the coefficients of self-diffusion of molecules of the first and second species respectively in the fluid mixture. They depend on n_a, n_b and T, but when the concentration $c_b = n_b/(n_a + n_b)$ of the second species approaches the value zero, D_a approaches the ordinary coefficient of self-diffusion for a fluid containing only molecules of the first species, and conversely.

Suppose that D_a and D_b are expressed as functions of the number densities n_a and n_b and the temperature. Then (3. 13) can be expressed in the form

$$(3.14) \quad \begin{cases} \mathbf{u}_a - \mathbf{u}_b = \tfrac{1}{3}\left\{ n_a \frac{\partial}{\partial n_a}(D_a - D_b) + D_a \right\} \frac{1}{n_a}\frac{\partial n_a}{\partial \mathbf{x}} + \\ \qquad + \tfrac{1}{3}\left\{ n_b \frac{\partial}{\partial n_b}(D_a - D_b) - D_b \right\} \frac{1}{n_b}\frac{\partial n_b}{\partial \mathbf{x}} + \tfrac{1}{3}\frac{\partial}{\partial T}(D_a - D_b)\frac{\partial T}{\partial \mathbf{x}}. \end{cases}$$

One has also

$$(3.15) \qquad \frac{\partial p}{\partial \mathbf{x}} = \frac{\partial p}{\partial n_a}\frac{\partial n_a}{\partial \mathbf{x}} + \frac{\partial p}{\partial n_b}\frac{\partial n_b}{\partial \mathbf{x}} + \frac{\partial p}{\partial T}\frac{\partial T}{\partial \mathbf{x}}.$$

From these last two equations, one can eliminate the variable

$$\frac{1}{n_a}\frac{\partial n_a}{\partial \mathbf{x}} + \frac{1}{n_b}\frac{\partial n_b}{\partial \mathbf{x}}$$

and obtain a relation of the form

$$(3.16) \quad \mathbf{u}_a - \mathbf{u}_b = -D_0\left(\frac{1}{n_a}\frac{\partial n_a}{\partial \mathbf{x}} - \frac{1}{n_b}\frac{\partial n_b}{\partial \mathbf{x}}\right) + \frac{(n_a+n_b)^2}{n_a n_b}\left(\frac{D_T}{T}\frac{\partial T}{\partial \mathbf{x}} + \frac{D_p}{p}\frac{\partial p}{\partial \mathbf{x}}\right),$$

where the coefficients D_0, D_T and D_p are somewhat complicated algebraic expressions whose values are easily written down if desired. For a gas, in which $p = (n_a + n_b)kT$, one obtains

$$(3.17) \quad \begin{cases} D_0 = \dfrac{n_a n_b}{3(n_a+n_b)^2}\left\{ \left(\dfrac{\partial}{\partial n_a} - \dfrac{\partial}{\partial n_b}\right)(D_a - D_b) + \dfrac{D_a}{n_a} + \dfrac{D_b}{n_b}\right\}, \\ D_T = \dfrac{n_a n_b}{3(n_a+n_b)^2}\left(1 + n_a\dfrac{\partial}{\partial n_a} + n_b\dfrac{\partial}{\partial n_b} - T\dfrac{\partial}{\partial T}\right)(D_a - D_b). \end{cases}$$

The coefficients D_a and D_b have not been determined except for rare gases. For a gas whose molecules can be represented as rigid spheres of mass m_a and diameter σ_a, Yang finds as a first approximation[1])

$$(3.18) \qquad D_a = \frac{0\cdot 8}{n_a \pi \sigma_a^2}\left(\frac{kT}{m_a}\right)^{\frac{1}{2}},$$

in substantial agreement with the formula obtained by Chapman &

---

[1]) This formula applies only when molecules of the second species are absent, and so cannot be used in (3. 17).

Pidduck, on the basis of the kinetic theory of gases. The interest of the formula (3.13) lies in the possibility of its application to liquids. It would probably be a fairly good approximation to substitute D_a and D_b into (3.13), neglecting the difference between (3.17) and the exact formula for a fluid mixture; this would allow at least the approximate evaluation of D_0 and D_T for condensed systems.

CHAPTER VII

FURTHER EQUILIBRIUM PROPERTIES

1. Elasticity

The phenomenon of elasticity is one which is most commonly associated with solid substances. It has, of course, always been recognized that fluids have elastic properties, and that, at least in respect of their compressibility, they are generally more elastic than solids. It has, however, sometimes been assumed that, because some fluids cannot support a static shearing stress, fluids in general cannot possess a shear modulus of elasticity. Such an assumption is easily seen to be incorrect if it is borne in mind that very viscous fluids, like the glasses, are at low temperatures almost indistinguishable from other solids in their elastic behaviour, and that no clearly defined distinction can be drawn between these and the much more mobile liquids and gases. This strongly suggests that it should be possible to define a shear modulus for all substances, irrespective of their state. It will naturally be an important feature of such a definition that it should permit the experimental measurement of the quantity defined.

If a stress of any kind is suddenly applied to a fluid at rest, a complicated re-adjustment of the molecular structure takes place. Within a short time — the so-called relaxation time measured by the parameter a in § 4 of chapter V — the distribution of molecular velocites, and also the radial distribution of the molecules about a given molecule, are changed; and the fluid as a whole is thus brought into a steady state which, macroscopically, is observed to be one of viscous flow. The flow is quite imperceptible in very viscous substances like the glasses, and there is no difficulty in measuring any elastic deformation of the fluid which may have resulted from the imposition of the stress. In mobile liquids and gases, however, it is almost impossible to detect the elastic response, which takes place as soon as the stress is applied, but is immediately effaced by the subsequent motion of the fluid. Under these conditions, it is indeed not easy to see how elasticity and viscosity can be distinguished.

From the theoretical point of view, elasticity and viscosity differ in the circumstance that, whereas viscous flow is an irreversible process, elastic deformation is ideally reversible. Viscosity dissipates the mechanical energy of the macroscopic motion, converting it irreversibly into thermal energy. The work done by the stress system in effecting an elastic deformation is, on the other hand, used to increase the mutual potential energy of the molecules in the fluid, and can, in principle, be recovered when the stress is relaxed.

The fact that elastic deformation, as opposed to viscous flow, is a reversible process, suggests an ideal experiment by which it might, in principle, be measured. One might suddenly impose a stress system, and, after a second or two, equally suddenly remove it, observing the variation in the strain during the interval. The permanent displacement of the fluid would then indicate the extent of the viscous flow which had taken place. Although the commencement and cessation of steady viscous flow is not quite instantaneous, the error in assuming that the viscous motion was steady during the period in which the stress was applied would probably be small. A reliable estimate of the elastic deformation could therefore be inferred directly from the results of such an ideal experiment, as illustrated by figure 1 below.

The times t_0, t_1, correspond to the instants of application and removal of the stress. The dotted curve shows the contribution to the strain from the action of viscosity alone. The ordinate e, at the intersection of the tangent to the straight portion of the curve with the vertical through t_0, measures the elastic strain.

A much more practical method of measuring the shearing elasticity of liquids has been developed recently by MASON, BAKER, McSKIMIN & HEISS [1949]. Instead of a constant stress applied for a short interval of time, a harmonically varying stress is applied, with a period of the order of the relaxation time. As the latter is often extremely short, ultrasonic frequencies need to be employed. The strain, which apart from a possible difference in phase, varies harmonically with the stress, loses its purely viscous character when a certain frequency is passed, and a non-dissipative component can be observed which must be elastic in origin.

It is clear from these experiments that a well defined shear modulus of elasticity exists even for the most mobile fluids. To understand how this is determined by the molecular structure, it is necessary to consider how the radial distribution of the molecules about a given molecule is deformed by the strain. When a stress is applied to a fluid, it is transmitted almost instantaneously, by the motion and mutual interactions of the molecules, throughout the fluid. The motion of the individual molecules is influenced by the change in the forces acting upon them, in a complicated way which can only be described statistically. The mean velocities change, and also the

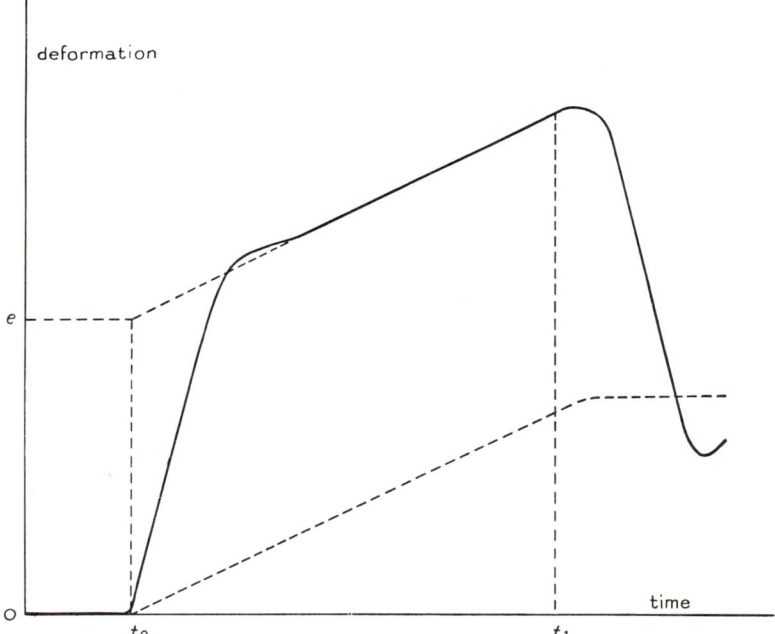

Fig. 1. The deformation of a fluid under a stress suddenly applied at time t_0 and removed at time t_1.

distribution of molecular velocities about the mean. More significant from the present standpoint, the mean configuration of the molecules relative to a given molecule is changed. This change in the molecular distribution may be regarded as due in part to the initial displacement of the whole fluid, and in part to the motion of different layers relative to one another. Only the first contribution, which is of an elastic nature, need be considered here; the viscous motion can be disregarded.

Leaving viscosity out of consideration, the effect of the stress is to displace each molecule by an average amount s depending on its position x. If a molecule at the point **x** is displaced by the amount **s(x)**, a second molecule situated at the point **x** + **r** will be displaced, on the average, by the amount

$$\mathbf{s}(\mathbf{x}+\mathbf{r}) = \mathbf{s}(\mathbf{x}) + \mathbf{r}\cdot\frac{\partial}{\partial\mathbf{x}}\mathbf{s}(\mathbf{x}) + \ldots,$$

and the final position of the molecule relative to the first will therefore be

(1. 1) $$\mathbf{r}^{\dagger} = \mathbf{r} + \mathbf{r}\cdot\frac{\partial}{\partial\mathbf{x}}\mathbf{s} + \ldots.$$

Neglecting terms quadratic in the tensor $\frac{\partial}{\partial\mathbf{x}}\mathbf{s}$, the distance between the two molecules will therefore be

(1. 2) $$r^{\dagger} = r + \mathbf{r}\cdot\left(\frac{\partial}{\partial\mathbf{x}}\mathbf{s}\right)\cdot\mathbf{r}/r$$

after the deformation. This depends on the direction of the vector **r** joining the two molecules; if one averages over all directions, keeping the distance r fixed, the result is

(1. 3) $$\overline{r^{\dagger}} = r\left(1 + \frac{1}{3}\frac{\partial}{\partial\mathbf{x}}\cdot\mathbf{s}\right).$$

From this one can infer the relation between the local number densities n and $n^{\dagger}$ before and after the deformation. For the same molecules which initially occupied a sphere of radius r in the fluid, will occupy an almost spherical ellipsoid of average radius $\overline{r^{\dagger}}$ in the strained fluid; and the number of such molecules is, on the average,

$$4\pi n r^3/3 = 4\pi n^{\dagger}(\overline{r^{\dagger}})^3/3.$$

It follows, with the help of (1. 3), that the number densities are related by

(1. 4) $$n = \left(1 + \frac{\partial}{\partial\mathbf{x}}\cdot\mathbf{s}\right)n^{\dagger}.$$

A similar consideration gives the molecular distribution function $n_2^{\dagger}(\mathbf{r}^{\dagger})$ after the deformation in terms of the radial distribution function $n_2(r)$ for the undeformed fluid. For the probability $n_2(r)\,d\mathbf{r}/n$ of finding a molecule in the volume element $d\mathbf{r}$ whose position relative to a given molecule is **r** before the deformation, must be the same as the probability $n_2^{\dagger}(\mathbf{r}^{\dagger})\,d\mathbf{r}^{\dagger}/n^{\dagger}$ of finding a molecule in the displaced volume element $d\mathbf{r}^{\dagger}$ after the deformation. Thus

(1. 5) $$n_2^{\dagger}(\mathbf{r}^{\dagger})\,d\mathbf{r}^{\dagger}/n^{\dagger} = n_2(r)\,d\mathbf{r}/n.$$

As the deformation is practically instantaneous, there is no time for the transfer of thermal energy to take place, and the change must be considered adiabatic. If there is a change in the density of the fluid, there will also be a change in the temperature, from the value T in the undeformed fluid, to a value $T^\dagger$ given by

$$(1.\ 6) \qquad T^\dagger = T + (n^\dagger - n) \left(\frac{\partial T}{\partial n}\right)_s,$$

where the coefficient $\left(\frac{\partial T}{\partial n}\right)_s$ is to be calculated from the adiabatic law.

It is now possible to evaluate the stress tensor in the deformed fluid, from the general formula (4. 7) of chapter II. The term arising from the thermal motion of the molecules reduces, in the usual way, to $n^\dagger k T^\dagger$, so the pressure tensor is

$$\boldsymbol{p}^\dagger = n^\dagger k T^\dagger \boldsymbol{\delta} - \tfrac{1}{2} \int n_2^\dagger(\mathbf{r}^\dagger)\, \phi'(r^\dagger)\, (r^\dagger)^{-1}\, \mathbf{r}^\dagger \mathbf{r}^\dagger\, d\mathbf{r}^\dagger.$$

This expression can be developed with the help of (1. 1), (1. 2) and (1. 5):

$$(1.\ 7) \quad \begin{cases} \boldsymbol{p}^\dagger = n^\dagger k T^\dagger \boldsymbol{\delta} - \tfrac{1}{2}(n^\dagger/n) \int n_2(r) \Big[\phi'(r)\, r^{-1} \mathbf{r}\mathbf{r} + \\ \qquad + \phi'(r)\, r^{-1} \left\{ \left(\mathbf{r}\cdot\frac{\partial}{\partial \mathbf{x}}\mathbf{s}\right)\mathbf{r} + \mathbf{r}\left(\mathbf{r}\cdot\frac{\partial}{\partial \mathbf{x}}\mathbf{s}\right) \right\} + \\ \qquad + \frac{d}{dr}\left\{\frac{\phi'(r)}{r}\right\} \frac{\mathbf{r}}{r}\cdot\left(\frac{\partial}{\partial \mathbf{x}}\mathbf{s}\right)\cdot \mathbf{r}\mathbf{r}\mathbf{r} \Big]\, d\mathbf{r}. \end{cases}$$

The integral can be expressed (*vide* § 4 of chapter I) in terms of the macroscopic strain tensor, which is defined by

$$(1.\ 8) \quad \begin{cases} \mathbf{e} = \overline{\left(\dfrac{\partial}{\partial \mathbf{x}}\mathbf{s}\right)}; \\ e_{11} = \dfrac{1}{3}\left(2\dfrac{\partial s_1}{\partial x_1} - \dfrac{\partial s_2}{\partial x_2} - \dfrac{\partial s_3}{\partial x_3}\right),\ \text{etc.}; \\ e_{23} = \dfrac{1}{2}\left(\dfrac{\partial s_2}{\partial x_3} + \dfrac{\partial s_3}{\partial x_2}\right),\ \text{etc.} \end{cases}$$

If $\boldsymbol{p}$ is the pressure tensor in the undeformed fluid, in which $\mathbf{e} = 0$, (1. 7) reduces to

$$(1.\ 9) \quad \begin{cases} \boldsymbol{p}^\dagger = \boldsymbol{p} + (n^\dagger - n)\left\{\boldsymbol{p}/n + n\,k\left(\dfrac{\partial T}{\partial n}\right)_s \boldsymbol{\delta}\right\} - \\ \quad - \left(\mathbf{e} + \tfrac{1}{3}\dfrac{\partial}{\partial \mathbf{x}}\cdot\mathbf{s}\,\boldsymbol{\delta}\right)\left[\tfrac{1}{3}\int n_2(r)\,\phi'(r)\,r\,d\mathbf{r} + \tfrac{1}{15}\int n_2(r)\,\dfrac{d}{dr}\left\{\dfrac{\phi'(r)}{r}\right\}r^3\,d\mathbf{r}\right]. \end{cases}$$

The shear modulus (μ) and bulk modulus (λ) of elasticity are defined by the macroscopic formula

(1.10) $$\mathbf{p}^\dagger - \mathbf{p} = -2\mu\mathbf{e} - \lambda\frac{\partial}{\partial \mathbf{x}}\cdot\mathbf{s}\,\boldsymbol{\delta},$$

which also expresses Hooke's law for fluids. By the comparison of (1.8) and (1.9), it can be seen immediately that

(1.11) $$\mu = \frac{4\pi}{30}\int_0^\infty n_2(r)\frac{d}{dr}\{r^4\phi'(r)\}\,dr$$

and

(1.12) $$\lambda = p + n^2 k\left(\frac{\partial T}{\partial n}\right)_s + \tfrac{2}{3}\mu.$$

Thus the elastic constants have been expressed in terms of the intermolecular forces, and a derivation of Hooke's law has been given on the basis of molecular theory. Elastic deviations from Hooke's law can be obtained by repeating this calculation, but retaining terms quadratic in the strain, wherever they appear. For deviations from the law due to the operation of viscosity, one has to refer to the general theory of fluids in motion.

The experimental data on the elasticity of liquids relate mostly to liquids whose molecular structure is so complicated that they do not make good theoretical models. The reason for choosing such liquids for experimental study was the fact that they have relaxation times which are comparatively long. It is, however, probably only a matter of time before the experimental technique is adapted to the investigation of liquids with shorter relaxation times, and correspondingly simpler structure.

2. Surface Tension

Molecular phenomena at the surface separating the liquid and the saturated vapour, or the liquid and the walls of its containing vessel, are appreciably more complex than those which occur inside the homogeneous fluid, and it is difficult to state much of a rigorous quantitative nature concerning them. The essential difficulty is that, whereas from the macroscopic standpoint there is always a well-defined surface of separation between the two phases, on the microscopic scale there is only a surface *zone*, in crossing which the structure of the fluid undergoes a progressive modification. It is in this surface zone that the dynamic equilibrium between the molecules of the vapour and those of the liquid is established. Owing to the attractive forces exerted by the molecules of the liquid proper on one another, only the faster moving members can penetrate the layer and escape into the vapour; in the process, they lose kinetic energy, and, on the average, attain the same velocity as the molecules in the vapour. Also, the number escaping cannot, on the average, exceed the number entering the liquid from the much rarer vapour.

From the statistical point of view, the density of the fluid is the most important variable in the surface layer; it does not, of course, suffer an abrupt change, but varies continuously in passing through the surface zone from its value in the liquid to the generally much lower value in the vapour. In consequence, it is possible to specify only rather arbitrarily where the liquid phase ends and the gaseous phase begins. It is convenient for some purposes to define the interface as a certain surface, *of constant density*, within the surface zone, such that, *if* each of the two phases remained homogeneous up to this surface, the total number of molecules would be the same. A better definition, however, can be founded on the thermodynamical theory which will now be explained.

It will be supposed that the surface of separation has been assigned in such a way as to divide the total volume V into a part V_A occupied by the liquid, and a part $V_B = V - V_A$ occupied by the saturated vapour; the interface must, of course, be a surface of constant density but is otherwise unspecified. If n_A and U_1^A represent the number density and internal energy per molecule in the liquid proper, and n_B and U_1^B represent the number density and internal energy per molecule in the vapour proper, the total energy calculated on the hypothesis that both liquid and vapour remain homogeneous up to the surface of separation is $n_A V_A U_1^A + n_B V_B U_1^B$. The difference (u) between this quantity and the actual total energy (U) of the fluid may be called the total surface energy, or the *superficial* internal energy:

(2. 1) $$u = U - (n_A V_A U_1^A + n_B V_B U_1^B).$$

If n and U_1 represent the actual number density and internal energy per molecule, this can be expressed in the form

$$u = \int_A (nU_1 - n_A U_1^A)\, d\mathbf{x} + \int_B (nU_1 - n_B U_1^B)\, d\mathbf{x},$$

where the suffixes A and B to the signs of integration indicate that the integrations are restricted to the liquid and vapour respectively.

Since, in the interior of the liquid, $n = n_A$ and $U_1 = U_1^A$, the effective contribution to the integral

$$\int_A (nU_1 - n_A U_1^A)\, d\mathbf{x}$$

comes only from the region very near the surface of separation. Since also the surface is nearly flat in a region of microscopic dimensions, this integral reduces to

(2. 2) $$A \int_0^a (nU_1 - n_A U_1^A)\, dz$$

where A is the total area of the interface, and the coordinate z measures the distance along the normal to the interface, shown in the figure. The distance a represents the depth of the surface zone in the liquid; as the upper limit of integration in (2. 2) it could, of course, be replaced by any larger value.

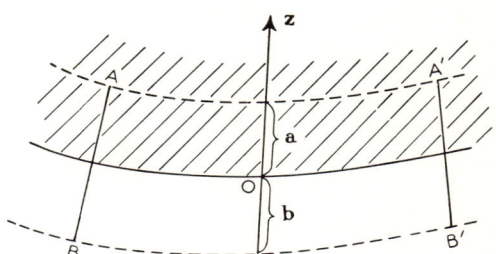

Fig. 2. The surface zone. On the microscopic scale there is no well-defined surface of separation between two phases.

Now the difference $nU_1 - n_A U_1^A$ can be written in the form

$$-\int_z^a \frac{\partial}{\partial z}(nU_1)\, dz;$$

hence

$$\int_A (nU_1 - n_A U_1^A)\, d\mathbf{x} = -A \int_0^a \int_z^a \frac{\partial}{\partial z}(nU_1)\, dz\, dz$$

$$= -A \int_0^a z \frac{\partial}{\partial z}(nU_1)\, dz,$$

integrating by parts. Similarly

$$\int_B (nU_1 - n_B U_1^B)\, d\mathbf{x} = -A \int_{-b}^{0} z \frac{\partial}{\partial z}(nU_1)\, dz$$

where b is the depth of the surface zone in the vapour. Hence one obtains finally

(2.3) $$u = -A \int_{-b}^{a} z \frac{\partial}{\partial z}(nU_1)\, dz.$$

The internal energy per molecule can be expressed by the usual formula (cf. § 4 of chapter II)

(2.4) $$U_1 = \tfrac{3}{2} kT + \tfrac{1}{2} n^{-1} \int n_2\, \phi(r)\, d\mathbf{r},$$

but it has to be remembered that the molecular distribution function n_2 in the surface zone will depend on the distance of the point $\mathbf{x}^{(1)}$ below the surface, and the angle made by the vector $\mathbf{r} = \mathbf{x}^{(2)} - \mathbf{x}^{(1)}$ with the z-direction, as well as the distance r. Also, the number density n depends on the distance z from the interface.

The superficial free energy (f) and surface entropy (s) are defined like u, and can be expressed by the formulae

(2.5) $$\begin{cases} f = -A \int_{-b}^{a} z \frac{\partial}{\partial z}(nF_1)\, dz, \\ s = -A \int_{-b}^{a} z \frac{\partial}{\partial z}(nS_1)\, dz, \end{cases}$$

where F_1 and S_1 represent the free energy per molecule and entropy per molecule respectively. There exists the relation

(2.6) $$u = f + Ts,$$

which follows from the ordinary thermodynamical relation $U_1 = F_1 + TS_1$. The quantity (γ) analogous to the pressure is called the surface tension, and satisfies

(2.7) $$\begin{cases} T\, ds = du - \gamma\, dA, \\ df = \gamma\, dA - s\, dT. \end{cases}$$

It follows from the second of these relations that, if the temperature is kept constant, the free energy increases with area at a rate proportional to the surface tension, which is itself independent of area;

hence $f = \gamma A$ — the surface tension is simply the superficial free energy per unit area of surface. This property, with the help of (2. 6) and (2. 7), enables one to relate the surface tension to the superficial internal energy, thus:

(2. 8) $$u = A\left(\gamma - T\frac{d\gamma}{dT}\right) = -AT^2\frac{d}{dT}\left(\frac{\gamma}{T}\right).$$

The last result is due in principle to LORD KELVIN [1911] [1]).

Now the surface tension will be obtained directly, by calculating the total stress across a surface (ABB'A' in figure 2) normal to the surface of separation. This is the method recently adopted by KIRKWOOD & BUFF [1949]. If r_j is the component of the vector **r** in a direction normal to the surface ABB'A', it follows from (4. 7) of chapter II that the stress per unit area across this surface is [2])

(2. 9) $$p_{11} = nkT - \tfrac{1}{2}\int n_2\,\phi'(r)\,r^{-1}\,r_1^2\,d\mathbf{r}$$

at any point $\mathbf{x}^{(1)}$. To obtain the surface tension, one simply subtracts from this the pressure $p_A = p_B$ in the uniform fluid, and integrates over a strip of the surface ABB'A', of unit length along the surface of separation, and sufficiently wide to cover the surface zone. This procedure is obviously in conformity with the experimental measurement of the surface tension, and leads to the formula

$$\gamma = \int_0^a (p_{11} - p_A)\,dz + \int_{-b}^0 (p_{11} - p_B)\,dz.$$

The latter can be transformed in exactly the same way as the formula for u, with the result

(2. 10) $$\gamma = \int_{-b}^a z\,\frac{\partial p_{11}}{\partial z}\,dz.$$

Because the pressure is the same in the liquid and gas, the final formula (2. 10) for the surface tension is independent of the choice of the surface of separation; one can replace z by $z + z_0$, and the additional integral $z_0\int_{-b}^a \frac{\partial p_{11}}{\partial z}\,dz$ obtained will vanish. The same is *not*

---

[1]) The date refers to the time of publication of some collected work, of earlier origin.

[2]) The notation indicates that the pressure in the boundary layer is not isotropic. The idea of the anisotropy of the pressure seems to originate with H. Hulshof, a pupil of van der Waals, and was used by G. Bakker in his book "Kapillarität."

true of the formula (2. 3) for the superficial internal energy; if, indeed, in this formula, the value of z is changed by the amount z_0, the value of u will be changed by the amount

$$A z_0 \int_{-b}^{a} \frac{\partial}{\partial z} (n U_1) \, dz = A z_0 (n_A U_1^A - n_B U_1^B).$$

Since the superficial internal energy is unique, and can be derived from the surface tension by means of the formula (2. 8), it is clear that the choice of the surface of separation is not arbitrary. In principle it is *necessary* to define the surface of separation as that which secures the agreement of the formulae (2. 3) and (2. 8).

The next question to be considered is the variation of the number density of molecules in the surface zone. This can be studied with the help of a relation which is derived from (4. 10) of chapter III simply by substituting $q = 1$:

(2. 11) $$\frac{\partial n}{\partial \mathbf{x}^{(1)}} + \beta \int n_2 \frac{\partial \phi}{\partial \mathbf{x}^{(1)}} d\mathbf{x}^{(2)} = 0.$$

If r_3 represents the component of the vector $\mathbf{r}$ in the z-direction, this gives

(2. 12) $$\frac{\partial n}{\partial z} = \beta \int n_2 \phi'(r) r^{-1} r_3 \, d\mathbf{r}.$$

The molecular distribution function n_2 is not known exactly, even when the radial distribution functions n_2^A and n_2^B for the uniform liquid and vapour are known; one is therefore compelled to use the approximation

(2. 13) $$\begin{cases} n_2 = n_A^2 g(r) & (\mathbf{x}^{(1)} \text{ and } \mathbf{x}^{(2)} \text{ both in the liquid}) \\ = 0 & (\mathbf{x}^{(1)} \text{ or } \mathbf{x}^{(2)} \text{ in the vapour}), \end{cases}$$

where $g(r) = n_2^A/n_A^2$. A less drastic approximation would be to assume, instead of $n_2 = 0$, $n_2 = n_B^2 g(r)$ when both $\mathbf{x}^{(1)}$ and $\mathbf{x}^{(2)}$ are in the vapour, and $n_2 = n_A n_B g(r)$ when one of $\mathbf{x}^{(1)}$ and $\mathbf{x}^{(2)}$ is in the vapour and the other in the liquid. This would give considerably improved results near the critical point. Away from the critical point, however, the density of the vapour is so much smaller than that of the liquid that n_2 is really negligible when either $\mathbf{x}^{(1)}$ or $\mathbf{x}^{(2)}$ lies in the vapour; and (2. 13) may be used in this region. Then (2. 12) gives

$$\frac{\partial n}{\partial z} = \beta n_A^2 \int_0^\infty \int_{-(r,z)}^{r} g(r) \phi'(r) r^{-1} r_3 \cdot 2\pi r \, dr_3 \, dr,$$

where (r, z) stands for the smaller of the two quantities r and z; or,

$$(2.14) \qquad \frac{\partial n}{\partial z} = \pi \beta n_A^2 \int_z^\infty g(r) \, \phi'(r) \, (r^2 - z^2) \, dr.$$

Further, one has from (2.9):

$$p_{11} = \beta^{-1} n - \tfrac{1}{2} n_A^2 \int_0^\infty \int_{-(r,z)}^r g(r) \, \phi'(r) \, r^{-1} \cdot \tfrac{1}{2} (r^2 - r_3^2) \cdot 2\pi r \, dr_3 \, dr,$$

so

$$\frac{\partial p_{11}}{\partial z} = \frac{1}{\beta} \frac{\partial n}{\partial z} - \tfrac{1}{2} \pi n_A^2 \int_z^\infty g(r) \, \phi'(r) \, (r^2 - z^2) \, dr$$

$$= \tfrac{1}{2} \pi n_A^2 \int_z^\infty g(r) \, \phi'(r) \, (r^2 - z^2) \, dr$$

with the help of (2.14). Hence, (2.10) leads finally to the formula

$$(2.15) \qquad \begin{cases} \gamma = \tfrac{1}{2} \pi n_A^2 \int_0^\infty z \int_z^\infty g(r) \, \phi'(r) \, (r^2 - z^2) \, dr \, dz \\ = \tfrac{1}{8} \pi n_A^2 \int_0^\infty g(r) \, \phi'(r) \, r^4 \, dr, \end{cases}$$

obtained previously by FOWLER [1937].

The superficial internal energy can be evaluated in a similar way. From (2.4) and (2.13) it follows that

$$nU_1 = \tfrac{3}{2} \beta^{-1} n + \tfrac{1}{2} n_A^2 \int_0^\infty \int_{-(r,z)}^r g(r) \, \phi(r) \cdot 2\pi r \, dr_3 \, dr,$$

$$\frac{\partial}{\partial z}(nU_1) = \frac{3}{2\beta} \frac{\partial n}{\partial z} + \pi n_A^2 \int_z^\infty g(r) \, \phi(r) \, r \, dr.$$

The formula (2.3) therefore reduces to

$$u = -\pi A \, n_A^2 \int_0^\infty z \int_z^\infty g(r) \, \{ \tfrac{3}{2} \phi'(r) \, (r^2 - z^2) + \phi(r) \, r \} \, dr \, dz$$

$$= -\tfrac{1}{8} \pi n_A^2 \int_0^\infty g(r) \, \{ 3 \phi'(r) \, r^4 + 4 \phi(r) \, r^3 \} \, dr \, A,$$

if one chooses a surface of separation coincident with the discontinuity of the radial distribution function. This is actually not a good choice, and the resulting formula is in error; failing the rigorous thermodynamic definition, it is better to define the surface of separation as that which makes

$$\int_{-b}^{a} z \frac{\partial n}{\partial z} \, dz = 0,$$

when z represents the distance of the point $\mathbf{x}^{(1)}$ from the true surface of separation. This leaves the total number of molecules unchanged when it is supposed that both phases remain homogeneous right up to the interface. Even this definition does not coincide exactly with the thermodynamical one; however, if one accepts it as a reasonable approximation, the formula (2.3) for u reduces to

(2.16)
$$\begin{cases} u = - \int_{-b}^{a} z \frac{\partial}{\partial z} \left(\tfrac{1}{2} \int n_2 \phi \, d\mathbf{r} \right) dz A \\ = - \tfrac{1}{2} \pi n_A^2 \int_0^\infty g(r) \phi(r) r^3 \, dr A. \end{cases}$$

In principle, of course, it would be better — in fact, correct — to determine the superficial internal energy by combining the formulae (2.8) and (2.15). Kirkwood & Buff have, however, made a direct comparison of the formulae (2.15) and (2.16) with the experimentally measured values in liquid argon at 90° K. They use the potential function

$$\phi(r) = -\mu r^{-6} + \nu r^{-11.4},$$
$$\mu = 1.11 \times 10^{-10} \text{ erg. A}^6, \quad \nu = 8.62 \times 10^{-8} \text{ erg. A}^{11.4}$$

given by RUSHBROOKE [1940], and the distribution function

$$g(r) = (a_1/r)^s \exp \{(a_m/a_1)^t - (a_m/r)^t\}, \qquad r < a_1$$
$$= 1, \qquad r > a_1$$
$$a_1 = 4.50 \text{ A}, \ a_m = 3.55 \text{ A}, \ s = 7, \ t = 14,$$

which is designed to fit the experimental curve of Eisenstein & Gingrich (cf. § 3 of chapter III) as well as possible, consistent with analytical simplicity and certain thermodynamical requirements which need not be related here.

The numerical results, for liquid argon at 90° K., are

$$\int_0^\infty g(r) \phi'(r) r^4 \, dr = 8.82 \times 10^{-12} \text{ erg. A}^4$$
$$\int_0^\infty g(r) \phi(r) r^3 \, dr = -4.03 \times 10^{-12} \text{ erg. A}^4$$

	Theory	Experiment	
γ	14.9	11.9	dyne/cm
u/A	27.2	35	dyne/cm.

The agreement is perfectly satisfactory in view of the approximations introduced into the calculations.

The variation of the density in the surface zone can be obtained by an integration by parts from the equation (2. 14), thus:

$$(2.\ 17) \qquad n = n_A - \tfrac{1}{3}\pi\beta n_A^2 \int_z^\infty g(r)\,\phi'(r)\,(2r^3 - 3zr^2 + z^3)\,dr.$$

Subject to the approximation represented by (2. 13), it can be seen quite readily from this formula that the depth of the surface zone does not exceed a few molecular diameters, and near the interface the density is a *cubic* function of the distance.

It has been remarked that, in the neighbourhood of the critical point, one may not neglect the density of the vapour, and must therefore assume $n_2 = n_B^2 g(r)$ when both of $\mathbf{x}^{(1)}$ and $\mathbf{x}^{(2)}$ are in the vapour, and $n_2 = n_A n_B g(r)$ when one of these points is in the vapour and the other in the liquid. The effect of this modification of the theory is to replace the coefficient n_A^2 in (2. 15) and (2. 16) by $(n_A - n_B)^2$. In consequence,

$$(2.\ 18) \qquad \begin{aligned} u &= c_1 A\,(n_A - n_B)^2, \\ c_1 &= \tfrac{1}{4}\int r\,n_{2A}\,\phi\,d\mathbf{r}/n_A^2. \end{aligned}$$

If, in the neighbourhood of the critical temperature T_c, the temperature is expanded in powers of the difference $n_A - n_B$ between the densities of the two phases, one has

$$(2.\ 19) \qquad T = T_c - c_2(n_A - n_B)^2 + \ldots.$$

This, indeed, merely expresses the fact that, in the pressure-specific volume diagram, the condensation zone has a horizontal tangent at the critical point. Substituting (2. 19), with (2. 18), into (2. 8), one obtains by integration

$$(2.\ 20) \qquad \begin{cases} \gamma \sim \dfrac{c_1 T}{c_2}\left(\dfrac{T_c}{T} - 1 - \ln\dfrac{T_c}{T}\right) \sim \dfrac{c_1}{c_2}\dfrac{(T_c - T)^2}{2T} \\ \qquad\qquad\qquad\qquad \sim c_1\,c_2\,T_c^{-1}(n_A - n_B)^4 \end{cases}$$

for the immediate neighbourhood of the critical point. This result furnishes some justification for Macleod's rule, that the surface tension varies as the fourth power of the difference in density between the two phases. In fact this rule is found to hold empirically for many fluids over a much wider temperature range than the approximations made in the present derivation could warrant.

The preceding considerations apply strictly only to non-polar fluids. There is an abundance of evidence that surface phenomena are very different in polar and non-polar fluids. In non-polar fluids, the depth of the surface zone is probably never much more than a few molecular diameters. In polar fluids, however, the depth of the surface zone is often great enough to be observed by macroscopic methods. The experimental evidence and probable explanation has been reviewed by HENNIKER [1949]. The principal effects observed are a considerably increased viscosity, and an abnormally great elastic strength of the fluid at its surface, especially near an interface with a solid substance. The suggested explanation is that the surface molecules orient themselves in the direction of their polarization, and this causes a similar alignment of the molecules inside the fluid which often extends to a distance of more than 1,000 A from the surface. The X-ray diffraction patterns tend to confirm that this explanation is in fact correct. Nevertheless, from the quantitative point of view, much remains to be done in the elaboration of the theory of surface phenomena.

3. The Brownian Motion

The most direct evidence available for the molecular structure of fluids is to be seen in the Brownian motion. If sufficiently small grains of any foreign material are immersed in a fluid, they are subject to fluctuating forces, arising from the impacts transmitted by the individual molecules by virtue of their thermal motion. Under these fluctuating forces, the Brownian particles, as they are called, execute perpetual irregular motions which can be seen without difficulty under the microscope. The existence of such motions was predicted by Einstein in 1905 at a time when he was still unfamiliar with the experimental facts, which had been discovered much earlier.

It should not be supposed that the effect of individual collisions between the Brownian particle and the molecules of the fluid is visible; it is only the fluctuation in the resultant force due to many collisions which is directly responsible for the Brownian motion. In the theory of the Brownian motion, it is customary to separate the instantaneous resultant force on the Brownian particle into two components. The first is a comparatively slowly varying force, very much like the viscous force experienced by macroscopic bodies, opposing the motion through the fluid. The second component ($M\mathbf{g}$), on the other hand, is rapidly fluctuating both in time and direction, and, though its average value vanishes, over a short interval of time it can produce an observable acceleration of the particle. LANGEVIN [1908] proposed an equation of the type

$$(3.1) \qquad \frac{d}{dt}\boldsymbol{\xi} = -\zeta\boldsymbol{\xi} + \mathbf{g}(t)$$

to describe the motion of a particle with velocity $\boldsymbol{\xi}$ under such conditions. The parameter ζ is called the friction constant, and is assumed to be almost independent of velocity; KIRKWOOD [1946] has given an approximate method to determine ζ from a knowledge of the forces exerted by the individual molecules, and their distribution about the Brownian particle (cf. § 6 of chapter V). At present consideration will be given only to the statistical consequences of Langevin's equation.

It will be supposed that at some initial time ($t = 0$) the Brownian particle, situated at the point $\mathbf{x}_0$, is observed to have the velocity $\boldsymbol{\xi}_0$. Owing to the fact that nothing can be said with certainty concerning the magnitude or direction of the fluctuating force $\mathbf{g}(t)$, it will be impossible to predict with certainty the position ($\mathbf{x}$) and velocity ($\boldsymbol{\xi}$) of the particle at any later time (t). One may hope, however, to determine the *probability* $f(\boldsymbol{\xi}, \mathbf{x}, t)d\boldsymbol{\xi}d\mathbf{x}$ of finding the particle in the volume element $d\mathbf{x}$, at the point $\mathbf{x}$, with velocity in the range $\boldsymbol{\xi}$, $d\boldsymbol{\xi}$, at time t. This is the task which will be attempted in the present section; when accomplished, it will furnish the answer to any question of a quantitative nature which one might wish to ask concerning the Brownian motion.

The form of the distribution function $f(\boldsymbol{\xi}, \mathbf{x}, t)$ at the initial time ($t = 0$), and also after a very long interval ($t = \infty$), can be stated immediately. For the initial time, one has by hypothesis

$$(3.2) \qquad f(\boldsymbol{\xi}, \mathbf{x}, 0) = \delta(\boldsymbol{\xi} - \boldsymbol{\xi}_0)\,\delta(\mathbf{x} - \mathbf{x}_0)$$

where $\delta(\mathbf{x})$ represents the δ-function, which vanishes everywhere except in an infinitesimally small region surrounding the point $\mathbf{x} = 0$, and is there so large that

$$\int \delta(\mathbf{x})\,d\mathbf{x} = 1.$$

After a very long time, the particle will be in statistical equilibrium with the fluid, and its velocity distribution function must therefore be the equilibrium velocity distribution function:

(3. 3) $$f(\boldsymbol{\xi}, \mathbf{x}, \infty) = A \exp(-\tfrac{1}{2}\beta M \boldsymbol{\xi}^2),$$

where M is the mass of the particle, and A is *a* constant.

To circumvent the difficulty that the fluctuating force $\mathbf{g}(t)$ is unknown, it is convenient to introduce a probability function $w(\mathbf{G}, \tau)$ such that

$$w(\mathbf{G}, \tau)\, d\mathbf{G}$$

measures the probability that the time integral

$$\mathbf{G} = \int_t^{t+\tau} \mathbf{g}(t)\, dt$$

of $\mathbf{g}(t)$ over a given time interval τ should have a value in the range $\mathbf{G}, d\mathbf{G}$. It is assumed that, if the interval τ is macroscopically short, yet long compared with the period of the fluctuations of $\mathbf{g}(t)$, the probability function w depends on $\mathbf{G}$ and τ, but not on the time t or the velocity $\boldsymbol{\xi}$ of the particle. Now, according to Langevin's equation (3. 1), the change of velocity $\boldsymbol{\rho}$ of the particle in the small time interval τ is

(3. 4) $$\boldsymbol{\rho} = -\zeta \boldsymbol{\xi} \tau + \mathbf{G}$$

hence the probability that $\boldsymbol{\rho}$ should lie in the range $\boldsymbol{\rho}, d\boldsymbol{\rho}$ is

$$w(\boldsymbol{\rho} + \zeta \boldsymbol{\xi} \tau, \tau)\, d\boldsymbol{\rho}$$

Further, the relative probability $f(\boldsymbol{\xi}, \mathbf{x}, t + \tau)$ of finding the particle at the point $\mathbf{x}$, with velocity $\boldsymbol{\xi}$, at time $t + \tau$, is obtained from the relative probability $f(\boldsymbol{\xi} - \boldsymbol{\rho}, \mathbf{x} - \boldsymbol{\xi}\tau, t)$ of finding the particle at the point $\mathbf{x} - \boldsymbol{\xi}\tau$, with velocity $\boldsymbol{\xi} - \boldsymbol{\rho}$, at time t, by multiplying with the probability $w\{\boldsymbol{\rho} + \zeta(\boldsymbol{\xi} - \boldsymbol{\rho})\tau, \tau\} d\boldsymbol{\rho}$ that the velocity of the particle changes by the amount $\boldsymbol{\rho}$ in the small interval τ, and summing over all values of $\boldsymbol{\rho}$, thus:

(3. 5) $$f(\boldsymbol{\xi}, \mathbf{x}, t + \tau) = \int f(\boldsymbol{\xi} - \boldsymbol{\rho}, \mathbf{x} - \boldsymbol{\xi}\tau, t)\, w\{\boldsymbol{\rho} + \zeta(\boldsymbol{\xi} - \boldsymbol{\rho})\tau, \tau\}\, d\boldsymbol{\rho}.$$

The probability function w can now be determined by considering the state of the system after a long time ($t = \infty$), and substituting (3. 3) into (3. 5). One obtains in this way the integral equation

(3. 6) $$\exp(-\tfrac{1}{2}\beta M \boldsymbol{\xi}^2) = \int \exp(-\tfrac{1}{2}\beta M \boldsymbol{\xi}'^2)\, w(\boldsymbol{\xi} - \boldsymbol{\xi}' + \zeta \boldsymbol{\xi}'\tau, \tau)\, d\boldsymbol{\xi}'.$$

This equation is readily solved for w by Fourier transformations; it is sufficient to give the solution, which, one may easily verify, is

(3. 7) $$w(\mathbf{G}, \tau) = \left(\frac{\beta M}{4\pi\zeta\tau}\right)^{3/2} \exp\left(-\frac{\beta M \mathbf{G}^2}{4\zeta\tau}\right).$$

Thus it has been established that, in a fluid in equilibrium, the accelerations of a Brownian particle in short equal intervals of time must be distributed according to a Gaussian law.

From (3. 7) the mean values of various quantities of physical interest are easily calculated. The mean value of $\boldsymbol{\rho}$ is

$$\int \boldsymbol{\rho} w(\boldsymbol{\rho} + \zeta\boldsymbol{\xi}\tau, \tau) \, d\boldsymbol{\rho} = -\zeta\boldsymbol{\xi}\tau$$

and the mean value of $\mathbf{G}^2$, i.e., the mean square deviation of $\boldsymbol{\rho}$ from the mean, is

$$\int \mathbf{G}^2 w(\mathbf{G}, \tau) \, d\mathbf{G} = 6\zeta\tau/(\beta M).$$

From this one sees that the root mean square deviation of $\boldsymbol{\rho}$ from the mean is proportional to $\tau^{\frac{1}{2}}$, and much greater than the magnitude $\zeta\boldsymbol{\xi}\tau$ of the mean, provided τ is very small compared with $(\beta M\zeta\xi^2)^{-1}$. For reasonably large values of ξ, the last condition can be satisfied with a macroscopically significant value of τ.

The next question to be considered is the determination of the velocity distribution function f of the Brownian particle, at an arbitrary time. A differential equation for f can be obtained from (3. 5) by expanding both sides in powers of τ, and neglecting squares and higher powers. If $\mathbf{z} = \{\boldsymbol{\rho} + \zeta(\boldsymbol{\xi} - \boldsymbol{\rho})\tau\}/\tau^{\frac{1}{2}}$ the right hand side gives

$$\left(\frac{\beta M}{4\pi\zeta}\right)^{3/2} \int f\{\boldsymbol{\xi} - (\mathbf{z}\tau^{\frac{1}{2}} - \zeta\boldsymbol{\xi}\tau), \mathbf{x} - \boldsymbol{\xi}\tau, t\} \exp\left(-\beta M \mathbf{z}^2/4\zeta\right) \left(\frac{1}{1-\zeta\tau}\right)^3 d\mathbf{z}$$
$$= (1 + 3\zeta\tau)f +$$
$$+ \left(\frac{\beta M}{4\pi\zeta}\right)^{3/2} \tau \int \left\{-\boldsymbol{\xi} \cdot \frac{\partial f}{\partial \mathbf{x}} + \zeta\boldsymbol{\xi} \cdot \frac{\partial f}{\partial \boldsymbol{\xi}} + \tfrac{1}{2} \mathbf{z} \cdot \frac{\partial^2 f}{\partial \boldsymbol{\xi} \partial \boldsymbol{\xi}} \cdot \mathbf{z}\right\} \exp\left(-\beta M \mathbf{z}^2/4\zeta\right) \, d\mathbf{z},$$

so the result is

(3. 8) $$\frac{\partial f}{\partial t} + \boldsymbol{\xi} \cdot \frac{\partial f}{\partial \mathbf{x}} = \zeta \frac{\partial}{\partial \boldsymbol{\xi}} \cdot \left(\boldsymbol{\xi} f + \frac{1}{\beta M} \frac{\partial f}{\partial \boldsymbol{\xi}}\right).$$

This is a generalization — given and solved by CHANDRASEKHAR [1943] — of a well-known equation due to Fokker and Planck; the Fokker–Planck equation can, in fact, be obtained by suppressing the dependence of f on $\mathbf{x}$. The result may be compared with Kirkwood's equation (6. 19) of chapter V, where the velocity distribution function relates to a molecule of the fluid rather than a Brownian particle. Kirkwood's argument is, however, easily generalized for a Brownian particle in an environment of ordinary molecules.

The remaining task is now to obtain a solution of the differential equation (3. 8) which satisfies the initial condition (3. 2). If one makes the substitutions

(3. 9)
$$\begin{cases} \boldsymbol{\xi}\, e^{\zeta t} - \boldsymbol{\xi}_0 = \mathbf{v}, \quad \mathbf{x} - \mathbf{x}_0 + (\boldsymbol{\xi} - \boldsymbol{\xi}_0)/\zeta = \mathbf{r} \\ \chi = f\, e^{-3\zeta t}, \end{cases}$$

the equation (3. 8) reduces to

(3. 10) $$\frac{\partial \chi}{\partial t} = \frac{\zeta}{\beta M} \left(e^{2\zeta t} \frac{\partial^2 \chi}{\partial \mathbf{v} \cdot \partial \mathbf{v}} + 2\, \frac{e^{\zeta t}}{\zeta} \frac{\partial^2 \chi}{\partial \mathbf{v} \cdot \partial \mathbf{r}} + \frac{1}{\zeta^2} \frac{\partial^2 \chi}{\partial \mathbf{r} \cdot \partial \mathbf{r}} \right).$$

To solve an equation of this type, one may substitute the trial solution

(3. 11)
$$\begin{cases} \chi = (2\pi \varDelta^{\frac{1}{2}})^{-3} \exp\{-(a\mathbf{v}^2 + 2h\mathbf{r} \cdot \mathbf{v} + b\mathbf{r}^2)/2\varDelta\}, \\ \varDelta = ab - h^2 \end{cases}$$

where a, b and h may be functions of the time only. If one imposes the initial conditions $a = b = h = 0$ when $t = 0$, this solution will reduce for $t \to 0$ to the form

$$\chi = \delta(\mathbf{v})\, \delta(\mathbf{r})$$

required by (3. 2); the initial condition for f will then be satisfied automatically. Now, when (3. 11) is substituted into (3. 10), this equation reduces to

(3. 12) $$\frac{da}{dt} = \frac{2}{\zeta \beta}, \quad \frac{db}{dt} = \frac{2\zeta}{\beta} e^{2\zeta t}, \quad \frac{dh}{dt} = -\frac{2}{\beta} e^{\zeta t};$$

hence

(3. 13)
$$\begin{cases} a = 2t/\zeta\beta, \quad b = \beta^{-1}(e^{2\zeta t} - 1), \\ h = -(2/\zeta\beta)(e^{\zeta t} - 1). \end{cases}$$

Collecting results, the solution of (3. 8) is

$$f = \{4\pi^2(ab-h^2)\}^{-3/2} \exp\{3\zeta t - \tfrac{1}{2}(a\mathbf{v}^2 + 2h\mathbf{r}\cdot\mathbf{v} + b\mathbf{r}^2)/(ab-h^2)\},$$

where $\mathbf{v}$ and $\mathbf{r}$ are defined in (3. 9), and a, b and h are given by (3. 13).

In this way the probability has been obtained of finding the particle with any position and velocity at any time later than the initial observation was made. Such experimental observations as have been made are in excellent agreement with the Gaussian probability law which has been obtained.

4. The Second Virial Coefficient

It has been seen already [(4. 8) of chapter II] that the hydrostatic pressure in a fluid in equilibrium is given by

$$p = nkT - \tfrac{1}{6} \int n_2(r)\,\phi'(r)\,r\,d\mathbf{r}$$

The second term on the right-hand side represents the deviation from the perfect gas law, $p = nkT$. If one substitutes for the radial distribution function $n_2(r)$ the first approximation

$$n_2(r) \sim n^2 \exp\{-\beta\phi(r)\},$$

where n is the number density of the molecules, and β is inversely proportional to the absolute temperature, one obtains

(4. 1)
$$\begin{cases} p - nkT = -\tfrac{1}{6} n^2 \int \exp\{-\beta\phi(r)\}\,\phi'(r)\,r\,d\mathbf{r} \\ \qquad\quad = -\tfrac{2}{3}\pi n^2 \int\limits_0^\infty \exp\{-\beta\phi(r)\}\,\phi'(r)\,r^3\,dr. \end{cases}$$

Integrating by parts, this reduces to

(4. 2)
$$\begin{cases} p - nkT = \dfrac{2\pi n^2}{3\beta} \int r^3\,d\,[\exp\{-\beta\phi(r)\} - 1] \\ \qquad\quad = -\tfrac{1}{2}(n^2/\beta)\int\limits_0^\infty [\exp\{-\beta\phi(r)\} - 1]\,4\pi r^2\,dr. \end{cases}$$

The approximation used for the derivation of (4. 2) has the consequence that it is correct to terms quadratic in the number density (n), but neglects terms proportional to cubic and higher powers $(n^3, n^4$ etc.). This indeed can be seen directly from the exact development, expressed by the formula (3. 26) of chapter IV. The expression

(4. 3) $$\beta_1 = \tfrac{1}{2}\int\limits_0^\infty [\exp\{-\beta\phi(r)\} - 1]\,4\pi r^2 dr$$

is called the second virial coefficient. It can obviously be determined experimentally by measuring the pressure exerted by a gas at various densities, but the same temperature, and exhibiting the experimental values as a power series in the density. If this is repeated for various temperatures, one obtains the second virial coefficient as a function $\beta_1(T)$ of the temperature.

By means of the formula (4. 3), it would be quite easy to determine the function $\beta_1(T)$ if the intermolecular potential energy $\phi(r)$ were known as a function of the distance r between a pair of molecules. In practice, the inverse procedure is most frequently followed; from the knowledge of the function obtained experimentally, the intermolecular force law is determined. This would be possible in principle even if one knew nothing concerning the function $\phi(r)$; one would simply regard (4. 3) as an integral equation from which the unknown function $\phi(r)$ had to be determined. However, the task is simplified by the circumstance that the form of the function is already known from other considerations. It was seen in § 7 of chapter III that the attractive part of the intermolecular field is always proportional to r^{-6}, and the repulsive part can be represented by $\exp(-r/\varrho)$; and that the potential energy between two molecules can be represented in any of the alternative forms

(4. 4) $$-\mu r^{-6} + N e^{-r/\varrho},$$

(4. 5) $$-\mu r^{-6} + \nu r^{-m},$$

(4. 6) $$-M e^{-r/\sigma} + N e^{-r/\varrho},$$

of which the first is the best, from the theoretical standpoint. Since the form of the potential energy may be regarded as known, only the constants μ, ν, M, N, ϱ and σ remain to be determined. The easiest way of doing this is to substitute (4. 4) — (4. 6) into (4. 3), and ascertain which values of the constants give the best agreement with the experimentally measured values of the first virial coefficient.

The form (4. 5) of the function $\phi(r)$ was adopted by LENNARD-JONES [1937], who found it was possible to evaluate the resulting integral, in the way now to be described. If one substitutes (4. 5) into (4. 3), and changes the variable of integration from r to $s = \beta \nu r^{-m}$, the result is

(4. 7) $$\beta_1 = 2\pi (\beta \nu)^{3/m}/m \int_0^\infty \{\exp(y s^{6/m} - s) - 1\} s^{-(3/m)-1} \, ds,$$

where

(4. 8) $$y = \beta \mu (\beta \nu)^{-6/m}.$$

To evaluate the integral in (4. 7), one uses

$$\exp(y s^{6/m}) = \sum_{k=0}^{\infty} (y s^{6/m})^k/k!,$$

$$\int_0^\infty (e^{-s} - 1) s^{-(3/m)-1} \, ds = -\tfrac{1}{3} m \int_0^\infty e^{-s} s^{-3/m} \, ds,$$

$$\int_0^\infty e^{-s} s^\lambda \, ds = \Gamma(1 + \lambda),$$

where Γ denotes the Γ-function, of which many tables exist. Then

$$(4.9) \quad \beta_1 = \tfrac{2}{3}\pi(\beta\nu)^{3/m}\left\{-\Gamma\left(\frac{m-3}{m}\right) + \frac{3}{m}\sum_{k=1}^{\infty}\frac{y^k}{k!}\Gamma\left(\frac{6k-3}{m}\right)\right\}.$$

The first few terms of the infinite series always suffice to provide a sufficiently accurate result, when m is chosen with a value greater than 6. Subject to this condition, the results are not very sensitive to the precise value of m which is chosen; by suitable adjustment of μ and ν, one can always obtain good agreement with the experimentally measured values of β_1. In the table below are the best values of μ and ν for various values of m, obtained for diatomic hydrogen and nitrogen by Lennard–Jones, and for helium, neon and argon by Buckingham. The units in which μ and ν have the values stated are 10^{-10} erg.Å^6 and 10^{-10} erg.Å^m respectively

m	8	9	10	12	14
He μ	0.0370	0.0237	0.0173	0.01085	0.0076
ν	0.304	0.510	0.948	3.56	13.9
Ne μ	0.207	0.145	0.114	0.0832	0.0678
ν	1.83	3.50	7.32	35.5	182
A μ	2.37	1.70	1.37	1.03	0.0867
ν	31.5	76.8	205	1620	13650
H μ		0.178	0.142	0.105	
ν		5.1	11.6	64.9	
N μ		2.27	1.85	1.40	
ν		132.0	393	3700	

The numerical data for the second virial coefficient, on which these values were based, were due to HOLBORN & OTTO [1924—26].

For helium, it is possible to compare the function $\phi(r)$ obtained by this method with that calculated directly from the known structure of the helium atom by SLATER & KIRKWOOD [1931]. The agreement is good; but it is noteworthy that at low temperatures the formula (4.3) is subject to corrections of a quantum-mechanical origin.

It has been seen that agreement with the experimental data can be obtained with a fairly wide variety of repulsive interactions of the form νr^{-m}. One concludes naturally that it should also be possible to fit the data with a repulsive interaction of the form $Ne^{-r/\varrho}$, which is theoretically much better justified. In this way one would obtain the constants associated with the intermolecular potential energy (4. 4). This was done by Buckingham, who obtained the results given in the table below. The constant μ is again in 10^{-10} erg.$Å^6$; N is in 10^{-10} erg, and ϱ in Å.

	μ	N	ϱ
He	0.0147	8.71	0.216
Ne	0.090	25.7	0.235
A	1.02	169	0.273

5. Gravity and Boundary Forces

Under normal experimental conditions, fluids are subject to the action of gravity, which is a particular instance of a body-force. The effect of gravity on a fluid has not yet been considered explicitly, and it is a matter of importance either to justify or to remedy this neglect.

In fact, most of the consequences of the action of external forces can be inferred from macroscopic hydrodynamics, without direct reference to the molecular constitution of the fluid. It is known that, in equilibrium, a conservative force system is balanced by the pressure distribution. The vertical equilibrium of the atmosphere, for example, is preserved by an upward decrease of pressure, which is so adjusted that the pressure gradient is everywhere the same as the gravitational force per unit volume. Thus, in any fluid in equilibrium which is under the influence of gravity, the pressure is not uniform. The only reason why this can usually be disregarded is that, unless the fluid is very dense or of great depth, the variation of the pressure within the fluid is small compared with its absolute value. However, in dense fluids like mercury, or where a deep column of fluid is under consideration, the variation in pressure may be relatively large.

Associated with the variation in pressure in a fluid under the action of a body force, there is normally a variation in density. This is necessary to satisfy the requirements of the equation of state, which imposes a relation between the density, temperature and pressure in the fluid. Assuming that the fluid is in thermal equilibrium, the density must therefore vary with the pressure according to the isothermal law. The simplest expression of these purely thermodynamical relations is the statement that, in any fluid, the sum of the thermodynamic potential (Gibbs' free energy) and the potential energy of the external forces is everywhere the same, under isothermal conditions. This is easily proved, in the following way. If G_1 is the thermodynamic potential per molecule, related to F_1, the free energy per molecule, by $G_1 = F_1 + p/n - \chi$, and S_1 is the entropy per molecule, one has in general

$$dG_1 = -S_1 dT + dp/n;$$

so

(5. 1) $$dG_1 = dp/n,$$

if the temperature is the same throughout the fluid. Representing the potential energy of a molecule at the point $\mathbf{x}$, due to the presence of the external forces, by $\chi(\mathbf{x})$, the hydrodynamical equation of motion of the fluid is

$$mn \frac{d}{dt} \mathbf{u} + \frac{\partial}{\partial \mathbf{x}} \cdot \mathbf{P} = -n \frac{\partial \chi}{\partial \mathbf{x}},$$

reducing to

(5. 2) $$\frac{\partial p}{\partial \mathbf{x}} = -n \frac{\partial \chi}{\partial \mathbf{x}}$$

in equilibrium. From (5. 1) and (5. 2) it follows that

(5. 3) $$\frac{\partial}{\partial \mathbf{x}} (G_1 + \chi) = 0,$$

which was to be proved.

The distribution function (f_N) which measures the probability of finding each of the N molecules with assigned positions and velocities is given by the appropriate adaptation of the formula (5. 15) of chapter II:

(5. 4) $$f_N = \exp \beta \, [F - \sum_{i=1}^{N} \{\tfrac{1}{2} m \boldsymbol{\xi}^{(i)2} + \chi(\mathbf{x}^{(i)})\} + \Phi],$$

where Φ is, as usual, the total mutual potential energy of the molecules, $\beta = 1/(kT)$, and F is a constant which may be interpreted as the free energy of the system. Since

$$\int \stackrel{(2N)}{\cdots} \int f_N \, d\boldsymbol{\xi}^{(1)} \, d\mathbf{x}^{(1)} \ldots d\boldsymbol{\xi}^{(N)} \, d\mathbf{x}^{(N)} = N!$$

one has

(5. 5) $$\begin{cases} \exp \beta F = \lambda^N \int \stackrel{(2N)}{\cdots} \int \exp \{-\beta (\chi + \Phi)\} \, d\boldsymbol{\xi}^{(1)} \, d\mathbf{x}^{(1)} \ldots d\boldsymbol{\xi}^{(N)} \, d\mathbf{x}^{(N)}/N!, \\ \chi = \sum \chi(\mathbf{x}^{(i)}), \qquad \lambda = (2\pi/\beta m)^{3/2}. \end{cases}$$

The force exerted by gravity on a molecule in a fluid may, under laboratory conditions, be represented by the constant vector $-m\mathbf{g}$, where $-\mathbf{g}$, the acceleration due to gravity, is directed vertically downwards, and has a magnitude of about 980 cm/sec². Thus one has

$$(5.\ 6) \qquad \chi(\mathbf{x}) = m\mathbf{g}\cdot\mathbf{x}.$$

According to (5. 3), the thermodynamic potential G_1 must therefore decrease vertically upwards at a rate given by the law

$$(5.\ 7) \qquad G_1 = -m\mathbf{g}\cdot\mathbf{x} + \text{const.},$$

where the constant of integration may depend on the temperature, as well as the origin of coordinates. *In a rare gas*,

$$(5.\ 8) \qquad \beta G_1 = \ln(n\beta^{3/2}) + \text{const.};$$

it follows, therefore, from (5. 7), that the density must vary according to the law

$$(5.\ 9) \qquad n = \text{const. exp.}(-\beta m\mathbf{g}\cdot\mathbf{x}).$$

In condensed fluids, the spatial variation of density is even smaller, and can be neglected for most purposes.

The theory of fluids under the action of conservative forces can also be used to examine phenomena at the boundary of a fluid, where it is in contact with a solid surface. Real solid surfaces are, of course, made up of molecules not essentially different in their nature from the molecules of the fluid. The interaction between a molecule of the fluid and a molecule of the boundary wall can be represented by a function $\chi(\mathbf{x}^{(i)}, \mathbf{x}^{(j)})$ of their respective positions $\mathbf{x}^{(i)}$ and $\mathbf{x}^{(j)}$. The position $\mathbf{x}^{(j)}$ of the molecule in the wall continually changes with time, as a result of the thermal motion. If the substance of the wall is crystalline, however, this motion consists of small oscillations about a mean position $\mathbf{x}_0^{(j)}$; and, even if the wall is itself a fluid, such as glass, the molecules of which it is made will not be so mobile as those of the fluid which it contains. It is therefore permissible for most purposes to regard the molecules of the wall as stationary at their mean positions $\mathbf{x}_0^{(j)}$, and to represent the potential energy of a molecule of the fluid, due to the presence of the molecules in the wall, as a function

$$\chi(\mathbf{x}^{(i)}) = \sum_j \chi(\mathbf{x}^{(i)}, \mathbf{x}_0^{(j)})$$

of its own position ($\mathbf{x}^{(i)}$) alone. The effect of the wall is thus the same as that of a conservative field of force. The equivalent field of force is notwithstanding very different in its nature from the gravitational field, for example; the function defined above is effectively zero everywhere inside the fluid, except in a very small region near the boundary wall. At the wall itself, $\chi(\mathbf{x})$ rises to an extremely large, effectively infinite, value, which prevents the molecules from escaping through the wall.

The behaviour of the fluid near the wall can be inferred from (5. 3). As χ rises from zero inside the fluid to a very large value near the wall, the thermodynamic potential G_1 decreases from its value inside the fluid to some very large negative value, which corresponds to a very small value of the density n. At the wall itself, and in the region immediately next to the wall, the density is so small that (5. 8) and (5. 9) are perfectly applicable. The pressure exerted by the fluid is also very slight, because of the small density.

At first sight it might appear rather paradoxical that, although the pressure exerted by a fluid is measured at the wall, the actual value of this pressure is quite different in the immediate neighbourhood of the wall from that in the interior of the fluid. The paradox is easily resolved, however. Let the equation (5. 2) be integrated over the region between a section of the wall and a surface S, parallel with the wall and well inside the fluid. The pressure is zero on the outer surface; on the inner surface it has the value p appropriate to the interior of the fluid.

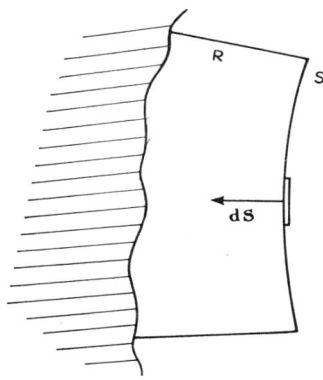

Fig. 3. The surface S is parallel to the wall, and sufficiently far from it to avoid the irregularities near the interface.

By an application of Gauss' theorem, one has, therefore,

$$\int_R \frac{\partial p}{\partial \mathbf{x}} d\mathbf{x} = - \int_S p \, \mathbf{dS}$$

where the suffix R indicates that the integration is limited to the region R between the two surfaces, and **dS** is a vector element of the surface S inside the fluid. The equation (5. 2) therefore leads to the result

(5. 10) $$p \int_S \mathbf{dS} = \int_R n \frac{\partial \chi}{\partial \mathbf{x}} d\mathbf{x},$$

Now, $-\int_R n \frac{\partial \chi}{\partial \mathbf{x}} d\mathbf{x}$ is the resultant force exerted by the molecules of the wall on the fluid in the region R, and this must be equal and opposite to the force exerted by the fluid on the wall. The relation (5. 10) shows that this force measures the pressure inside the fluid, and *not* the pressure exerted by the molecules just inside the wall.

Of course, the transition layer in which the variation of density and pressure takes place is deep only by molecular standards, and could not be detected by macroscopic means. Even so, the representation of boundary walls in this way is rather highly idealized. Walls which appear to be perfectly smooth, even under the microscope, are actually very rough on the molecular scale, and possess holes and protuberances which impede the motion of the fluid parallel to the wall. In addition to forces of the van der Waals' type, chemical or polar forces are often in operation between the molecules of the fluid and those of the wall. The thermal motion of the molecules of the wall cannot always be neglected. All these factors lead to complications of the simple theory described above. The representation of a wall by a conservative field of force does, however, enable one to investigate the essential characteristics of boundary phenomena in the simplest possible way.

6. The Dielectric Constant

In the theory of the dielectric properties of a fluid, the principal object is to obtain an expression for the dielectric constant in terms of the electrostatic properties of the individual molecules. Some molecules, including those which constitute what are called *polar* fluids, are permanent dipoles in the sense that they have a dipole moment even in the absence of an impressed electrostatic field. Polar fluids present special problems from the theoretical point of view, which are touched upon in the theory of complex fluids presented in the previous chapter; it is not permissible, for example, to represent the mutual interactions of the molecules in such fluids by central forces. Attention will therefore be restricted here to those non-polar fluids for which the interactions between the molecules are of the simple type usually assumed. In non-polar fluids, the molecules are not permanent dipoles, but can be polarized under the influence of an electrostatic field. If the impressed electrostatic field **E** at the mean centre of an isolated molecule is not too large, the relation between the dipole moment of the molecule and the field can be represented by

$$\mathbf{P} = a\mathbf{E}$$

where a is the polarizability of the molecule. A more exact expression would include a term proportional to the spatial gradient of the field **E** at the mean centre of the molecule, and it would not be difficult to take account of such a term; however, its effect is usually negligible in comparison with that of the term $a\mathbf{E}$, and it will therefore be disregarded.

The detailed effect of the polarizability of the molecules in a fluid under the influence of an electrostatic field is rather complex. In the first place, the structure of the fluid, as described by the molecular distribution functions, is changed, due both to the interaction between the molecules of the fluid and the field, and to the additional mutual interactions between the molecules of the fluid arising from their polarization. The change is, however, small when, as is assumed, the impressed field is weak, and it has no influence on the value of the dielectric constant. It is, of course, essential to consider the direct effect of the external field in polarizing the molecules; and there are, in addition, secondary effects arising from the action of the induced polarization of the molecules on one another. As a result of such secondary effects, the dipole moment of a molecule in a fluid under the influence of an external field **E** may be quite different from the dipole moment of an isolated molecule in the same electric field. Owing to the motion of the molecules, the dipole moment of any molecule in the fluid will, in fact, be a fluctuating quantity. Even the mean value $\mathbf{P}_1$ of the dipole moment will differ from the value **P** which would obtain in the absence of the other molecules.

The quantitative treatment which follows is a modification of the method adopted by Yvon [1937]. The fluid considered is supposed to be bounded by a surface S, and exposed to an external electrostatic field **E** which may vary from point to point. The mean resultant field acting on a molecule situated at the point $\mathbf{x}^{(1)}$ is denoted by $\mathbf{E}_1^{(1)}$; it is the sum of the value $\mathbf{E}^{(1)}$ of the external field at the point $\mathbf{x}^{(1)}$, and the mean resultant field due to the polarization of the other molecules. If $\mathbf{P}_2^{(2)}$ is the mean dipole moment of a second molecule which occupies the volume element $d\mathbf{x}^{(2)}$ at $\mathbf{x}^{(2)}$, the mean field due to this molecule at the point $\mathbf{x}^{(1)}$ is

$$r^{-5}(3\,\mathbf{r}\,\mathbf{r}\cdot\mathbf{P}_2^{(2)} - r^2\,\mathbf{P}_2^{(2)}) = \mathbf{T}^{(12)}\cdot\mathbf{P}_2^{(2)}$$

where $\mathbf{r} = \mathbf{x}^{(2)} - \mathbf{x}^{(1)}$, and

(6. 1) $$\mathbf{T}^{(12)} = r^{-5}(3\,\mathbf{r}\,\mathbf{r} - r^2\,\boldsymbol{\delta}).$$

Also, the probability of finding a molecule in the volume element $d\mathbf{x}^{(2)}$, when it is already known that there is a molecule at $\mathbf{x}^{(1)}$, is $n_2 d\mathbf{x}^{(2)}/n$, where n is the value of the radial distribution function for the distance r. Hence the mean field acting on the molecule at $\mathbf{x}^{(1)}$ is

(6. 2) $$\mathbf{E}_1^{(1)} = \mathbf{E}^{(1)} + \int \mathbf{T}^{(12)}\cdot\mathbf{P}_2^{(2)}\, n_2\, d\mathbf{x}^{(2)}/n;$$

and the mean dipole moment of this molecule is

(6. 3) $$\mathbf{P}_1^{(1)} = a\,(\mathbf{E}^{(1)} + \int \mathbf{T}^{(12)}\cdot\mathbf{P}_2^{(2)}\, n_2\, d\mathbf{x}^{(2)}/n).$$

Of course, the value $\mathbf{P}_2^{(2)}$ of the mean dipole moment of the second molecule when it is known that a molecule is situated at $\mathbf{x}^{(1)}$ differs from the value $\mathbf{P}_1^{(2)}$ expected when nothing is known concerning the configuration of the surrounding molecules. For, in the former instance, the mean field acting on the molecule at $\mathbf{x}^{(2)}$ is

(6. 4) $$\mathbf{E}_2^{(2)} = \mathbf{E}^{(2)} + \mathbf{T}^{(21)}\cdot\mathbf{P}_2^{(1)} + \int \mathbf{T}^{(23)}\cdot\mathbf{P}_3^{(3)}\, n_3\, d\mathbf{x}^{(3)}/n,$$

where $\mathbf{P}_3^{(3)}$ is the mean dipole moment of a third molecule when it is known that other molecules are situated at $\mathbf{x}^{(1)}$ and $\mathbf{x}^{(2)}$. One has, therefore,

(6. 5) $$\mathbf{P}_2^{(2)} = a\,(\mathbf{E}^{(2)} + \mathbf{T}^{(21)}\cdot\mathbf{P}_2^{(1)} + \int \mathbf{T}^{(23)}\cdot\mathbf{P}_3^{(3)}\, n_3\, d\mathbf{x}^{(3)}/n_2).$$

Similarly one has in general

(6. 6) $$\mathbf{P}_q^{(q)} = a\,(\mathbf{E}^{(q)} + \sum_{i=1}^{q-1} \mathbf{T}^{(qi)}\cdot\mathbf{P}_q^{(i)} + \int \mathbf{T}^{(q,q+1)}\cdot\mathbf{P}_{q+1}^{(q+1)}\, n_{q+1}\, d\mathbf{x}^{(q+1)}/n_q).$$

The integrals appearing in (6. 3), (6. 5) and (6. 6) are not in a suitable form for computation, because they are so slowly convergent. They include a term representing the field due to the charge distribution induced in the fluid by the external field; this has to be separated from the remaining term, which is a true displacement field. If one writes

$$\int \boldsymbol{T}^{(q-1,q)} \cdot \mathbf{P}_q^{(q)} \, n_q \, d\mathbf{x}^{(q)} = \int \boldsymbol{T}^{(q-1,q)} \cdot (\mathbf{P}_q^{(q)} \, n_q - n n_{q-1} \, \mathbf{P}_1^{(q)}) \, d\mathbf{x}^{(q)} + \\ + n n_{q-1} \int \boldsymbol{T}^{(q-1,q)} \cdot \mathbf{P}_1^{(q)} \, d\mathbf{x}^{(q)},$$

the first integral on the right-hand side is now quickly convergent, and the second is readily evaluated in terms of recognizable physical quantities. For, since $\boldsymbol{T}^{(q-1,q)}$, given by (6.1), can be expressed in the form

$$\boldsymbol{T}^{(q-1,q)} = -\frac{\partial}{\partial \mathbf{x}^{(q)}} (s^{-3} \mathbf{s}), \quad \mathbf{s} = \mathbf{x}^{(q)} - \mathbf{x}^{(q-1)},$$

one obtains by an application of Gauss' theorem to the whole fluid, with the exception of a tiny sphere surrounding the point $\mathbf{x}^{(1)}$,

$$\int \{\mathbf{P}_1^{(q)} \cdot \boldsymbol{T}^{(q-1,q)} - s^{-3} \mathbf{s} \, \frac{\partial}{\partial \mathbf{x}^{(q)}} \cdot \mathbf{P}_1^{(q)} \} \, d\mathbf{x}^{(q)} \\ = -\int s^{-3} \mathbf{s} \, \mathbf{P}_1^{(q)} \cdot d\mathbf{S}^{(q)} + \tfrac{4}{3} \pi \mathbf{P}_1^{(q-1)},$$

where $d\mathbf{S}^{(q)}$ is a typical element of the boundary surface of the fluid. Hence

(6. 7)
$$\begin{aligned} & \int \boldsymbol{T}^{(q-1,q)} \cdot \mathbf{P}_q^{(q)} \, n_q \, d\mathbf{x}^{(q)}/n_{q-1} \\ & = \tfrac{4}{3} \pi n \mathbf{P}_1^{(q-1)} + \int \boldsymbol{T}^{(q-1,q)} \cdot (n_q \, \mathbf{P}_q^{(q)}/n_{q-1} - n \, \mathbf{P}_1^{(q)}) \, d\mathbf{x}^{(q)} + \\ & + n \, \{\int s^{-3} \mathbf{s} \, \frac{\partial}{\partial \mathbf{x}^{(q)}} \cdot \mathbf{P}_1^{(q)} \, d\mathbf{x}^{(q)} - \int s^{-3} \mathbf{s} \, \mathbf{P}_1^{(q)} \cdot d\mathbf{S}^{(q)} \}. \end{aligned}$$

The last two integrals have an obvious physical significance. The first represents the electric field at the point $\mathbf{x}^{(q-1)}$ due to the charge distribution, of density $n \frac{\partial}{\partial \mathbf{x}^{(q)}} \cdot \mathbf{P}_1^{(q)}$, which is induced in the fluid by the external field. The second represents the electric field at the point $\mathbf{x}^{(q-1)}$ due to the surface charge distribution, of density $n\mathbf{P}_1^{(q)} \cdot \mathbf{a}^{(q)}$, where $\mathbf{a}^{(q)}$ is the unit normal to the surface at the point $\mathbf{x}^{(q)}$, induced at the boundary of the fluid by the external field. These two terms are properly part of the electric intensity, as distinct from the electric displacement, at the point $\mathbf{x}^{(q-1)}$ in the fluid. If one defines the total electric intensity $\mathbf{E}_0^{(q-1)}$ at the point $\mathbf{x}^{(q-1)}$ by

(6. 8) $\quad \mathbf{E}_0^{(q-1)} = \mathbf{E}^{(q-1)} + n \{\int s^{-3} \mathbf{s} \, \frac{\partial}{\partial \mathbf{x}^{(q)}} \cdot \mathbf{P}_1^{(q)} \, d\mathbf{x}^{(q)} - \int s^{-3} \mathbf{s} \, \mathbf{P}_1^{(q)} \cdot d\mathbf{S}^{(q)} \},$

the general formula (6.6) reduces to

(6.9) $$\begin{cases} \mathbf{P}_q^{(q)} = a\,\{\mathbf{E}_0^{(q)} + \sum_{i=1}^{q-1} \mathbf{T}^{(qi)} \cdot \mathbf{P}_q^{(i)} + \\ \quad + \int \mathbf{T}^{(q,q+1)} \cdot (n_{q+1}\,\mathbf{P}_{q+1}^{(q+1)}/n_q - n\mathbf{P}_1^{(q+1)})\,d\mathbf{x}^{(q+1)} + \tfrac{4}{3}\pi n \mathbf{P}_1^{(q)}\}. \end{cases}$$

The value of $\mathbf{P}^{(1)}$ as a power series in a is readily obtained from this equation by successive substitutions. The result is

(6.10) $$\begin{cases} (1 - \tfrac{4}{3}\pi n a)\,\mathbf{P}_1^{(1)} = \\ \quad = a\mathbf{E}_0^{(1)} + a^2(1 + \tfrac{4}{3}\pi n a) \int \mathbf{T}^{(12)} \cdot \mathbf{E}_0^{(2)} (n_2/n - n)\,d\mathbf{x}^{(2)} + \\ \quad + a^3 \int \mathbf{T}^{(12)} \cdot \mathbf{T}^{(21)} \cdot \mathbf{E}_0^{(1)}\,n_2/n\,d\mathbf{x}^{(2)} + \\ \quad + a^3 \iint \mathbf{T}^{(12)} \cdot \mathbf{T}^{(23)} \cdot \mathbf{E}_0^{(3)} (n_3/n - n_2^{(12)} - n_2^{(23)} + n^2)\,d\mathbf{x}^{(2)}\,d\mathbf{x}^{(3)}, \end{cases}$$

with neglect of a^4 and higher powers of a. All the integrals contained in this formula converge quickly, so that only the value of the field $\mathbf{E}_0$ in the immediate neighbourhood of the point $\mathbf{x}^{(1)}$ is important. In such a small region, the variation of $\mathbf{E}_0$ from point to point can be neglected, and since

$\int \mathbf{T}^{(12)} (n_2/n - n)\,d\mathbf{x}^{(2)} = 0,$

$\int \mathbf{T}^{(12)} \cdot \mathbf{T}^{(21)} (n_2/n)\,d\mathbf{x}^{(2)} = 2 \int r^{-6}\,n_2(r)\,d\mathbf{r}/n,$

$\iint \mathbf{T}^{(12)} \cdot \mathbf{T}^{(23)} (n_3/n - n_2^{(12)} - n_2^{(23)} + n^2)\,d\mathbf{x}^{(2)}\,d\mathbf{x}^{(3)}$
$\quad = \iint (rs)^{-5} \{3(\mathbf{r}\cdot\mathbf{s})^2 - r^2 s^2\}\,\{n_3(\mathbf{r},\mathbf{s})/n - n_2(r) - n_2(s) + n^2\}\,d\mathbf{r}\,d\mathbf{s},$

the equation (6.10) reduces to

(6.11) $$\begin{cases} (1 - \tfrac{4}{3}\pi n a)\,\mathbf{P}_1 = a\mathbf{E}_0\,(1 + \lambda a^2), \\ n\lambda = \int r^{-6}\,n_2(r)\,d\mathbf{r} + \\ \quad + \iint (rs)^{-5} \{3(\mathbf{r}\cdot\mathbf{s})^2 - r^2 s^2\}\,\{n_3(\mathbf{r},\mathbf{s}) - nn_2(r) - nn_2(s) + n^3\}\,d\mathbf{r}\,d\mathbf{s}. \end{cases}$$

The macroscopically observed dielectric constant is defined by

(6.12) $$\mathbf{E}_0 + 4\pi n \mathbf{P}_1 = \varkappa \mathbf{E}_0.$$

Hence

(6.13) $$(\varkappa - 1)(1 - \tfrac{4}{3}\pi n a) = 4\pi n a\,(1 + \lambda a^2),$$

with the value of λ given by (6.11). This corrects the formula given by Yvon [1937], who did not discuss sufficiently the effect of the induced charge distribution. Since λ is in general a function of temperature, as well as density, $\varkappa$ also depends on the temperature. Owing to the small value of a, the temperature variation is, however, generally small in non-polar fluids of the type considered. Much larger variations can occur in polar substances, but these are primarily a consequence of the variation with temperature of the degree of association of the molecules in the fluid.

So far only the behaviour of the fluid in electrostatic fields has been considered. It is important, however, to take into consideration the effect of oscillatory electromagnetic fields, because, although the dielectric constant is never measured directly in such fields, it is easily measured indirectly through the well-known relation between the dielectric constant and the refractive index. One can easily measure the refractive index of the fluid for electromagnetic radiation of any given frequency, and infer the value of the dielectric constant for this frequency.

The theoretical determination of the dielectric constant $\varkappa(p)$ for electromagnetic radiation of frequency $p/2\pi$ does not differ in principle from the electrostatic determination already considered. Only the electrodynamical relations are changed. The external electric field can be represented by the *real part* of a *complex* vector **E** given by

$$\mathbf{E} = \mathbf{e} \exp ipt$$

where **e** is a complex vector depending on position but not on time. The dipole moment of an isolated molecule at the point $\mathbf{x}^{(1)}$ is then the real part of a complex vector $\mathbf{P}^{(1)} = a\mathbf{E}^{(1)}$, where the factor a depends on the frequency, and is in general complex, since the polarization of the molecule will not usually be in phase with the impressed radiation. Just as in the electrostatic case, the precise value of a can be inferred from a knowledge of the structure of the molecule.

To determine the electromagnetic radiation emitted by the molecule, one requires not only the dipole moment of the molecule, but also its first and second derivatives with respect to time; these are the real parts of the complex vectors $ip\mathbf{P}^{(1)}$ and $-p^2\mathbf{P}^{(1)}$ respectively. Then, according to elementary electrodynamics, the electric field experienced at the point $\mathbf{x}^{(2)}$ due to the polarization of the molecule at $\mathbf{x}^{(1)}$ is the real part of the vector $\mathbf{T}^{(21)} \cdot \mathbf{P}^{(1)}$, where

(6.14) $$\mathbf{T} = p^2(r^2\boldsymbol{\delta} - \mathbf{rr})/(c^2 r^3)$$

and $\boldsymbol{\delta}$ is the unit tensor.

From this point the derivation of (6.10) is formally unchanged. The formulae (6. 11) — (6. 13) are also correct if, as in the instance of optical phenomena, the wave-length of the radiation is large compared with intermolecular distances, and also the relativistic effects, represented by the terms proportional to higher powers of pr/c in T, are neglected. Under these conditions, the results derived for electrostatic fields can be immediately adapted to determine the optical behaviour of the fluid.

It is noteworthy that, whereas (6. 1) is applicable to fluids on which a purely electrostatic field is acting, (6. 14) above is applicable for fluids exposed to monochromatic radiation of frequency $p/2\pi$. In the case where electrostatic and radiation fields are both present, the effects are additive. When the radiation is not monochromatic, however, typical interference effects may appear.

The electromagnetic properties of fluids which have just been investigated are remarkable inasmuch as they cannot apparently be described completely in terms of the radial distribution function n_2, but the distribution functions relating to groups of more than two molecules are involved. Fortunately, the influence of these complex distributions is small; nevertheless it is interesting to find some properties of a fluid depending on the more delicate internal structure, to which little attention has yet been given.

CHAPTER VIII

THE KINETIC THEORY OF FLUIDS

1. Introductory

The modern kinetic theory of fluids has its origin in the work of Maxwell and Boltzmann, who first studied the behaviour of the velocity distribution function for non-uniform gases. It was not, however, till 1916—17 that CHAPMAN [1916, 1917] and ENSKOG [1917] succeeded independently in determining the exact form of the velocity distribution function, by methods which are best summarized in the elegant treatise of CHAPMAN & COWLING [1939]. These methods will be considered in this chapter, principally with regard to their extension to dense gases and liquids. They enable one to derive exact formulae for such important quantities as the coefficients of viscosity and thermal conduction, which are amenable to numerical development.

The characteristic assumption made in the kinetic theory of gases is that the molecules suffer only 'binary encounters'. One may imagine that each molecule is surrounded by a sphere, with a certain radius r_0 depending on the intermolecular law of force, outside which other molecules are practically unaffected by its attractive field. In a gas, a molecule spends most of its time outside the range of interaction of the other molecules, and the number of molecules within this sphere is therefore generally zero, and only rarely more than one. The postulate of binary encounters strictly amounts to neglecting the small probability of finding two or more molecules simultaneously within the sphere of interaction of any given molecule. This neglect is obviously perfectly well justified in relation to rare gases, not so well in relation to dense gases, and not at all in relation to liquids. However, in practice it is used in such a way that its consequences probably have a rather wider range of validity than the assumption of binary encounters itself. The errors entailed in the application of the theory of Chapman and Enskog to dense fluids are certainly due less to this assumption than to another approximation concealed in the usual derivation of Boltzmann's equation.

This second assumption can be described in the following way. To determine the mean force exerted on a molecule with a definite position ($\mathbf{x}^{(1)}$) and velocity ($\boldsymbol{\xi}^{(1)}$), by reason of the collisions which it suffers with the other molecules, one requires to know the probability that it will encounter a second molecule with a given velocity, and moving along a given line of incidence relative to the molecule under consideration. In assessing the mean force, Boltzmann assumed that this probability would be that appropriate to the position of the first molecule, rather than that appropriate to the position which the incident molecule actually occupied at the instant of the encounter. These two probabilities differ by terms which are proportional to the local gradients of density, temperature, and macroscopic velocity, and are therefore highly relevant to the theory of non-uniform fluids. Fortunately the neglected terms are also proportional to the square of the density, and so may be legitimately neglected in the application to rare gases.

In seeking to generalize the theory of Chapman and Enskog to dense fluids, one naturally considers whether either or both of these approximative assumptions can be eliminated. It is soon evident that the second approximation is more easily dispensed with than the first. The calculation of the additional terms proportional to the gradients of the macroscopic parameters is precisely the kind of task which the methods developed in this book were designed to accomplish; but the difficulty in taking exact account of clusters of an arbitrary number of molecules promises to be at least as fundamental in the theory of non-uniform fluids as it has proved in the theory of fluids in the equilibrium state. To extend the theory of gases to take account even of ternary encounters would involve using the solution of the three-body problem; and, in a liquid, the very idea of molecular encounters as required by the theory of gases is quite inapplicable. The path followed by an individual molecule in a dense fluid is exceedingly intricate, and impossible to describe accurately except perhaps in a statistical sense. The best approximation to the mean trajectory of one molecule relative to another appears to be that described under the influence of the mean force, which is the gradient of the potential $-\beta^{-1} \ln \bar{n}_2^0(r)$, where $\bar{n}_2^0(r)$ is the equilibrium value of the radial distribution function.

From the general standpoint, there is a difficulty of principle which has not yet been overcome. This is connected with the question of irreversibility, and the way in which it should be introduced into the fundamental equations of the theory. The equations (2. 9) of chapter V which determine the distribution functions do not, as they stand, determine the essential characteristics of irreversible phenomena. This can be seen from the fact that the sign of the *time* is not uniquely determined; corresponding to any solution, there exists another with the sign of the time and velocities reversed. Nor does the assumption of the existence of 'normal' solutions, specified by the distribution of density, temperature and macroscopic velocity in the fluid, completely remove this difficulty. For guidance one may turn to the theory of gases, where the element of irreversibility is introduced through the application of a very fundamental tenet of the theory of probability. One uses the principle that if two molecules have not been in interaction *in the past*, their velocities must be independent of one another. In this way the distinction between past and future is insinuated. It is, however, more difficult to formulate this principle in its application to liquids.

2. Boltzmann's Equation

Attempts to formulate the fundamental equation of the kinetic theory of gases were first made by Maxwell and Boltzmann in the last century. Their methods will not be followed here, because the result which they obtained is not quite exact, even under the assumption, which they made, of binary encounters between the molecules. Instead, the equation (2. 7) of chapter V, namely

$$(2.\,1) \quad \frac{\partial f_1}{\partial t} + \boldsymbol{\xi}^{(1)} \cdot \frac{\partial f_1}{\partial \mathbf{x}^{(1)}} + \frac{1}{m} \mathbf{F} \cdot \frac{\partial f_1}{\partial \boldsymbol{\xi}^{(1)}} = \frac{1}{m} \int\int \frac{\partial f_2}{\partial \boldsymbol{\xi}^{(1)}} \cdot \frac{\partial \phi}{\partial \mathbf{x}^{(1)}} \, d\boldsymbol{\xi}^{(2)} \, d\mathbf{x}^{(2)},$$

will be used as a starting point. This method enables one to give a precise specification of the approximations involved, and also to determine the correction which must be applied to Boltzmann's equation in its original form.

It must first be noticed that the potential function $\phi(r)$ and its derivative $\phi'(r)$ decrease very rapidly with r, beyond a certain value r_0; the contribution to the value of the integral of (2. 1) from positions of $\mathbf{x}^{(2)}$ outside a sphere of radius r_0 surrounding the point $\mathbf{x}^{(1)}$ can therefore be neglected. It will be supposed, then, that the domain of integration with respect to $\mathbf{x}^{(2)}$ is restricted to such a sphere. Obviously one can make the error involved as small as one pleases by taking a sufficiently large value of r_0. In practice, however, it is not necessary for r_0 to exceed a few molecular diameters; and it will be supposed that r_0 is sufficiently small to allow one to neglect the probability of finding two or more molecules within the distance r_0 of a given molecule. In a rare gas, there is a wide range of values of r_0 which satisfy these requirements. The assumption of binary encounters is then perfectly valid.

One makes use of the concept of binary encounters to express the distribution function f_2 in terms of f_1. Let $\mathbf{x}^{(1)}(t)$, $\mathbf{x}^{(2)}(t)$ and $\mathbf{\xi}^{(1)}(t)$, $\mathbf{\xi}^{(2)}(t)$ refer to the positions and velocities at time t of two molecules moving only under the influence of their mutual interaction. It will be supposed that the distance between $\mathbf{x}^{(1)}(t)$ and $\mathbf{x}^{(2)}(t)$ is less than r_0. Then, because there is no interaction with any other molecule, the configurational probability

(2.2) $\quad f_2 \{\mathbf{\xi}^{(1)}(t), \mathbf{\xi}^{(2)}(t), \mathbf{x}^{(1)}(t), \mathbf{x}^{(2)}(t), t\} \, d\mathbf{\xi}^{(1)}(t) \, d\mathbf{\xi}^{(2)}(t) \, d\mathbf{x}^{(1)}(t) \, d\mathbf{x}^{(2)}(t)$

of finding two molecules with velocities in the ranges $\mathbf{\xi}^{(1)}(t)$, $d\mathbf{\xi}^{(1)}(t)$; $\mathbf{\xi}^{(2)}(t)$, $d\mathbf{\xi}^{(2)}(t)$ in the volume elements $d\mathbf{x}^{(1)}(t)$, $d\mathbf{x}^{(2)}(t)$ respectively at time t, will remain unchanged as t varies and the molecules pursue their trajectories relative to one another. This (cf. § 2 of chapter V) is a direct consequence of the fact that the configurations of the pair of molecules at any two instants are causally related to one another. Also, according to Liouville's theorem, the volume

(2.3) $\qquad\qquad d\mathbf{x}^{(1)}(t) \, d\mathbf{x}^{(2)}(t) \, d\mathbf{\xi}^{(1)}(t) \, d\mathbf{\xi}^{(2)}(t)$

of phase space 'occupied' by the two molecules does not change with time. A proof of Liouville's theorem appropriate to this simple example is given in the following subsection; the main argument is resumed in § 2.2 below.

2.1 Liouville's Theorem for Gases

The expression $\partial(a_1, \ldots a_k)/\partial(b_1, \ldots b_k)$ represents the *Jacobian*, or the determinant of the partial differential coefficients $\partial a_m/\partial b_n$. With this notation, one has

(2.4) $\quad\begin{cases} \dfrac{d\mathbf{x}^{(1)}(t+dt) \, d\mathbf{x}^{(2)}(t+dt) \, d\mathbf{\xi}^{(1)}(t+dt) \, d\mathbf{\xi}^{(2)}(t+dt)}{d\mathbf{x}^{(1)}(t) \, d\mathbf{x}^{(2)}(t) \, d\mathbf{\xi}^{(1)}(t) \, d\mathbf{\xi}^{(2)}(t)} \\ = \dfrac{\partial(\mathbf{x}^{(1)}+\dot{\mathbf{x}}^{(1)}dt, \, \mathbf{x}^{(2)}+\dot{\mathbf{x}}^{(2)}dt, \, \mathbf{\xi}^{(1)}+\dot{\mathbf{\xi}}^{(1)}dt, \, \mathbf{\xi}^{(2)}+\dot{\mathbf{\xi}}^{(2)}dt)}{\partial(\mathbf{x}^{(1)}, \, \mathbf{x}^{(2)}, \, \mathbf{\xi}^{(1)}, \, \mathbf{\xi}^{(2)})}, \end{cases}$

since $\dot{\mathbf{x}}^{(1)}$, $\dot{\mathbf{x}}^{(2)}$, $\dot{\boldsymbol{\xi}}^{(1)}$, $\dot{\boldsymbol{\xi}}^{(2)}$ are the time derivatives at time t, given by

$$\dot{\mathbf{x}}^{(1)} = \boldsymbol{\xi}^{(1)} \quad , \quad \dot{\mathbf{x}}^{(2)} = \boldsymbol{\xi}^{(2)},$$
$$m\dot{\boldsymbol{\xi}}^{(1)} = \phi'(r)\,\mathbf{r}/r \quad , \quad m\dot{\boldsymbol{\xi}}^{(2)} = -\phi'(r)\,\mathbf{r}/r.$$

On account of the identity $\dot{\boldsymbol{\xi}}^{(1)} + \dot{\boldsymbol{\xi}}^{(2)} = 0$, the Jacobian which is the coefficient of dt in (2. 4) must vanish. Therefore

$$\frac{d}{dt}\{d\mathbf{x}^{(1)}(t)\,d\mathbf{x}^{(2)}(t)\,d\boldsymbol{\xi}^{(1)}(t)\,d\boldsymbol{\xi}^{(2)}(t)\}$$

vanishes: the volume of phase space occupied by the two molecules must remain constant in time.

2. 2 Derivation of Boltzmann's Equation

Since (2. 2) and (2. 3) are both independent of the time t,

$$f_2\{\boldsymbol{\xi}^{(1)}(t),\,\boldsymbol{\xi}^{(2)}(t),\,\mathbf{x}^{(1)}(t),\,\mathbf{x}^{(2)}(t),\,t\}$$

must itself remain unchanged as t varies.

Now, it has been supposed that $|\mathbf{x}^{(2)}(t) - \mathbf{x}^{(1)}(t)| < r_0$; so, assuming for the time being that the two molecules are not forever circling round one another in a closed trajectory relative to their mean centre, there will have been some time t_0, earlier than t, at which the distance $|\mathbf{x}^{(2)}(t_0) - \mathbf{x}^{(1)}(t_0)|$ between the molecules was exactly r_0. At this time t_0, the two molecules will have had no appreciable interaction in the past, and their velocities will therefore be completely independent of one another. It follows that

(2. 5)
$$\begin{cases} f_2\{\boldsymbol{\xi}^{(1)}(t),\,\boldsymbol{\xi}^{(2)}(t),\,\mathbf{x}^{(1)}(t),\,\mathbf{x}^{(2)}(t),\,t\} \\ = f_2\{\boldsymbol{\xi}^{(1)}(t_0).\,\boldsymbol{\xi}^{(2)}(t_0),\,\mathbf{x}^{(1)}(t_0),\,\mathbf{x}^{(2)}(t_0),\,t_0\} \\ = f_1\{\boldsymbol{\xi}^{(1)}(t_0),\,\mathbf{x}^{(1)}(t_0),\,t_0\}\,f_1\{\boldsymbol{\xi}^{(2)}(t_0),\,\mathbf{x}^{(2)}(t_0),\,t_0\}. \end{cases}$$

One can now specify the second of the approximations implicit in Boltzmann's derivation of his equation for gases. This consists in the neglect of the difference

(2. 6)
$$\begin{cases} \varDelta = f_1\{\boldsymbol{\xi}^{(1)}(t_0),\,\mathbf{x}^{(1)}(t_0),\,t_0\}\,f_1\{\boldsymbol{\xi}^{(2)}(t_0),\,\mathbf{x}^{(2)}(t_0),\,t_0\} - \\ \quad - f_1\{\boldsymbol{\xi}^{(1)}(t_0),\,\mathbf{x}^{(1)}(t),\,t\}\,f_1\{\boldsymbol{\xi}^{(2)}(t_0),\,\mathbf{x}^{(1)}(t),\,t\}. \end{cases}$$

As the differences $t_0 - t$, $\mathbf{x}^{(1)}(t_0) - \mathbf{x}^{(1)}(t)$, and $\mathbf{x}^{(2)}(t_0) - \mathbf{x}^{(1)}(t)$ are all small by macroscopic standards, it may be hoped that the error involved in this approximation will not be large. Its actual value is determined in the next section. For brevity, $\boldsymbol{\xi}^{(1)}(t)$, $\boldsymbol{\xi}^{(2)}(t)$, $\mathbf{x}^{(1)}(t)$, $\mathbf{x}^{(2)}(t)$

will now be represented simply by $\boldsymbol{\xi}^{(1)}$, $\boldsymbol{\xi}^{(2)}$, $\mathbf{x}^{(1)}$, $\mathbf{x}^{(2)}$, and $\boldsymbol{\xi}^{(1)}(t_0)$, $\boldsymbol{\xi}^{(2)}(t_0)$ by $\boldsymbol{\xi}_0^{(1)}$, $\boldsymbol{\xi}_0^{(2)}$; also, $f_1(\boldsymbol{\xi}, \mathbf{x}^{(1)}, t)$ will be contracted to $f_1(\boldsymbol{\xi})$. Then it follows from (2. 5) and (2. 6) that

$$(2.7) \quad \begin{cases} f_2 = \bar{f}_2 + \Delta, \\ \bar{f}_2 = f_1(\boldsymbol{\xi}_0^{(1)}) f_1(\boldsymbol{\xi}_0^{(2)}). \end{cases}$$

The initial velocities $\boldsymbol{\xi}_0^{(1)}$, $\boldsymbol{\xi}_0^{(2)}$ are actually functions of $\boldsymbol{\xi}^{(1)}$, $\boldsymbol{\xi}^{(2)}$ and $\mathbf{r} = \mathbf{x}^{(2)} - \mathbf{x}^{(1)}$. The form of these functions can be determined by solving the equations of motion

$$(2.8) \quad \frac{d\mathbf{r}}{\boldsymbol{\xi}^{(2)} - \boldsymbol{\xi}^{(1)}} = \frac{m\, d\boldsymbol{\xi}^{(1)}}{(\partial \phi/\partial \mathbf{r})} = \frac{m\, d\boldsymbol{\xi}^{(2)}}{(-\partial \phi/\partial \mathbf{r})} \; (= dt)$$

of the two molecules to obtain the eight constants of the motion. The latter will be represented by

$$I_l(\boldsymbol{\xi}^{(1)}, \boldsymbol{\xi}^{(2)}, \mathbf{r}) \qquad (l = 1, \ldots 8).$$

Of these, seven may be selected from components of the total momentum vector, angular momentum vector, and energy, thus:

$$(2.9) \quad \begin{cases} (I_1, I_2, I_3) = m(\boldsymbol{\xi}^{(1)} + \boldsymbol{\xi}^{(2)}), \; (I_4, I_5, I_6) = m\, \mathbf{r} \wedge (\boldsymbol{\xi}^{(2)} - \boldsymbol{\xi}^{(1)}), \\ I_7 = \tfrac{1}{2} m\, (\boldsymbol{\xi}^{(1)2} + \boldsymbol{\xi}^{(2)2}) + \phi(r). \end{cases}$$

The remaining constant is a more complicated function which might, for example, be specified as the cosine of the angle between the direction of the initial relative velocity $\boldsymbol{\xi}_0^{(2)} - \boldsymbol{\xi}_0^{(1)}$ and some arbitrary direction in the plane of the relative motion. Its precise form is unimportant, as only the differential equation

$$(2.10) \quad (\boldsymbol{\xi}^{(2)} - \boldsymbol{\xi}^{(1)}) \cdot \frac{\partial I_l}{\partial \mathbf{r}} + \frac{1}{m} \frac{\partial \phi}{\partial \mathbf{r}} \cdot \left(\frac{\partial I_l}{\partial \boldsymbol{\xi}^{(1)}} - \frac{\partial I_l}{\partial \boldsymbol{\xi}^{(2)}} \right) = 0$$

satisfied by *all* the I_l will be required for the present purpose. The values of $\boldsymbol{\xi}_0^{(1)}$ and $\boldsymbol{\xi}_0^{(2)}$ (and also the value $\mathbf{r}_0$ of $\mathbf{r}$ at time t_0, if needed) can be obtained as functions of the I_l and r_0 by solving the set of nine simultaneous equations

$$(2.11) \quad \begin{cases} I_l(\boldsymbol{\xi}_0^{(1)}, \boldsymbol{\xi}_0^{(2)}, \mathbf{r}_0) = I_l, \\ |\mathbf{r}_0| = r_0. \end{cases}$$

As $\boldsymbol{\xi}_0^{(1)}$ and $\boldsymbol{\xi}_0^{(2)}$ are functions of the constants of the motion I_l, they must also satisfy the equation (2. 10); and since $\bar{f}_2$, defined by the equation (2. 7), is a function of $\boldsymbol{\xi}_0^{(1)}$ and $\boldsymbol{\xi}_0^{(2)}$ (and $\mathbf{x}^{(1)}$ and t in addition).

it must also satisfy the equation (2. 10). Thus

(2. 12) $\quad (\boldsymbol{\xi}^{(2)} - \boldsymbol{\xi}^{(1)}) \cdot \dfrac{\partial \bar{f}_2}{\partial \mathbf{r}} + \dfrac{1}{m} \dfrac{\partial \phi}{\partial \mathbf{r}} \cdot \left(\dfrac{\partial \bar{f}_2}{\partial \boldsymbol{\xi}^{(1)}} - \dfrac{\partial \bar{f}_2}{\partial \boldsymbol{\xi}^{(2)}} \right) = 0.$

Now, if one substitutes in (2. 1) the value of

$$\dfrac{\partial \bar{f}_2}{\partial \boldsymbol{\xi}^{(1)}} \cdot \dfrac{\partial \phi}{\partial \mathbf{x}^{(1)}} = - \dfrac{\partial \phi}{\partial \mathbf{r}} \cdot \left(\dfrac{\partial \bar{f}_2}{\partial \boldsymbol{\xi}^{(1)}} + \dfrac{\partial \varDelta}{\partial \boldsymbol{\xi}^{(1)}} \right)$$

found from this equation, the right-hand side becomes

(2. 13) $\quad \displaystyle\iint \left\{ -\dfrac{1}{m} \dfrac{\partial \phi}{\partial \mathbf{r}} \cdot \dfrac{\partial \bar{f}_2}{\partial \boldsymbol{\xi}^{(2)}} + (\boldsymbol{\xi}^{(2)} - \boldsymbol{\xi}^{(1)}) \cdot \dfrac{\partial \bar{f}_2}{\partial \mathbf{r}} - \dfrac{1}{m} \dfrac{\partial \phi}{\partial \mathbf{r}} \cdot \dfrac{\partial \varDelta}{\partial \boldsymbol{\xi}^{(1)}} \right\} d\mathbf{x}^{(2)} d\boldsymbol{\xi}^{(2)}.$

By an application of Gauss' theorem to the velocity space of $\boldsymbol{\xi}^{(2)}$, the first term vanishes. The second term does not vanish, but can be converted to a surface integral on the sphere of radius r_0 which, it will be remembered, bounds the region of integration. If $d\mathbf{b}$ represents the projection of the element $d\mathbf{S}$ of the surface of the sphere on to the plane normal to the vector $\boldsymbol{\rho} = \boldsymbol{\xi}^{(2)} - \boldsymbol{\xi}^{(1)}$ this surface integral can be written

$$\iint \varrho \, \{\bar{f}_2 (d\mathbf{S} \cdot \boldsymbol{\rho} > 0) - \bar{f}_2 (d\mathbf{S} \cdot \boldsymbol{\rho} < 0)\} \, d\mathbf{b} \, d\boldsymbol{\xi}^{(2)},$$

the first term arising from the hemisphere where $\mathbf{r} \cdot \boldsymbol{\rho} > 0$, and the second from the hemisphere with $\mathbf{r} \cdot \boldsymbol{\rho} < 0$. Now, where $d\mathbf{S} \cdot \boldsymbol{\rho} < 0$, the two molecules are entering for the first time one another's sphere of influence; hence $\boldsymbol{\xi}_0^{(1)} = \boldsymbol{\xi}^{(1)}$, $\boldsymbol{\xi}_0^{(2)} = \boldsymbol{\xi}^{(2)}$, and

$$\bar{f}_2 (d\mathbf{S} \cdot \boldsymbol{\rho} < 0) = f_1(\boldsymbol{\xi}^{(1)}) \, f_1(\boldsymbol{\xi}^{(2)}).$$

On the other hand, where $d\mathbf{S} \cdot \boldsymbol{\rho} > 0$, the two molecules are passing out of one another's sphere of influence, and their initial velocities $\boldsymbol{\xi}_0^{(1)}$, $\boldsymbol{\xi}_0^{(2)}$ have to be calculated from the final velocites $\boldsymbol{\xi}^{(1)}$, $\boldsymbol{\xi}^{(2)}$ by integrating the equations of motion. On this understanding, one may however write

$$\bar{f}_2 (d\mathbf{S} \cdot \boldsymbol{\rho} > 0) = f_1(\boldsymbol{\xi}_0^{(1)}) \, f_1(\boldsymbol{\xi}_0^{(2)});$$

then the equation (2. 1) becomes

(2. 14) $\quad \begin{cases} \dfrac{\partial f_1}{\partial t} + \boldsymbol{\xi}^{(1)} \cdot \dfrac{\partial f_1}{\partial \mathbf{x}^{(1)}} + \dfrac{1}{m} \mathbf{F} \cdot \dfrac{\partial f_1}{\partial \boldsymbol{\xi}^{(1)}} = \\ = \displaystyle\iint \varrho \, \{f_1(\boldsymbol{\xi}_0^{(1)}) f_1(\boldsymbol{\xi}_0^{(2)}) - f_1(\boldsymbol{\xi}^{(1)}) f_1(\boldsymbol{\xi}^{(2)})\} \, d\mathbf{b} \, d\boldsymbol{\xi}^{(2)} - \\ \qquad - \dfrac{1}{m} \displaystyle\iint \dfrac{\partial \phi}{\partial \mathbf{r}} \cdot \dfrac{\partial \varDelta}{\partial \boldsymbol{\xi}^{(1)}} \, d\mathbf{x}^{(2)} \, d\boldsymbol{\xi}^{(2)}, \end{cases}$

which, apart from the last term, is Boltzmann's equation in its original form. This is the fundamental equation of the kinetic theory of gases.

In the derivation of (2.14) from (2.1), the effect of those configurations in which the molecules are not actually in collision, but circling round one another under the influence of their mutual attraction, has been disregarded. Such configurations are not common in gases at high temperatures, but it is of interest to verify that they do not modify the result. For, in such configurations, there are no unique 'initial' velocities, and the latter may be chosen arbitrarily from among those assumed simultaneously by the two molecules. Also, the 'final' velocities will be exactly the same as the initial velocities, since the corresponding trajectories are closed. Such trajectories therefore contribute nothing to the 'collision' integral in (2.14), so long as it is assumed that no third molecule perturbs the motion. The latter assumption should be regarded as implied by the assumption of binary encounters between the molecules which has already been made.

Apart from the assumption of binary encounters, an independent statistical principle has been introduced in the derivation of (2.14) from (2.1). This principle, which involves a distinction between past and future, is expressed by the relation (2.5). The fact that (2.5) is correct when $\xi_0^{(1)}, \xi_0^{(2)}$ represent initial velocities, and would be incorrect if $\xi_0^{(1)}, \xi_0^{(2)}$ represented final velocities, is responsible for the implication of irreversibility carried by the equation (2.14), though not by (2.1). It has the effect of selecting those configurations which correspond to physical reality from a wider class of configurations, which are compatible with the mechanical laws, but most of which are never experimentally realized.

3. Corrections to Boltzmann's Equation

To assess the errors implicit in Boltzmann's original equation, one has first to evaluate the final integral in (2.14). The correction Δ, defined by (2.6), can be split into two parts:

$$\Delta = \Delta_1 + \Delta_2,$$

(3.1) $\begin{cases} \Delta_1 = f_1(\xi_0^{(1)}, \mathbf{x}_0^{(1)}, t_0) f_1(\xi_0^{(2)}, \mathbf{x}_0^{(2)}, t_0) - f_1(\xi_0^{(1)}, \mathbf{x}, t) f_1(\xi_0^{(2)}, \mathbf{x}, t), \\ \Delta_2 = f_1(\xi_0^{(1)}, \mathbf{x}, t) f_1(\xi_0^{(2)}, \mathbf{x}, t) - f_1(\xi_0^{(1)}, \mathbf{x}^{(1)}, t) f_1(\xi_0^{(2)}, \mathbf{x}^{(1)}, t), \end{cases}$

where $\mathbf{x}$ is the mean centre $\frac{1}{2}(\mathbf{x}^{(1)} + \mathbf{x}^{(2)})$ at time t.

For simplicity, it will be supposed that there is no external force.

Then, since, on the assumption of binary encounters, the distribution function f_3 is negligible when $|\mathbf{x}^{(2)} - \mathbf{x}^{(1)}| \leqslant r_0$, and either $|\mathbf{x}^{(3)} - \mathbf{x}^{(2)}| \leqslant r_0$ or $|\mathbf{x}^{(1)} - \mathbf{x}^{(3)}| \leqslant r_0$, one has from (2. 9) of chapter V, with $q = 2$,

$$(3.2) \quad \left(\frac{\partial}{\partial t_0} + \boldsymbol{\xi}_0^{(1)} \cdot \frac{\partial}{\partial \mathbf{x}_0^{(1)}} + \boldsymbol{\xi}_0^{(2)} \cdot \frac{\partial}{\partial \mathbf{x}_0^{(2)}}\right) f_2(\boldsymbol{\xi}_0^{(1)}, \boldsymbol{\xi}_0^{(2)}, \mathbf{x}_0^{(1)}, \mathbf{x}_0^{(2)}, t_0) = 0.$$

On the other hand, it follows from (3. 1) that

$$(3.3) \quad \begin{cases} \Delta_1 = (t_0 - t)\left(\frac{\partial}{\partial t_0} + \boldsymbol{\xi}_0^{(1)} \cdot \frac{\partial}{\partial \mathbf{x}_0^{(1)}} + \boldsymbol{\xi}_0^{(2)} \cdot \frac{\partial}{\partial \mathbf{x}_0^{(2)}}\right) f_2 + \\ + \{(t - t_0)\boldsymbol{\xi}_0^{(1)} + \mathbf{x}_0^{(1)} - \mathbf{x}\} \cdot \left\{\frac{\partial f_1}{\partial \mathbf{x}}(\boldsymbol{\xi}_0^{(1)}) f_1(\boldsymbol{\xi}_0^{(2)})\right\} + \\ + \{(t - t_0)\boldsymbol{\xi}_0^{(2)} + \mathbf{x}_0^{(2)} - \mathbf{x}\} \cdot \left\{f_1(\boldsymbol{\xi}_0^{(1)}) \frac{\partial f_1}{\partial \mathbf{x}}(\boldsymbol{\xi}_0^{(2)})\right\}, \end{cases}$$

with neglect of second derivatives with respect to time and the coordinates of the mean centre $\mathbf{x}$. The first term vanishes, because of (3. 2); and, since

$$\mathbf{x} = \tfrac{1}{2}(\mathbf{x}_0^{(1)} + \mathbf{x}_0^{(2)}) + \tfrac{1}{2}(\boldsymbol{\xi}_0^{(1)} + \boldsymbol{\xi}_0^{(2)})(t - t_0),$$

(3. 3) reduces to

$$(3.4) \quad \begin{cases} \Delta_1 = \tfrac{1}{2} \mathbf{r}' \cdot \left\{\frac{\partial f_1}{\partial \mathbf{x}}(\boldsymbol{\xi}_0^{(2)}) f_1(\boldsymbol{\xi}_0^{(1)}) - \frac{\partial f_1}{\partial \mathbf{x}}(\boldsymbol{\xi}_0^{(1)}) f_1(\boldsymbol{\xi}_0^{(2)})\right\}, \\ \mathbf{r}' = (\mathbf{x}_0^{(2)} - \mathbf{x}_0^{(1)}) + (\boldsymbol{\xi}_0^{(2)} - \boldsymbol{\xi}_0^{(1)})(t - t_0). \end{cases}$$

As in chapter V, the parts of f_1 and f_2 which are independent of the gradients of density, temperature and macroscopic velocity will be denoted by

$$(3.5) \quad \begin{cases} \bar{f}_1^0 = n\lambda^{-1} \exp\{-\tfrac{1}{2}\beta m \mathbf{v}^{(1)2}\}, \\ \bar{f}_2^0 = \bar{n}_2^0 \lambda^{-2} \exp\{-\tfrac{1}{2}\beta m (\mathbf{v}^{(1)2} + \mathbf{v}^{(2)2})\}, \\ \lambda = (2\pi/\beta m)^{3/2}, \quad \mathbf{v}^{(i)} = \boldsymbol{\xi}^{(i)} - \mathbf{u}. \end{cases}$$

Then (3. 4) reduces further to

$$(3.6) \quad \begin{cases} \Delta_1 = \tfrac{1}{2} \mathbf{r}' \cdot \left\{-\tfrac{1}{2} m (\mathbf{v}_0^{(2)2} - \mathbf{v}_0^{(1)2}) \frac{\partial \beta}{\partial \mathbf{x}} + \right. \\ \left. + m\beta \left(\frac{\partial}{\partial \mathbf{x}}\mathbf{u}\right) \cdot (\mathbf{v}_0^{(2)} - \mathbf{v}_0^{(1)})\right\} \bar{f}_1^0(\mathbf{v}_0^{(1)}) \bar{f}_1^0(\mathbf{v}_0^{(2)}), \quad \mathbf{v}_0^{(i)} = \boldsymbol{\xi}_0^{(i)} - \mathbf{u}. \end{cases}$$

The vector $\mathbf{r}'$ is obviously independent, as it should be, of the precise value of t_0 and therefore of r_0. Like $\boldsymbol{\xi}_0^{(1)}$ and $\boldsymbol{\xi}_0^{(2)}$, it can be determined without difficulty when the potential function $\phi(r)$ is known. The value of

$$\frac{1}{m} \iint \frac{\partial \Delta_1}{\partial \boldsymbol{\xi}^{(1)}} \cdot \frac{\partial \phi}{\partial \mathbf{r}} \, d\mathbf{r} \, d\boldsymbol{\xi}^{(2)}$$

may therefore be regarded as determined.

The corresponding term involving Δ_2 can be completely evaluated. One has

$$\Delta_2 = \tfrac{1}{2} \mathbf{r} \cdot \frac{\partial \bar{f}_2^0}{\partial \mathbf{x}},$$

correct to terms linear in the gradients; hence

(3.7) $\begin{cases} \dfrac{1}{m} \iint \dfrac{\partial \Delta_2}{\partial \boldsymbol{\xi}^{(1)}} \cdot \dfrac{\partial \phi}{\partial \mathbf{r}} \, d\mathbf{r} \, d\boldsymbol{\xi}^{(2)} = -\dfrac{1}{2} \dfrac{\partial}{\partial \mathbf{x}} \cdot \iint \beta \bar{f}_2^0 \, \mathbf{r} \, \mathbf{v}^{(1)} \cdot \dfrac{\partial \phi}{\partial \mathbf{r}} \, d\mathbf{r} \, d\mathbf{v}^{(2)} \\ \qquad\qquad\qquad\qquad\qquad = \dfrac{\partial}{\partial \mathbf{x}} \cdot \{(p^0 - nkT)\, \beta \mathbf{v}^{(1)} \bar{f}_1^0\}. \end{cases}$

The left-hand side of (2.14) is expressed in terms of the macroscopic parameters and their gradients by the equation (5.17) of chapter V. The corrected form of Boltzmann's equation therefore reduces to

(3.8) $\begin{cases} \dfrac{\beta p^0 \bar{f}_1^0}{n} \left\{ (\tfrac{1}{2}\beta m \mathbf{v}^{(1)2} - \tfrac{5}{2})\, \mathbf{v}^{(1)} \cdot \left(\dfrac{1}{T}\dfrac{\partial T}{\partial \mathbf{x}}\right) + \beta m \mathbf{v}^{(1)} \cdot \overline{\left(\dfrac{\partial}{\partial \mathbf{x}} \mathbf{u}\right)} \cdot \mathbf{v}^{(1)} \right\} + \\ + (\gamma + \tfrac{2}{3}\beta p^0/n)\, \bar{f}_1^0 (\tfrac{1}{2}\beta m \mathbf{v}^{(1)2} - \tfrac{3}{2}) \dfrac{\partial}{\partial \mathbf{x}} \cdot \mathbf{u} + \dfrac{1}{m} \iint \dfrac{\partial \Delta_1}{\partial \mathbf{v}^{(1)}} \cdot \dfrac{\partial \phi}{\partial \mathbf{r}} \, d\mathbf{r} \, d\mathbf{v}^{(2)} \\ = \iint \varrho\, \{\bar{f}_1(\mathbf{v}_0^{(1)})\, \bar{f}_1(\mathbf{v}_0^{(2)}) - \bar{f}_1(\mathbf{v}^{(1)})\, \bar{f}_1(\mathbf{v}^{(2)})\}'\, d\mathbf{b}\, d\mathbf{v}^{(2)}. \end{cases}$

This equation, which determines $f_1(\mathbf{v})$ in non-uniform conditions, is exact to terms linear in the gradients, apart from the approximation of binary encounters which was introduced in the previous section. The question now arises whether it is possible to eliminate this approximation also, and thus obtain a generalization which can be applied with confidence to condensed states of the fluid. The remainder of this section is devoted to some formal considerations leading towards this end.

The probability that two molecules, with velocities in the ranges $\boldsymbol{\xi}^{(1)}, d\boldsymbol{\xi}^{(1)}$; $\boldsymbol{\xi}^{(2)}, d\boldsymbol{\xi}^{(2)}$ respectively, are situated in the volume elements $d\mathbf{x}^{(1)}, d\mathbf{x}^{(2)}$ at time t, is

$$f_2(\boldsymbol{\xi}^{(1)}, \boldsymbol{\xi}^{(2)}, \mathbf{x}^{(1)}, \mathbf{x}^{(2)}, t)\, d\boldsymbol{\xi}^{(1)}\, d\boldsymbol{\xi}^{(2)}\, d\mathbf{x}^{(1)}\, d\mathbf{x}^{(2)}.$$

The probability that at time $t_0 = t - \tau$ the same molecules occupied volume elements $ds^{(1)}$, $ds^{(2)}$ at $\mathbf{x}_0^{(1)} = \mathbf{x}^{(1)} - \mathbf{s}^{(1)}$ and $\mathbf{x}_0^{(2)} = \mathbf{x}^{(2)} - \mathbf{s}^{(2)}$, and had velocities in the ranges $\boldsymbol{\xi}_0^{(1)}$, $d\boldsymbol{\xi}_0^{(1)}$; $\boldsymbol{\xi}_0^{(2)}$, $d\boldsymbol{\xi}_0^{(2)}$ respectively, will be the product of the *a priori* probability

$$f_2(\boldsymbol{\xi}_0^{(1)}, \boldsymbol{\xi}_0^{(2)}, \mathbf{x}_0^{(1)}, \mathbf{x}_0^{(2)}, t_0) \, d\boldsymbol{\xi}_0^{(1)} \, d\boldsymbol{\xi}_0^{(2)} \, ds^{(1)} \, ds^{(2)}$$

and the probability

$$p(\mathbf{s}^{(1)}, \mathbf{s}^{(2)}, \boldsymbol{\xi}_0^{(1)}, \boldsymbol{\xi}_0^{(2)}; \mathbf{r}, \boldsymbol{\xi}^{(1)}, \boldsymbol{\xi}^{(2)}) \, d\mathbf{x}^{(1)} \, d\mathbf{x}^{(2)} \, d\boldsymbol{\xi}^{(1)} \, d\boldsymbol{\xi}^{(2)}$$

that the molecules originally at $\mathbf{x}_0^{(1)}$ and $\mathbf{x}_0^{(2)}$ will, in the time τ, suffer displacements in the ranges $\mathbf{s}^{(1)}$, $d\mathbf{x}^{(1)}$; $\mathbf{s}^{(2)}$, $d\mathbf{x}^{(2)}$, and have their velocities changed by amounts in the ranges $\boldsymbol{\xi}^{(1)} - \boldsymbol{\xi}_0^{(1)}$, $d\boldsymbol{\xi}^{(1)}$ and $\boldsymbol{\xi}^{(2)} - \boldsymbol{\xi}_0^{(2)}$, $d\boldsymbol{\xi}^{(2)}$ respectively. Hence

$$(3.9) \quad \begin{cases} f_2(\boldsymbol{\xi}^{(1)}, \boldsymbol{\xi}^{(2)}, \mathbf{x}^{(1)}, \mathbf{x}^{(2)}, t) = \iiiint f_2(\boldsymbol{\xi}_0^{(1)}, \boldsymbol{\xi}_0^{(2)}, \mathbf{x}^{(1)} - \mathbf{s}^{(1)}, \mathbf{x}^{(2)} - \mathbf{s}^{(2)}, t - \tau) \times \\ \qquad \times p(\mathbf{s}^{(1)}, \mathbf{s}^{(2)}, \boldsymbol{\xi}_0^{(1)}, \boldsymbol{\xi}_0^{(2)}; \mathbf{r}, \boldsymbol{\xi}^{(1)}, \boldsymbol{\xi}^{(2)}) \, d\boldsymbol{\xi}_0^{(1)} \, d\boldsymbol{\xi}_0^{(2)} \, ds^{(1)} \, ds^{(2)}. \end{cases}$$

This purely formal result is exact. The difficulty, however, in the exact determination of the probability function p makes it desirable to simplify the above relation by means of some approximation. The simplest assumption which suggests itself may be expressed by the formula

$$(3.10) \qquad f_2(\boldsymbol{\xi}^{(1)}, \boldsymbol{\xi}^{(2)}, \mathbf{x}^{(1)}, \mathbf{x}^{(2)}, t) = f_2(\boldsymbol{\xi}_0^{(1)}, \boldsymbol{\xi}_0^{(2)}, \mathbf{x}_0^{(1)}, \mathbf{x}_0^{(2)}, t_0)$$

where $\mathbf{x}_0^{(1)}$, $\mathbf{x}_0^{(2)}$, $\boldsymbol{\xi}_0^{(1)}$, $\boldsymbol{\xi}_0^{(2)}$ have the *most probable* values of the positions and velocities of the molecules at time t_0 in the condensed fluid. These can be obtained by computing the trajectories described by each molecule with an acceleration at every point equal to its most probable value:

$$(3.11) \qquad \frac{d\boldsymbol{\xi}^{(2)}}{dt} = \frac{kT}{m} \frac{\partial}{\partial \mathbf{r}} \{\ln n_2(\mathbf{r})/n^2\} = -\frac{d\boldsymbol{\xi}^{(1)}}{dt}.$$

If the time interval τ is sufficiently long, the initial velocities of the molecules may be assumed to be independent of one another, so that

$$(3.12) \quad \begin{cases} f_2(\boldsymbol{\xi}_0^{(1)}, \boldsymbol{\xi}_0^{(2)}, \mathbf{x}_0^{(1)}, \mathbf{x}_0^{(2)}, t_0) = f_1(\boldsymbol{\xi}_0^{(1)}, \mathbf{x}_0^{(1)}, t_0) f_1(\boldsymbol{\xi}_0^{(2)}, \mathbf{x}_0^{(2)}, t_0) \\ \qquad = \varDelta + f_1(\boldsymbol{\xi}_0^{(1)}, \mathbf{x}^{(1)}, t) f_1(\boldsymbol{\xi}_0^{(2)}, \mathbf{x}^{(1)}, t). \end{cases}$$

In evaluating the correction $\varDelta$, it is not necessary to modify the argument of the early part of this section; for, although the presence of many molecules will induce an acceleration of the mean centre of the pair considered, the average value of this acceleration is proportional to the macroscopic gradients and can thus be neglected.

With the help of (3.10) and (3.12), the fundamental integral equation (2.1) is reduced to one involving f_1 alone; the latter is then soluble in principle, in spite of the fact that it cannot be transformed to a form resembling Boltzmann's equation. In view of the fact that the accelerations specified by (3.11) do not differ very much in a dense fluid from their values in the vapour, it seems quite probable, however, that the corrected form (3.8) of Boltzmann's equation is fairly reliable even at high densities.

4. Solution of Boltzmann's Equation

The most general solution of the corrected form (3.8) of Boltzmann's equation is necessarily of the form

$$(4.1) \quad \overline{f}_1(\mathbf{v}) = \overline{f}_1^0(\mathbf{v}) \left\{ 1 + \varphi_0(\tfrac{1}{2}\beta m \mathbf{v}^2) \frac{\partial}{\partial \mathbf{x}} \cdot \mathbf{u} + \varphi_1(\tfrac{1}{2}\beta m \mathbf{v}^2) \mathbf{v} \cdot \left(\frac{1}{T}\frac{\partial T}{\partial \mathbf{x}}\right) + \varphi_2(\tfrac{1}{2}\beta m \mathbf{v}^2) \beta m \mathbf{v} \cdot \overline{\left(\frac{\partial}{\partial \mathbf{x}}\mathbf{u}\right)} \cdot \mathbf{v} \right\},$$

where φ_0, φ_1, φ_2 are functions of the dimensionless scalar quantity $\tfrac{1}{2}\beta m \mathbf{v}^2$. Writing

$$(4.2) \quad \begin{cases} T\{A(\mathbf{v})\} = \iint \overline{f}_1^0(\mathbf{v}^{(2)}) \{A(\mathbf{v}_0^{(1)}) + A(\mathbf{v}_0^{(2)}) - A(\mathbf{v}^{(1)}) - A(\mathbf{v}^{(2)})\} \varrho \, d\mathbf{b} \, d\mathbf{v}^{(2)}, \\ \varrho = |\mathbf{v}^{(2)} - \mathbf{v}^{(1)}| = |\mathbf{v}_0^{(2)} - \mathbf{v}_0^{(1)}|, \end{cases}$$

where A is any scalar, vector or tensor function of $\mathbf{v}$, the equation (3.8) reduces to

$$(4.3) \quad \begin{cases} T\{\varphi_0(\tfrac{1}{2}\beta m \mathbf{v}^2)\} = \psi_0(\tfrac{1}{2}\beta m \mathbf{v}^{(1)2}), \\ T\{\varphi_1(\tfrac{1}{2}\beta m \mathbf{v}^2) \mathbf{v}\} = \psi_1(\tfrac{1}{2}\beta m \mathbf{v}^{(1)2}) \mathbf{v}^{(1)}, \\ T\{\varphi_2(\tfrac{1}{2}\beta m \mathbf{v}^2) \overline{\mathbf{v}\mathbf{v}}\} = \psi_2(\tfrac{1}{2}\beta m \mathbf{v}^{(1)2}) \overline{\mathbf{v}^{(1)}\mathbf{v}^{(1)}}, \end{cases}$$

where

$$(4.4) \quad \begin{cases} \psi_0(x) = (\gamma + \tfrac{2}{3}\beta p^0/n)(x - \tfrac{3}{2}) + K_0(x), \\ \psi_1(x) = (\beta p^0/n)(x - \tfrac{5}{2}) + K_1(x), \\ \psi_2(x) = (\beta p^0/n) + K_2(x), \end{cases}$$

and $K_0(x)$, $K_1(x)$, $K_2(x)$ are easily determined contributions from the integral on the left-hand side of (3.8). The detailed solution of the integral equations (4.3) is described in the standard work of CHAPMAN & COWLING [1939]; only a summary of the method will be given here.

It is convenient to expand the functions $\varphi_m(x)$, $\psi_m(x)$ in terms of the Sonine polynomials [1] $(-1)^k (k + m + \tfrac{1}{2})! S^{(k)}_{m+\frac{1}{2}}(x)$, defined by the identity

(4.5) $$(1 - s)^{-(n+1)} \exp\{-xs/(1-s)\} \equiv \sum_{k=0}^{\infty} S^{(k)}_n(x) s^k.$$

They satisfy the condition of orthogonality

(4.6) $$\int_0^{\infty} e^{-x} S^{(k)}_n(x) S^{(l)}_n(x) x^n dx = (k+n)!/k! \, \delta_{kl},$$

where δ_{kl} is 1 when $k = l$, and 0 when $k \neq l$. Then, if

(4.7) $$\begin{cases} \varphi_m(x) = \sum \varphi_m^{(k)} S^{(k)}_{m+1/2}(x); \\ \psi_m(x) = \sum \psi_m^{(k)} S^{(k)}_{m+1/2}(x), \end{cases}$$

it follows from (4.6) that

(4.8) $$\begin{cases} (k + m + \tfrac{1}{2})! \, \varphi_m^{(k)} = k! \int_0^{\infty} e^{-x} \varphi_m(x) S^{(k)}_{m+1/2}(x) x^{m+1/2} dx \\ (k + m + \tfrac{1}{2})! \, \psi_m^{(k)} = k! \int_0^{\infty} e^{-x} \psi_m(x) S^{(k)}_{m+1/2}(x) x^{m+1/2} dx. \end{cases}$$

The equations (4.3) then reduce to

(4.9) $$\sum_{l=0}^{\infty} a_m^{(kl)} \varphi_m^{(l)} = (k + m + \tfrac{1}{2})! \, \psi_m^{(k)}/k!$$

where

(4.10) $$\begin{cases} a_0^{(kl)} = \tfrac{1}{2}! \, n^{-1} \int_0^{\infty} \bar{f}_1^0(\mathbf{v}^{(1)}) S^{(k)}_{1/2}(\tfrac{1}{2}\beta m \mathbf{v}^{(1)2}) \, T\{S^{(l)}_{1/2}(\tfrac{1}{2}\beta m \mathbf{v}^2)\} d\mathbf{v}^{(1)}, \\ a_1^{(kl)} = \tfrac{1}{2}! \, n^{-1}(\tfrac{1}{2}\beta m) \int_0^{\infty} \bar{f}_1^0(\mathbf{v}^{(1)}) S^{(k)}_{3/2}(\tfrac{1}{2}\beta m \mathbf{v}^{(1)2}) \mathbf{v}^{(1)} \cdot T\{S^{(l)}_{3/2}(\tfrac{1}{2}\beta m \mathbf{v}^2) \mathbf{v}\} d\mathbf{v}^{(1)}, \\ a_2^{(kl)} = \tfrac{3}{2}! \, n^{-1}(\tfrac{1}{2}\beta m)^2 \int_0^{\infty} \bar{f}_1^0(\mathbf{v}^{(1)}) S^{(k)}_{5/2}(\tfrac{1}{2}\beta m \mathbf{v}^{(1)2}) \mathbf{v}^{(1)} \cdot T\{S^{(l)}_{5/2}(\tfrac{1}{2}\beta m \mathbf{v}^2) \overline{\mathbf{vv}}\} \cdot \mathbf{v}^{(1)} d\mathbf{v}^{(1)} \end{cases}$$

Now, $S_n^{(0)}(x) = 1$ and $S_n^{(1)}(x) = (n+1) - x$; so one has from (4.4) and (4.7)

(4.11) $$\begin{cases} \psi_0^{(k)} = (\gamma + \tfrac{2}{3} \beta p^0/n) \, \delta_{k1} + K_0^{(k)}, \\ \psi_1^{(k)} = (\beta p^0/n) \, \delta_{k1} + K_1^{(k)}, \\ \psi_2^{(k)} = (\beta p^0/n) \, \delta_{k0} + K_2^{(k)}. \end{cases}$$

The $\psi_m^{(k)}$ are therefore small when k is large, and the same is most probably true of the $\varphi_m^{(k)}$; one may therefore replace the infinite set

---

[1] Note that $\tfrac{1}{2}! = \sqrt{\pi}/2$, $\tfrac{3}{2}! = \tfrac{3}{2}(\tfrac{1}{2}!)$, etc.

of equations (4.9) by the finite set obtained by assuming $\psi_m^{(k)} = \varphi_m^{(k)} = 0$ when $k \geqslant k_0$, with an error which can be made as small as one pleases by taking k_0 sufficiently large. In practice, k_0 need not exceed 2 or 3. Then it is easy to obtain the $\varphi_m^{(k)}$ from (4.9), once the coefficients $a_m^{(kl)}$ have been evaluated.

The evaluation of the $a_m^{(kl)}$ is rather tedious in practice; formulae for $a_1^{(kl)}$ and $a_2^{(kl)}$ are derived in detail in the work by Chapman & Cowling (loc. cit.). To illustrate the method involved, the other coefficients $a_0^{(kl)}$ will be treated in a similar way. From (4.10) and (4.2) one has

$$(4.12) \quad \begin{cases} a_0^{(kl)} = 2(\tfrac{1}{2}!)\, n^{-1} \iiint \bar{f}_1^0(\mathbf{v}^{(1)})\, \bar{f}_1^0(\mathbf{v}^{(2)})\, S_{1/2}^{(k)}(\tfrac{1}{2}\beta m \mathbf{v}^{(1)2}) \times \\ \qquad \times \{S_{1/2}^{(l)}(\tfrac{1}{2}\beta m \mathbf{v}_0^{(2)2}) - S_{1/2}^{(l)}(\tfrac{1}{2}\beta m \mathbf{v}^{(2)2})\} \times \varrho\, d\mathbf{b}\, d\mathbf{v}^{(1)}\, d\mathbf{v}^{(2)}. \end{cases}$$

By the definition (4.5) of the Sonine polynomials, this is the coefficient of $s^k t^l$ in the expression

$$(4.13) \quad \begin{cases} A_0 = \dfrac{2(\tfrac{1}{2}!)\, n^{-1}}{\{(1-s)(1-t)\}^{3/2}} \iiint f_1^0(\mathbf{v}^{(1)})\, \bar{f}_1^0(\mathbf{v}^{(2)}) \{\exp[-\tfrac{1}{2}\beta m (\sigma \mathbf{v}^{(1)2} + \tau \mathbf{v}_0^{(2)2})] - \\ \qquad - \exp[-\tfrac{1}{2}\beta m (\sigma \mathbf{v}^{(1)2} + \tau \mathbf{v}^{(2)2})]\}\, \varrho\, d\mathbf{b}\, d\mathbf{v}^{(1)}\, d\mathbf{v}^{(2)}, \\ \sigma = s/(1-s), \quad \tau = t/(1-t). \end{cases}$$

In this expression, one substitutes $\mathbf{v}^{(1)} = \mathbf{v} - \tfrac{1}{2}\boldsymbol{\varrho}$, $\mathbf{v}^{(2)} = \mathbf{v} + \tfrac{1}{2}\boldsymbol{\varrho}$, and $\mathbf{v}_0^{(2)} = \mathbf{v} + \tfrac{1}{2}\boldsymbol{\varrho}_0$; then, since $\varrho_c = \varrho$,

$$(4.14) \quad \begin{cases} A_0 = \dfrac{2(\tfrac{1}{2}!)\, n}{\{(1-s)(1-t)\}^{3/2}} \iiint \exp[-\tfrac{1}{2}\beta m (2+\sigma+\tau)(\mathbf{v}^2 + \tfrac{1}{4}\boldsymbol{\varrho}^2)] \times \\ \times \{\exp[-\tfrac{1}{2}\beta m (\tau \boldsymbol{\varrho}_0 - \sigma\boldsymbol{\varrho})\cdot\mathbf{v}] - \exp[-\tfrac{1}{2}\beta m (\tau-\sigma)\boldsymbol{\varrho}\cdot\mathbf{v}]\}\, \varrho\, d\mathbf{b}\, (m\beta/2\pi)^3\, d\boldsymbol{\varrho}\, d\mathbf{v} \\ = \dfrac{2(\tfrac{1}{2}!)\, n}{(2-s-t)^{3/2}} \iint \exp\!\left[-\tfrac{1}{2}\beta m \left\{\dfrac{(1+\sigma)(1+\tau)}{(2+\sigma+\tau)}\right\} \boldsymbol{\varrho}^2\right] \times \\ \qquad \times \left[\exp\!\left(-\tfrac{1}{2}\beta m \left\{\dfrac{\sigma\tau(1-z)}{2(2+\sigma+\tau)}\right\}\boldsymbol{\varrho}^2\right) - 1\right] \varrho\, d\mathbf{b}\, (m\beta/2\pi)^{3/2}\, d\boldsymbol{\varrho}, \end{cases}$$

where

$$(4.15) \quad z = \boldsymbol{\varrho}\cdot\boldsymbol{\varrho}_0/\varrho^2 = \cos\left[2 b\varrho \int_{r_0}^{\infty} \frac{dr}{r^2 \{\varrho^2(1-b^2/r^2) - 4\phi(r)/m\}^{\frac{1}{2}}}\right],$$

and r_0, a function of b and ϱ, is the zero of the denominator

$$\varrho^2(1 - b^2/r^2) - 4\phi(r)/m$$

in the integrand. The individual coefficients $a_0^{(kl)}$ may now be obtained by expanding (4.14) as a power series in s and t; with the help of the formula

$$(4.16) \begin{cases} (2-s-t)^{-3/2} \exp\left[-\tfrac{1}{2}\beta m \rho^2 \left\{ \frac{(1+\sigma)(1+\tau)+\sigma\tau(1-z)}{(2+\sigma+\tau)} \right\} \right] \\ = \exp(-\tfrac{1}{4}\beta m \rho^2) \sum_{r=0}^{\infty} (-\tfrac{1}{4}\beta m \rho^2)^r \{(s+t)+st(1-z)\}^r / \{r!\,(2-s-t)^{r+3/2}\} \end{cases}$$

the required coefficients are readily obtained, expressed in terms of integrals of the type

$$\int_0^\infty \exp(-\tfrac{1}{2}\beta m \rho^2) \cdot \int_0^\infty (1-z^a)\, b\, db\, \rho^c\, d\rho.$$

The final integrations must be performed numerically, except when the potential function $\phi(r)$ is very simple in form.

4.1 Viscosity and Thermal Conduction

It is unnecessary to determine the complete solution of Boltzmann's equation in order to compute the coefficients of viscosity and thermal conduction in a gas.

According to (4.1), the part of the pressure tensor arising from the thermal motion of the molecules is

$$(4.17)\quad \mathbf{P}_a = m \int \bar{f}_1(\mathbf{v})\mathbf{v}\mathbf{v}\,d\mathbf{v} = nkT\left[\left\{1 + \frac{4}{3\sqrt{\pi}}\int_0^\infty e^{-x}\varphi_0(x)\,x^{3/2}dx\left(\frac{\partial}{\partial \mathbf{x}}\cdot\mathbf{u}\right)\right\}\boldsymbol{\delta} \right. \\ \left. + \frac{8}{15\sqrt{\pi}}\int_0^\infty e^{-x}\varphi_2(x)\,x^{5/2}\,dx\,\overline{\left(\frac{\partial}{\partial \mathbf{x}}\mathbf{u}\right)}\right].$$

The part of the coefficient of viscosity η_a arising from this source is therefore

$$\eta_a = -\frac{4}{15\sqrt{\pi}}\int_0^\infty e^{-x}\varphi_2(x)\,x^{5/2}\,dx.$$

Substituting from (4.7) into this formula, with $S_{5/2}^{(0)}(x) = 1$, one has, by the use of the orthogonality relation (4.6),

$$(4.18)\qquad \eta_a = -\tfrac{1}{2}\varphi_2^{(0)}.$$

In rare gases, the viscosity is due almost entirely to the thermal motion, and (4.18) is a valid formula for the coefficient of viscosity.

In more condensed systems, however, one would have to take account of the contribution

(4. 19) $$\mathbf{p}_b = -\tfrac{1}{2} \iiint \bar{f}_2 \, r^{-1} \, \phi'(r) \, \mathbf{rr} \, d\mathbf{r} \, d\mathbf{v}^{(1)} \, d\mathbf{v}^{(2)}$$

of the intermolecular forces to the pressure tensor. The value of $\bar{f}_2$ can be found from the relations

$$f_2 = \bar{f}_2 + \tfrac{1}{2} r \cdot \frac{\partial}{\partial \mathbf{x}} \bar{f}_2^0$$

and (2. 7). Splitting the correction term $\varDelta$ into two parts $\varDelta_1$ and $\bar{f}_2$, as in (3. 1), one obtains, in fact,

(4. 20) $$\bar{f}_2 = \bar{f}_1(\mathbf{v}_0^{(1)}) \, \bar{f}_1(\mathbf{v}_0^{(2)}) + \varDelta_1,$$

correct to terms linear in the macroscopic gradients. The right-hand side of (4. 19) may therefore be evaluated when $\bar{f}_1(\mathbf{v})$ has been determined; but no calculations have yet been made.

The coefficient of thermal conduction may be dealt with in a similar way. The part of the thermal flux arising from the thermal motion is

(4. 21) $$\begin{cases} \mathbf{q}_a = \tfrac{1}{2} m \int \bar{f}_1(\mathbf{v}) \, v^2 \, \mathbf{v} \, d\mathbf{v} \\ = \dfrac{4n}{3\sqrt{\pi} m \beta^2} \int_0^\infty e^{-x} \varphi_1(x) \, x^{5/2} \, dx \left(\dfrac{1}{T} \dfrac{\partial T}{\partial \mathbf{x}} \right) \end{cases}$$

Again substituting from (4. 7), with $x = \tfrac{5}{2} S_{3/2}^{(0)}(x) - S_{3/2}^{(1)}(x)$, one has, by the use of (4. 6),

(4. 22) $$\lambda_a = -\frac{4 n k^2 T}{3 \sqrt{\pi} m} \int_0^\infty e^{-x} \, \varphi_1(x) \, x^{5/2} \, dx.$$

According to (3. 21) of chapter V, the part of the thermal flux arising from the action of the intermolecular forces is

(4. 23) $$\mathbf{q}_b = -\tfrac{1}{4} \iiint \bar{f}_2 \, \mathbf{r} \cdot (\mathbf{v}^{(1)} + \mathbf{v}^{(2)}) \, r^{-1} \, \phi'(r) \, \mathbf{r} \, d\mathbf{v}^{(1)} \, d\mathbf{v}^{(2)} \, d\mathbf{r}$$

and this could be evaluated by the substitution of (4. 20).

The only calculations yet made of the coefficients of viscosity and thermal conduction are those relating to rare gases. The results have been summarized by Chapman & Cowling. The viscosity and thermal conduction of a gas whose molecules were rigid spheres of diameter σ would be

(4.24)
$$\begin{cases} \eta = \frac{5}{16\,\sigma^2}\left(\frac{m}{\pi\beta}\right)^{\frac{1}{2}} \left\{ 1 + \frac{3}{202} + \frac{(347)^2}{808 \times 145043} + \cdots \right\} \\ \quad = 1\cdot 016 \times \frac{5}{16\,\sigma^2}\left(\frac{m}{\pi\beta}\right)^{\frac{1}{2}}, \\ \lambda = \frac{75\,k}{64\,\sigma^2 m}\left(\frac{m}{\pi\beta}\right)^{\frac{1}{2}} \left\{ 1 + \frac{1}{44} + \frac{(111)^2}{88 \times 66951} + \cdots \right\} \\ \quad = 1\cdot 025 \times \frac{75\,k}{64\,\sigma^2 m}\left(\frac{m}{\pi\beta}\right)^{\frac{1}{2}}. \end{cases}$$

If any two molecules repel one another with a force proportional to $r^{-\nu}$, the coefficients of viscosity and thermal conduction are proportional to T^a, where $a = \frac{1}{2} + 2/(\nu - 1)$; if, in addition, they attract one another with a force proportional to r^{-7}, one has to add a term proportional to T^b, where $b = \frac{1}{2} - (\nu - 9)/(\nu - 1)$.

The coefficient of thermal conduction in a rare gas should always be related approximately to the coefficient of viscosity by the formula

(4.25)
$$\lambda = \frac{15}{4}\frac{k}{m}\,\eta = \frac{5}{2}\,c_v\,\eta$$

where c_v is the specific heat at constant volume. This formula is actually verified by the experimental data for the monatomic gases. In diatomic and polyatomic gases, the values of λ and c_v are both considerably affected by the existence of the internal degrees of freedom of the molecules; for such gases, EUCKEN [1913] proposed the formula

$$\lambda = \tfrac{1}{4}(9\gamma - 5)\,c_v\eta$$

where γ is the ratio of the specific heats at constant pressure and constant volume. This law is closely obeyed by most simple gases.

CHAPTER IX

THE QUANTUM THEORY OF FLUIDS

1. General Principles

The theory developed in the foregoing chapters has been founded on the concept of a fluid as an assembly of similar molecules, all with well-defined positions and velocities. That this concept is not perfectly correct in a strict sense is now universally accepted, as a result of discoveries made in the field of quantum mechanics. It has been discovered that it is impossible in principle to measure exactly both the position and the velocity of a molecule, or any other ultra-microscopic particle, at the same time. The physical reason for this, is that, in order to observe the position of a particle with any precision it is necessary to illuminate it with radiation of correspondingly short wave-length; and then its velocity becomes indeterminate by virtue of the powerful reaction of the illuminating radiation. This is not merely an experimental difficulty, but a very fundamental aspect of the nature of the real world in the ultra-microscopic domain. Consequently it is not permissible to extrapolate one's experience of macroscopic 'particles' by supposing that definite positions and velocities may be assigned simultaneously to atoms or molecules. This raises the further difficulty that the concepts of Newtonian mechanics cannot be applied to molecular phenomena; instead, the concepts of *quantum* mechanics have to be introduced.

In quantum mechanics it would not be meaningful to ask the probability that a single molecule should have the position **x** and the velocity ξ at the same time.

A particle with a definite velocity cannot be regarded as located at a point, but must be imagined rather as a wave, filling the whole of space. The wave-length (λ) associated with such a particle is given by de Broglie's formula

$$\lambda = h/m\xi,$$

where h is Planck's constant (6.624×10^{-27} erg. sec), and m represents as usual the molecular mass. The wave has a direction of propagation parallel to the velocity of the molecule; at any given time it may therefore be represented by the phase factor

$$\exp 2\pi i m \boldsymbol{\xi} \cdot \mathbf{r}/h,$$

where $\mathbf{r}$ is the relative coordinate associated with the wave.

Although one may not speak of the probability that a molecule should have simultaneously a definite position $\mathbf{x}$ and a definite velocity $\boldsymbol{\xi}$, it is possible to define a function $f_1(\boldsymbol{\xi}, \mathbf{x}, t)$ which is closely analogous to the velocity distribution function of the classical theory of fluids. For the *mean value* $\overline{A}(\mathbf{x}, t)$ of any function $A(\boldsymbol{\xi})$ of the velocity of a molecule at the point $\mathbf{x}$ is, in principle, a measurable quantity, and the number density $n(\mathbf{x}, t)$ can be defined exactly as in the classical theory. A 'velocity distribution function' may therefore be defined as the solution of the integral equation

(1. 1) $$n(\mathbf{x}, t)\overline{A}(\mathbf{x}, t) = \int f_1(\boldsymbol{\xi}, \mathbf{x}, t) A(\boldsymbol{\xi}) \, d\boldsymbol{\xi},$$

where $A(\boldsymbol{\xi})$ is arbitrary. It should not be supposed that f_1, so defined, will have all the properties which one normally associates with a probability; there is no reason, for example, why it should be positive for all values of $\boldsymbol{\xi}$ and $\mathbf{x}$. It will, however, provide correct values for the averages of all quantities which are measurable, even in principle.

In quantum theory the fundamental quantity is not actually the velocity distribution function f_1 defined by (1. 1), but the function

$$\int f_1(\boldsymbol{\xi}, \mathbf{x}, t) \exp (2\pi i m \boldsymbol{\xi} \cdot \mathbf{r}/ih) \, d\boldsymbol{\xi}$$

which is by definition the product of the number density and the average value of the phase factor $\exp (2\pi i m \boldsymbol{\xi} \cdot \mathbf{r}/ih)$ for the point $\mathbf{x}$. It is usually represented as a function of two coordinates $\mathbf{x}'$ and $\mathbf{x}''$ whose mean centre is at the point $\mathbf{x}$, and whose relative position vector is $\mathbf{r}$, thus:

(1. 2) $$\begin{cases} \mathbf{x} = \tfrac{1}{2}(\mathbf{x}' + \mathbf{x}'') \, , & \mathbf{r} = \mathbf{x}'' - \mathbf{x}'; \\ \mathbf{x}' = \mathbf{x} - \tfrac{1}{2}\mathbf{r} \, , & \mathbf{x}'' = \mathbf{x} + \tfrac{1}{2}\mathbf{r}. \end{cases}$$

It is customary, also, to write

(1. 3) $$\hbar = h/2\pi$$

Then the function

(1. 4) $$\varrho_1(\mathbf{x}', \mathbf{x}'', t) = \int f_1(\boldsymbol{\xi}, \mathbf{x}, t) \exp (m \boldsymbol{\xi} \cdot \mathbf{r}/i\hbar) \, d\boldsymbol{\xi}$$

is called the *density matrix* of a molecule in the fluid.

This function of two coordinate variables has been found a useful tool in handling the quantum mechanics of fluids. By putting $\mathbf{x}' = \mathbf{x}''$ in the relation (1. 4), one has, with the help of (1. 1),

(1. 5) $$\varrho_1(\mathbf{x}, \mathbf{x}, t) = \int f_1(\boldsymbol{\xi}, \mathbf{x}, t)\, d\boldsymbol{\xi} = n(\mathbf{x}, t),$$

which agrees with (4. 2) of chapter III. This result might be described by saying that the diagonal element of the density matrix is the number density.

By differentiating (1. 4) with respect to $\mathbf{r}$, and then putting $\mathbf{x}' = \mathbf{x}''$, one obtains

(1. 6) $$\left[\frac{i\hbar}{2m}\left(\frac{\partial}{\partial \mathbf{x}''} - \frac{\partial}{\partial \mathbf{x}'}\right)\varrho_1(\mathbf{x}', \mathbf{x}'')\right]_{\mathbf{x}'=\mathbf{x}''} = \int f_1(\boldsymbol{\xi}, \mathbf{x}, t)\,\boldsymbol{\xi}\, d\boldsymbol{\xi}$$
$$= n(\mathbf{x}, t)\,\mathbf{u}(\mathbf{x}, t),$$

where $\mathbf{u}$ is the mean velocity of a molecule at the point $\mathbf{x}$, or, what is the same thing, the macroscopic velocity of flow at the point $\mathbf{x}$. One may proceed still further by differentiating (1. 4) twice with respect to $\mathbf{r}$, afterwards setting $\mathbf{x}' = \mathbf{x}''$; the result is

(1. 7) $$\left[-\frac{\hbar^2}{4m}\left(\frac{\partial}{\partial \mathbf{x}''} - \frac{\partial}{\partial \mathbf{x}'}\right)^2 \varrho_1(\mathbf{x}', \mathbf{x}'')\right]_{\mathbf{x}'=\mathbf{x}''} = \int f_1 \cdot m\boldsymbol{\xi}^2\, d\boldsymbol{\xi}$$
$$= n(3\,kT_1 + m\mathbf{u}^2).$$

Here T_1 is proportional to the mean kinetic energy of a molecule as measured by an observer moving with the fluid; but, in quantum theory, the classical demonstration that this is the same as the thermodynamical temperature (T) on the absolute scale cannot be carried through, and the two quantities are in fact appreciably different at low temperatures. To secure the necessary distinction, T_1 will be called the kinetic temperature, and T the thermodynamic temperature.

The preceding considerations are easily extended to clusters of q molecules in the fluid. The density matrix for such a cluster is a function of two sets of coordinates $\mathbf{x}^{(1)'}, \ldots \mathbf{x}^{(q)'}$ and $\mathbf{x}^{(1)''}, \ldots \mathbf{x}^{(q)''}$, as well as of the time, and is given by

(1. 8) $$\varrho_q = \int \overset{(q)}{\cdots} \int f_q \exp(m \sum \boldsymbol{\xi}^{(i)} \cdot \mathbf{r}^{(i)}/i\hbar).$$

One has, by the generalization of (1. 5), (1. 6), and (1. 7) respectively,

(1. 9) $$n_q = [\varrho_q]_{\mathbf{x}'=\mathbf{x}''},$$

(1. 10) $$n_q \mathbf{u}_q^{(i)} = \left[\frac{i\hbar}{2m}\left(\frac{\partial}{\partial \mathbf{x}^{(i)''}} - \frac{\partial}{\partial \mathbf{x}^{(i)'}}\right)\varrho_q\right]_{\mathbf{x}'=\mathbf{x}''},$$

and

(1. 11) $\quad n_q(3T_q^{(i)} + m\mathbf{u}_q^{(i)2}) = \left[-\dfrac{\hbar^2}{4m}\left(\dfrac{\partial}{\partial \mathbf{x}^{(i)''}} - \dfrac{\partial}{\partial \mathbf{x}^{(i)'}}\right)^2 \varrho_q\right]_{\mathbf{x}'=\mathbf{x}''}.$

The differences between the quantum theory and the classical theory of fluids all arise from the fact that quantum mechanics has to be applied instead of Newtonian mechanics. The classical equations (2. 7) and (2. 9) of chapter V are therefore not exact in quantum theory, but will appear as approximations valid for all except very low temperatures.

Instead of the classical equations of motion, quantum theory proceeds from the single equation of motion

(1. 12) $\qquad \dfrac{\partial \varrho_N}{\partial t} + \dfrac{i}{\hbar}(H' - H'')\varrho_N = 0,$

where H' and H'' are the operators

(1. 13) $\begin{cases} H' = -\dfrac{\hbar^2}{2m}\sum\limits_{i=1}^{N}\left(\dfrac{\partial}{\partial \mathbf{x}^{(i)'}}\right)^2 + \varPhi(\mathbf{x}'), \\ H'' = -\dfrac{\hbar^2}{2m}\sum\limits_{i=1}^{N}\left(\dfrac{\partial}{\partial \mathbf{x}^{(i)''}}\right)^2 + \varPhi(\mathbf{x}''). \end{cases}$

Here $\varPhi(\mathbf{x})$ represents as usual the potential energy of the system, expressed as a function of the coordinates $\mathbf{x}^{(i)}$ of the molecules; $\varPhi(\mathbf{x}')$ and $\varPhi(\mathbf{x}'')$ are naturally the same functions of $\mathbf{x}^{(i)'}$ and $\mathbf{x}^{(i)''}$ respectively.

The equation (1. 12) for the determination of the density matrix actually applies to any quantum mechanical system. In general the Hamiltonian operator H is obtained by expressing the total energy of the system as a function of the coordinates $(\mathbf{x}^{(i)})$ and momenta $(\mathbf{p}^{(i)})$ of the individual particles, and replacing the momenta $\mathbf{p}^{(i)}$ everywhere by the corresponding differential operators $-i\hbar\dfrac{\partial}{\partial \mathbf{x}^{(i)}}$. In the present instance, (1. 13) is obtained by applying this procedure to the total energy

$$\sum_{i=1}^{N} \mathbf{p}^{(i)2}/2m + \varPhi(\mathbf{x})$$

of the system of molecules considered.

The whole of the quantum theory of fluids can be deduced from the quantum equation of motion (1. 12), which determines the density matrix ϱ_N appropriate to the complete system of molecules. It will be noticed by those familiar with wave mechanics that the density matrix ϱ_N is related to the wave function Ψ for the whole system, thus:

(1. 14) $\begin{cases} \varrho_N(\mathbf{x}', \mathbf{x}'') = \Psi(\mathbf{x}')\,\Psi^*(\mathbf{x}''), \\ i\hbar\dfrac{\partial \Psi}{\partial t} = H\Psi, \end{cases}$

provided that the system is in what is known as a pure state. To postulate a pure state, however, is to impose a rather drastic restriction on the system; and when the state of the system is *impure*, no wave function Ψ exists. For example, no wave-function exists for any number of molecules ($q < N$) which are in statistical interaction with some other molecules, because the interaction energy of the latter cannot be expressed in terms of the coordinates of the former alone. The density matrix, on the other hand, always exists, whether the system under consideration is in statistical interaction with another system or not.

1. 1 The Classical Theory as a Limit

From the quantum equation of motion (1. 12), it is not difficult to derive the equation satisfied by the velocity distribution function f_N; by this means a direct comparison of the classical and quantum formulae can be made which enables one to assess the importance of the errors implicit in the classical theory.

The operator $H' - H''$ must first be expressed in terms of the variables $\mathbf{r}^{(i)}$ and $\mathbf{x}^{(i)}$ defined as in (1. 2). From (1. 13) it follows that

(1. 15) $\quad H' - H'' = \dfrac{\hbar^2}{m}\sum_{i=1}^{N}\left(\dfrac{\partial}{\partial \mathbf{x}^{(i)}}\cdot\dfrac{\partial}{\partial \mathbf{r}^{(i)}}\right) + \Phi(\mathbf{x} - \tfrac{1}{2}\mathbf{r}) - \Phi(\mathbf{x} + \tfrac{1}{2}\mathbf{r}).$

Hence, by substituting (1. 8), with $q = N$, into (1. 12), one obtains

(1. 16) $\begin{cases} \displaystyle\int^{(N)}\int \left[\dfrac{\partial f_N}{\partial t} + \sum_{i=1}^{N}\boldsymbol{\xi}^{(i)}\cdot\dfrac{\partial f_N}{\partial \mathbf{x}^{(i)}} + \dfrac{i}{\hbar}\left\{\Phi(\mathbf{x} - \tfrac{1}{2}\mathbf{r}) - \Phi(\mathbf{x} + \tfrac{1}{2}\mathbf{r})\right\}f_N\right]\times \\ \qquad\qquad \times \exp\left(m\sum\boldsymbol{\xi}^{(i)}\cdot\mathbf{r}^{(i)}/i\hbar\right)d\boldsymbol{\xi}^{(1)}\ldots d\boldsymbol{\xi}^{(N)} = 0. \end{cases}$

The equation satisfied by f_N can now be obtained by multiplying (1. 16) by $\exp(-m\Sigma\boldsymbol{\xi}_0^{(i)}\cdot\mathbf{r}^{(i)}/i\hbar)$ and integrating with respect to the relative coordinates $\mathbf{r}^{(i)}$. Since, by Fourier's theorem,

$\int^{(2N)}\int f_N(\boldsymbol{\xi})\exp\left\{m\sum(\boldsymbol{\xi}^{(i)} - \boldsymbol{\xi}_0^{(i)})\cdot\mathbf{r}^{(i)}/i\hbar\right\}d\boldsymbol{\xi}^{(1)}\ldots d\boldsymbol{\xi}^{(N)}\cdot d\mathbf{r}^{(1)}\ldots d\mathbf{r}^{(N)}$
$= (2\pi\hbar/m)^N f_N(\boldsymbol{\xi}_0),$

and also

$$\int \overset{(N)}{\cdots} \int \exp\{m \sum (\boldsymbol{\xi}^{(i)} - \boldsymbol{\xi}_0^{(i)}) \cdot \mathbf{r}^{(i)}/i\hbar\} \, \Phi(\mathbf{x} \pm \tfrac{1}{2}\mathbf{r}) \, d\mathbf{r}^{(1)} \ldots d\mathbf{r}^{(N)}$$
$$= (2\pi\hbar/m)^N \exp \pm \{2m \sum (\boldsymbol{\xi}^{(i)} - \boldsymbol{\xi}_0^{(i)}) \cdot \mathbf{x}^{(i)}/i\hbar\} \, X(\boldsymbol{\xi} - \boldsymbol{\xi}_0)$$

where

(1.17) $\quad X(\boldsymbol{\xi}) = \left(\dfrac{m}{2\pi\hbar}\right)^N \int \overset{(N)}{\cdots} \int \Phi(\tfrac{1}{2}\mathbf{r}) \cos(m \sum \boldsymbol{\xi}^{(i)} \cdot \mathbf{r}^{(i)}/\hbar) \, d\mathbf{r}^{(1)} \ldots d\mathbf{r}^{(N)},$

one thus obtains, after replacing $\boldsymbol{\xi}_0^{(i)}$ and $\boldsymbol{\xi}^{(i)}$ simultaneously by $\boldsymbol{\xi}^{(i)}$ and $\boldsymbol{\xi}_0^{(i)} + \boldsymbol{\xi}^{(i)}$ respectively,

(1.18) $\quad\begin{cases} \dfrac{\partial f_N}{\partial t} + \sum\limits_{i=1}^{N} \boldsymbol{\xi}^{(i)} \cdot \dfrac{\partial f_N}{\partial \mathbf{x}^{(i)}} + \\ \quad + \dfrac{2}{\hbar} \int \overset{(N)}{\cdots} \int X(\boldsymbol{\xi}_0) \sin(2m \sum \boldsymbol{\xi}_0^{(i)} \cdot \mathbf{x}^{(i)}/\hbar) f_N(\boldsymbol{\xi}_0 + \boldsymbol{\xi}) \, d\boldsymbol{\xi}_0^{(1)} \ldots d\boldsymbol{\xi}_0^{(N)} = 0. \end{cases}$

This is the fundamental integro-differential equation for the velocity distribution function f_N.

To make a direct comparison with the classical theory, one uses Taylor's theorem to expand $f_N(\boldsymbol{\xi}_0 + \boldsymbol{\xi})$ in powers of $\boldsymbol{\xi}_0$, thus:

(1.19) $\quad\begin{cases} f_N(\boldsymbol{\xi}_0 + \boldsymbol{\xi}) = f_N(\boldsymbol{\xi}) + \left(\sum \boldsymbol{\xi}_0^{(i)} \cdot \dfrac{\partial}{\partial \boldsymbol{\xi}^{(i)}}\right) f_N(\boldsymbol{\xi}) + \\ \quad + \dfrac{1}{2!} \left(\sum \boldsymbol{\xi}_0^{(i)} \cdot \dfrac{\partial}{\partial \boldsymbol{\xi}^{(i)}}\right)^2 f_N(\boldsymbol{\xi}) + \ldots; \end{cases}$

the last term in (1.18) then becomes

(1.20) $\quad\begin{cases} \dfrac{2}{\hbar} \int \overset{(N)}{\cdots} \int X(\boldsymbol{\xi}_0) \sin(2m \sum \boldsymbol{\xi}_0^{(i)} \cdot \mathbf{x}^{(i)}/\hbar) \left\{ \sum \boldsymbol{\xi}_0^{(i)} \cdot \dfrac{\partial f_N}{\partial \boldsymbol{\xi}^{(i)}} + \right. \\ \quad \left. + \dfrac{1}{3!} \left(\sum \boldsymbol{\xi}_0^{(i)} \cdot \dfrac{\partial}{\partial \boldsymbol{\xi}^{(i)}}\right)^3 f_N + \ldots \right\} d\boldsymbol{\xi}_0^{(1)} \ldots d\boldsymbol{\xi}_0^{(N)}. \end{cases}$

It follows, however, by Fourier's theorem from (1.17) that

(1.21) $\quad \Phi(\mathbf{x}) = \int \overset{(N)}{\cdots} \int X(\boldsymbol{\xi}_0) \cos(2m \sum \boldsymbol{\xi}_0^{(i)} \cdot \mathbf{x}^{(i)}/\hbar) \, d\boldsymbol{\xi}_0^{(1)} \ldots d\boldsymbol{\xi}_0^{(N)};$

hence (1.20) reduces to

(1.22) $\quad -\dfrac{1}{m} \sum\limits_i \dfrac{\partial \Phi}{\partial \mathbf{x}^{(i)}} \cdot \dfrac{\partial f_N}{\partial \boldsymbol{\xi}^{(i)}} + \dfrac{1}{3!} \dfrac{(\tfrac{1}{2}\hbar)^2}{m^3} \sum\limits_{ijk} \dfrac{\partial^3 \Phi}{\partial \mathbf{x}^{(i)} \partial \mathbf{x}^{(j)} \partial \mathbf{x}^{(k)}} \vdots \dfrac{\partial^3 f_N}{\partial \boldsymbol{\xi}^{(i)} \partial \boldsymbol{\xi}^{(j)} \partial \boldsymbol{\xi}^{(k)}} + \ldots,$

which is a power series in $\hbar^2$.

If one were to neglect all terms in (1.22) but the first, substitution of the resulting expression in (1.18) would yield exactly the classical equation

$$\dfrac{\partial f_N}{\partial t} + \sum\limits_i \left(\boldsymbol{\xi}^{(i)} \cdot \dfrac{\partial f_N}{\partial \mathbf{x}^{(i)}} - \dfrac{1}{m} \dfrac{\partial \Phi}{\partial \mathbf{x}^{(i)}} \cdot \dfrac{\partial f_N}{\partial \boldsymbol{\xi}^{(i)}}\right) = 0.$$

Hence *the classical theory corresponds to taking for Planck's constant h the value zero*, instead of its actual small value of 6.624×10^{-27} erg. sec. One may infer also that at sufficiently high temperatures the error involved in the classical theory is negligible. For, at high temperatures the contribution to all physically observable quantities from molecules with very small velocities can be neglected. But, in the last term of (1.18), the supposition that $\xi^{(i)}$ is large compared with $\xi_0^{(i)}$ allows one to neglect all but the early terms of the Taylor's series (1.19), and thus to obtain again the *classical* equation for the determination of the velocity distribution function f_N. This is, of course, only a qualitative conclusion, but such quantitative calculations as have been made indicate that the classical theory is adequate at all ordinary temperatures. The quantum theory is required, therefore, only for those fluids, like liquid hydrogen and liquid helium, which exist at the lowest temperatures.

1.2 Quantum Mechanical States and Operators

Special examples of quantum mechanical operators have already been encountered, such as the Hamiltonian operator, defined in (1.13), and the momentum operators

$$(1.23) \qquad \mathbf{p}^{(i)} = -i\hbar \frac{\partial}{\partial \mathbf{x}^{(i)}}$$

which it contains. The operator representing the total momentum is

$$(1.24) \qquad \mathbf{P} = \sum_{i=1}^{N} \mathbf{p}^{(i)} = \sum_i \left(-i\hbar \frac{\partial}{\partial \mathbf{x}^{(i)}} \right),$$

and other quantum mechanical operators representing physical variables can be derived in a similar way, simply by replacing the momenta everywhere by the operators defined in (1.23). As a further example, the operator representing the total angular momentum is

$$(1.25) \qquad \mathbf{A} = \sum_i \mathbf{x}^{(i)} \wedge \mathbf{p}^{(i)} = \sum \left(-i\hbar\, \mathbf{x}^{(i)} \wedge \frac{\partial}{\partial \mathbf{x}^{(i)}} \right).$$

It has been found that the values which an experiment can give for many observable quantities can be determined mathematically as the eigenvalues of certain functional or differential equations. To each observable quantity there corresponds a particular functional equation, and the mathematical operator occurring in that equation is related in this way to the observable quantity. This has led to the general conviction, that with each observable quantity there is associated a mathematical operator; the latter may be just a common multiplicative factor, but is usually a differential operator of the kind described above. For example, one might — in principle if not in practice — obtain a fluid in a state in which the Hamiltonian energy could be measured exactly and had the numerical value H_L. In such a state, the density matrix ϱ_N would satisfy the differential equations

$$H' \varrho_N = H_L \varrho_N, \quad H'' \varrho_N = H_L \varrho_N.$$

Usually, of course, the Hamiltonian energy cannot be measured exactly in this way; it is possible, however, to measure the total internal energy of the fluid, and add to this the total energy of the macroscopic motion, but the result is only a *mean value* of the Hamiltonian energy.

Two operators X and Y are said to *commute*, if the order in which they are applied to the density matrix is immaterial, i.e., if

(1. 26) $$X(Y \varrho_N) = Y(X \varrho_N).$$

In order that it should be possible to measure each of the physical quantities represented by X and Y without altering the state of the system, it is necessary that the operators should commute. For, if X_L and Y_L are the measured values, one has

$$X(Y \varrho_N) = Y_L X \varrho_N = Y_L X_L \varrho_N$$

and

$$Y(X \varrho_N) = X_L Y \varrho_N = X_L Y_L \varrho_N,$$

so (1. 26) must be satisfied.

The operator **P** representing the total momentum commutes with that (H) representing the Hamiltonian energy; so, also, does the operator **A** representing the total angular momentum. Such operators, which commute with H, represent what are known as constants of the motion. As an example of two operators which do *not* commute, one may cite the components p_1 and x_1, of the momentum and position vector of a molecule respectively in the same direction. One has in fact

$$x_1(p_1 \varrho_N) = - i\hbar\, x_1 \frac{\partial \varrho_N}{\partial x_1},$$

$$p_1(x_1 \varrho_N) = - i\hbar \left(x_1 \frac{\partial \varrho_N}{\partial x_1} + \varrho_N \right),$$

so one may write

(1.27) $$x_1 p_1 - p_1 x_1 = i\hbar$$

meaning

$$(x_1 p_1 - p_1 x_1)\varrho_N = i\hbar \varrho_N.$$

The relation (1.27) was first obtained by Heisenberg, Born and Jordan. It shows that it must be impossible to measure simultaneously the position **x** and velocity **p**/m of the same molecule. As a second example, although the components of the total angular momentum vector **A** commute with H, they do not commute with one another; it is, in fact, readily verified that

$$A_2 A_3 - A_3 A_2 = i\hbar\, A_1$$
$$A_3 A_1 - A_1 A_3 = i\hbar\, A_2$$
$$A_1 A_2 - A_2 A_1 = i\hbar\, A_3$$

However, any component of **A** commutes with the square of the resultant vector:

$$A_1(A_1^2 + A_2^2 + A_3^2) = (A_1^2 + A_2^2 + A_3^2)A_1;$$

it is therefore possible in principle to measure simultaneously the resultant angular momentum, and its component in a given direction.

A fundamental problem of quantum mechanics is to determine a complete set of operators which are constants of the motion, i.e., commute with H, and also commute with one another. For a system of two molecules, H, P_1, P_2, P_3, A_1 and $(A_1^2 + A_2^2 + A_3^2)$ constitute such a complete set. When there are more than two molecules, however, it is in general necessary to supplement these by other, more complicated operators, in order to obtain a complete set. The importance of such complete commuting sets of operators is that they determine a complete set of states of the system, namely, those states in which H, P_1, P_2, P_3, A_1 etc. can, in principle, all be measured without changing the state of the system. The density matrix for the system in such a state (L) is given by

(1.28) $$\varrho_N(L; \mathbf{x}', \mathbf{x}'') = \Psi(L, \mathbf{x}')\,\Psi^*(L, \mathbf{x}''),$$

where

$$H\Psi(L, \mathbf{x}) = H_L \Psi(L, \mathbf{x}),$$
$$P_1 \Psi(L, \mathbf{x}) = (P_1)_L \Psi(L, \mathbf{x}), \quad \text{etc.,}$$

and $\Psi^*(L, \mathbf{x})$ represents the complex conjugate of $\Psi(L, \mathbf{x})$. The functions $\Psi(L, \mathbf{x})$ constitute what is known as an *orthogonal* system of functions, satisfying the relation

(1.29) $\quad \int \underset{}{\overset{(N)}{\cdots}} \int \Psi(L', \mathbf{x}) \Psi^*(L'', \mathbf{x}) \, d\mathbf{x}^{(1)} \ldots d\mathbf{x}^{(N)} = N! \, \delta(L', L'')$,

where $\delta(L', L'')$ vanishes when L' and L'' represent different states. It should be remarked that the density matrix as defined by (1.28) corresponds only to a state L, in which H, P_1, P_2 etc. all have definite numerical values H_L, $(P_1)_L$, $(P_2)_L$ etc. In general, the density matrix has to be represented as a quadratic function

(1.30) $\quad \varrho_N(\mathbf{x}', \mathbf{x}'') = \sum_{L'L''} \varrho(L', L'') \Psi(L', \mathbf{x}') \Psi^*(L'', \mathbf{x}'')$

of the state functions $\Psi(L, \mathbf{x})$. In that case, the diagonal element $\varrho(L, L)$ of the matrix $\varrho(L', L'')$ represents the *probability* of finding the system in the state L.

1.3 Transition Probabilities

The equation (1.12) for the density matrix is correct on the assumption that the interaction of the system of molecules with external systems can be neglected. For some purposes, it is necessary to take into account the influence of the environment of the fluid, which may consist of a vapour, or the walls of the containing vessel. Then (1.12) must be replaced by

(1.31) $\quad \dfrac{\partial \varrho_N}{\partial t} + \dfrac{i}{\hbar} \{(H' + V') \varrho_N - (H'' + V'') \varrho_N\} = 0$,

where $V(\mathbf{x})$ is the potential energy of the fluid due to its interaction with the environment, and $V' = V(\mathbf{x}')$, $V'' = V(\mathbf{x}'')$. If (1.30) is substituted in (1.31), and

$$H' \Psi(L', \mathbf{x}') = H'_L \Psi(L', \mathbf{x}'),$$
$$H'' \Psi^*(L'', \mathbf{x}'') = H''_L \Psi^*(L'', \mathbf{x}''),$$

then one obtains

$$\sum_{L'L''} \left\{ \frac{\hbar}{i} \frac{\partial \varrho(L', L'')}{\partial t} + (H'_L - H''_L) \varrho(L', L'') + (V' - V'') \varrho(L', L'') \right\} \times$$
$$\times \Psi(L', \mathbf{x}') \Psi^*(L'', \mathbf{x}'') = 0,$$

where H'_L and H''_L are now mere numbers. Multiplying this result by $\Psi^*(L', \mathbf{x}') \Psi(L'', \mathbf{x}'')$, and integrating with respect to the coordinates

$\mathbf{x}^{(i)\prime}$ and $\mathbf{x}^{(i)\prime\prime}$, one has further, with the help of (1.29),

(1.32) $\quad \begin{cases} \dfrac{\hbar}{i} \dfrac{\partial \varrho(L', L'')}{\partial t} + (H'_L - H''_L) \varrho(L', L'') \\ \qquad + \sum\limits_L \{V(L', L) \varrho(L, L'') - \varrho(L', L) V(L, L'')\} = 0, \end{cases}$

where

(1.33) $\quad \begin{cases} N! \, V(L', L'') = \int^{(N)} \int \Psi^*(L', \mathbf{x}) \, V(\mathbf{x}) \, \Psi(L'', \mathbf{x}) \, d\mathbf{x}^{(1)} \ldots d\mathbf{x}^{(N)}, \\ V(L', L'') = V^*(L'', L'), \end{cases}$

It will be supposed that, at some initial time t_0, $\varrho(L', L'')$ has the value $\varrho_0(L', L'')$. The solution of (1.32) can then be written in the form

(1.34) $\quad \varrho(L', L'') = \sum\limits_{L_1 L_2} U(L', L_1) \, U^*(L'', L_2) \, \varrho_0(L_1, L_2)$

provided

(1.35) $\quad \begin{cases} \dfrac{\hbar}{i} \dfrac{\partial U(L', L_1)}{\partial t} + H'_L \, U(L', L_1) + \sum\limits_L V(L', L) \, U(L, L_1) = 0, \\ -\dfrac{\hbar}{i} \dfrac{\partial U^*(L'', L_2)}{\partial t} + H''_L \, U^*(L'', L_2) + \sum\limits_L V^*(L'', L) \, U^*(L, L_2) = 0. \end{cases}$

It follows from (1.35) and (1.33) that

(1.36) $\quad \dfrac{\partial}{\partial t} \{\sum\limits_L U^*(L, L_2) \, U(L, L_1)\} = 0.$

Also, since $\varrho(L', L'')$ must reduce to $\varrho_0(L', L'')$ at time t_0, (1.34) requires that, *at time* t_0,

$$U(L', L_1) = \delta(L', L_1) \quad U(L'', L_2) = \delta(L'', L_2).$$

As a consequence,

(1.37) $\quad \sum\limits_L U^*(L, L_2) \, U(L, L_1) = \delta(L_2, L_1)$

holds at time t_0; and, according to (1.36), the same must be true for all time. By virtue of the relation (1.37), is called a *unitary* matrix.

Suppose now that the system is definitely in the state L at time t_0; then

$$\varrho_0(L_1, L_2) = \delta(L, L_1) \, \delta(L, L_2).$$

According to (1.34), therefore,

$$\varrho(L', L'') = U(L', L) \, U^*(L'', L),$$

and the probability of finding the system in the state L_1 at time t must be

$$\varrho(L_1, L_1) = U(L_1, L) \, U^*(L_1, L)$$

Hence the probability that the system will make a transition from the state L to the state L_1 between time t_0 and time t is

(1. 38) $$P(L, L_1) = |U(L_1, L)|^2$$

It follows also from the unitary condition (1. 37) that

$$\sum_{L_1} P(L, L_1) = \sum_{L_1} |U(L_1, L)|^2 = 1,$$

as required by the theory of probability.

As indicated by the equation (1. 33), the matrix $V(L', L'')$ has the property

$$V(L', L'') = V^*(L'', L')$$

and is, in consequence, called *hermitian*. It can be seen from the equations (1. 35), which define $U(L, L_1)$, that U also has the hermitian property. Therefore

(1. 39) $$P(L, L_1) = |U(L, L_1)|^2 = P(L_1, L)$$

The probability that the system will, in the time $t - t_0$, make a transition between the states L and L_1, is therefore the same as the probability that it will make a transition between the states L_1 and L.

2. Equilibrium in the Quantum Theory

The derivation of Boltzmann's law given in §§ 5. 1 of chapter II is quite unaffected by the presupposition that quantum mechanics, rather than classical mechanics, must be employed. Indeed, in quantum theory the proof is rather more straightforward, inasmuch as the probability of a transition from one state to another is easily derived from the considerations advanced in the previous section. One may therefore conclude immediately that, in thermal and mechanical equilibrium, the probability of finding the system in a state (L) of energy $E = H_L$ must be $a \exp(-\beta E)$. It also follows, without change of argument, from §§ 5. 2 of chapter II that $\beta = 1/(kT)$, where T is the temperature thermodynamically defined, and $a = \exp \beta F$, where F is the total free energy of the fluid. On the other hand, the determination of the number of states $[g(E)]$ with energy E is a matter of some difficulty in quantum theory; it cannot be obtained directly, except by the determination of a complete commuting set of constants of the motion, and the various numerical values which they may assume. Indirect methods, however, enable one to avoid this difficult and tedious process.

One makes use of the formula

$$(2.1) \qquad \sum_L \Psi(L, \mathbf{x}') \Psi^*(L, \mathbf{x}'') = \delta_N(\mathbf{x}' - \mathbf{x}''),$$

where $\delta_N(\mathbf{x}' - \mathbf{x}'')$ denotes the δ-function which vanishes when any of the vectors $\mathbf{x}^{(i)'} - \mathbf{x}^{(i)''}$ differs appreciably from zero, but nevertheless satisfies

$$\int \overset{(N)}{\cdots} \int \delta_N(\mathbf{x}' - \mathbf{x}'') \, d\mathbf{x}^{(1)'} \ldots d\mathbf{x}^{(N)'} = 1.$$

The formula (2. 1) can be verified by multiplying with $\Psi^*(L, \mathbf{x}')$ and integrating with respect to the coordinates $\mathbf{x}^{(i)'}$; the result is an identity, by virtue of (1. 29).

Now, if one applies the operator

$$\exp(-\beta H') = 1 - \beta H' + \tfrac{1}{2}(\beta H')^2 + \ldots$$

to (2. 1), one obtains

$$(2.2) \qquad \sum_L \exp(-\beta H_L) \, \Psi(L, \mathbf{x}') \, \Psi^*(L, \mathbf{x}'') = \exp(-\beta H') \, \delta_N(\mathbf{x}' - \mathbf{x}'').$$

Thus, putting $\mathbf{x}' = \mathbf{x}'' = \mathbf{x}$, and integrating with respect to the coordinates $\mathbf{x}^{(i)}$, one has with the help of (1. 29)

$$(2.3) \quad N! \sum_L \exp(-\beta H_L) = \int \overset{(N)}{\cdots} \int [\exp(-\beta H') \, \delta_N(\mathbf{x}' - \mathbf{x}'')]_{\mathbf{x}' = \mathbf{x}''} \, d\mathbf{x}^{(1)} \ldots d\mathbf{x}^{(N)}.$$

The left-hand side is $\sum_E g(E) \exp(-\beta E)$, and by expressing the right-hand side as a function of β in this form, the coefficients $g(E)$ might in principle be determined. Such a procedure is, however, unnecessary for the present purpose.

Since

$$\sum_L \exp \beta(F - H_L) = 1,$$

it follows from (2. 3) that the free energy F is given by

$$(2.4) \quad \exp(-\beta F) = \frac{1}{N!} \int \overset{(N)}{\cdots} \int [\exp(-\beta H') \, \delta_N(\mathbf{x}' - \mathbf{x}'')]_{\mathbf{x}' = \mathbf{x}''} \, d\mathbf{x}^{(1)} \ldots d\mathbf{x}^{(N)}.$$

Further, as $\exp \beta(F - H_N)$ is the probability, in equilibrium, of finding the fluid in the state L, a comparison of (2. 2) with (1. 30) shows that the density matrix must be

(2. 5) $$\varrho_N^0(\mathbf{x}', \mathbf{x}'') = \exp \beta(F - H') \, \delta_N(\mathbf{x}' - \mathbf{x}'')$$

in equilibrium. This is further confirmed by observing that, since

$$H'\delta_N(\mathbf{x}' - \mathbf{x}'') = H''\delta_N(\mathbf{x}' - \mathbf{x}''),$$

the right-hand side of (2. 5) satisfies the equation (1. 12) for the density matrix.

The equations (2. 4) and (2. 5) are strictly correct only when the interaction of the fluid with its environment is neglected. When this interaction is taken into account, the density matrix satisfies (1. 31) instead of (1. 12), so the operator H' has to be replaced by $H' + V'$ in (2. 4) and (2. 5).

In the quantum theory of fluids in equilibrium, the fundamental problem is the evaluation of the free energy as a function of the number density (n) and the temperature (T). It is very difficult to obtain a useful result which is valid over the whole temperature scale, but approximations can be obtained valid for all high temperatures, and moderately low temperatures, by developing the phase integral (2. 4) as a power series in $\hbar^2$. Even so, the multiple integrals can be evaluated exactly only under the same conditions as the classical phase integral, at sufficiently low densities. The appropriate calculations have been made by UHLENBECK & BETH [1936—37] and DE BOER [1940] among others. An alternative method, however, was suggested by WIGNER [1932], who attempted to determine the velocity distribution function which satisfies the equation (1. 18). This method has many advantages, including the fact that it may be modified to give results valid near absolute zero.

First the development as a power series in $\hbar^2$ will be examined. This can be obtained most easily by the solution of the equation

(2. 6) $$\left\{ \sum_i \left(\boldsymbol{\xi}^{(i)} \cdot \frac{\partial f_N}{\partial \mathbf{x}^{(i)}} - \frac{1}{m} \frac{\partial \Phi}{\partial \mathbf{x}^{(i)}} \cdot \frac{\partial f_N}{\partial \boldsymbol{\xi}^{(i)}} \right) + \right.$$
$$\left. + \frac{1}{3!} \frac{(\tfrac{1}{2}\hbar)^2}{m^3} \sum_{i,j,k} \frac{\partial^3 \Phi}{\partial \mathbf{x}^{(i)} \, \partial \mathbf{x}^{(j)} \, \partial \mathbf{x}^{(k)}} \vdots \frac{\partial^3 f_N}{\partial \boldsymbol{\xi}^{(i)} \, \partial \boldsymbol{\xi}^{(j)} \, \partial \boldsymbol{\xi}^{(k)}} + \ldots = 0,$$

to which (1. 18) reduces in equilibrium, when the integral expression is replaced by (1. 22). Since the solution must reduce to the classical solution when $\hbar = 0$, it may be assumed to be of the form

(2.7) $\quad f_N = \exp \beta (F - \Phi - \sum_i \tfrac{1}{2} m \boldsymbol{\xi}^{(i)2}) \cdot (1 + C_1 \hbar^2 + C_2 \hbar^4 + \ldots)$,

where C_1, C_2 etc. are functions of the $\mathbf{x}^{(i)}$ and $\boldsymbol{\xi}^{(i)}$, but do not involve $\hbar$. If this is substituted into (2.6), and the coefficient of $\hbar^2$ selected from each term, the following equation is obtained for C_1:

$$(2.8) \quad \begin{cases} \sum_i \left(\boldsymbol{\xi}^{(i)} \cdot \dfrac{\partial C_1}{\partial \mathbf{x}^{(i)}} - \dfrac{1}{m} \dfrac{\partial \Phi}{\partial \mathbf{x}^{(i)}} \cdot \dfrac{\partial C_1}{\partial \boldsymbol{\xi}^{(i)}} \right) = \\ = \dfrac{\beta^3}{24} \left(\sum_i \boldsymbol{\xi}^{(i)} \cdot \dfrac{\partial}{\partial \mathbf{x}^{(i)}} \right)^3 \Phi - \dfrac{\beta^2}{8m} \left(\sum_i \boldsymbol{\xi}^{(i)} \cdot \dfrac{\partial}{\partial \mathbf{x}^{(i)}} \right) \sum_j \dfrac{\partial^2 \Phi}{\partial \mathbf{x}^{(j)} \cdot \partial \mathbf{x}^{(j)}}. \end{cases}$$

The solution of this equation is obviously

$$(2.9) \quad C_1 = \dfrac{\beta^3}{24} \left(\sum_i \boldsymbol{\xi}^{(i)} \cdot \dfrac{\partial}{\partial \mathbf{x}^{(i)}} \right)^2 \Phi - \dfrac{\beta^2}{8m} \sum_i \dfrac{\partial^2 \Phi}{\partial \mathbf{x}^{(i)} \cdot \partial \mathbf{x}^{(i)}} + \dfrac{\beta^3}{24m} \sum_i \dfrac{\partial \Phi}{\partial \mathbf{x}^{(i)}} \cdot \dfrac{\partial \Phi}{\partial \mathbf{x}^{(i)}}.$$

Similar, but rather more complicated expressions can be obtained in the same way for C_2, C_3 etc.

By integrating (2.7) with respect to the velocities $\boldsymbol{\xi}^{(i)}$, one obtains

$$(2.10) \quad \begin{cases} n_N = \lambda^N \exp \beta (F - \Phi) \left\{ 1 + \dfrac{\hbar^2}{m} \left(\dfrac{\beta^3}{24} \sum_i \dfrac{\partial \Phi}{\partial \mathbf{x}^{(i)}} \cdot \dfrac{\partial \Phi}{\partial \mathbf{x}^{(i)}} - \right. \right. \\ \left. \left. - \dfrac{\beta^2}{12} \sum_i \dfrac{\partial^2 \Phi}{\partial \mathbf{x}^{(i)} \cdot \partial \mathbf{x}^{(i)}} \right) + \ldots \right\}, \quad \lambda = (2\pi/\beta m)^{3/2}. \end{cases}$$

This is the molecular distribution function appropriate to the entire molecular assembly. By further integration of (2.10) over the coordinates $\mathbf{x}^{(i)}$, one has

$$\exp(-\beta F) = \exp(-\beta F^c) \left[1 + \dfrac{\hbar^2 \beta^2 V}{24 m} \left\{ \int n_2^c \left(\beta \dfrac{\partial \phi}{\partial \mathbf{r}} \cdot \dfrac{\partial \phi}{\partial \mathbf{r}} - 2 \dfrac{\partial^2 \phi}{\partial \mathbf{r} \cdot \partial \mathbf{r}} \right) d\mathbf{r} \right. \right.$$
$$\left. \left. + \iint n_3^c \dfrac{\partial \phi^{(12)}}{\partial \mathbf{x}^{(1)}} \cdot \dfrac{\partial \phi^{(13)}}{\partial \mathbf{x}^{(1)}} d\mathbf{x}^{(2)} d\mathbf{x}^{(3)} \right\} + \ldots \right],$$

where F^c is the classical free energy of the fluid, given by (6.3) of chapter II, and n_2^c is the classical molecular distribution function, given by (6.11) of chapter II. The classical distribution function n_3^c for a trio of molecules which appears in the last term can be eliminated by means of (6.4) of chapter III, and one then obtains

$$(2.11) \quad \begin{cases} F = F^c + \dfrac{\hbar^2 \beta V}{24 m} \int \left(\dfrac{\partial n_2^c}{\partial \mathbf{r}} \cdot \dfrac{\partial \phi}{\partial \mathbf{r}} + 2 n_2^c \dfrac{\partial^2 \phi}{\partial \mathbf{r} \cdot \partial \mathbf{r}} \right) d\mathbf{r} + \ldots \\ = F^c + \dfrac{\hbar^2 \beta V}{24 m} \int n_2^c \dfrac{\partial^2 \phi}{\partial \mathbf{r} \cdot \partial \mathbf{r}} d\mathbf{r} + \ldots. \end{cases}$$

The pressure is therefore

$$(2.12) \qquad p = p^c + \frac{\hbar^2 \beta}{24 m} \int \left(n \frac{\partial n_2^c}{\partial n} - n_2^c \right) \frac{\partial^2 \phi}{\partial \mathbf{r} \cdot \partial \mathbf{r}} \, d\mathbf{r} + \dots,$$

where p^c is the classical expression, and the internal energy is

$$(2.13) \qquad U = U^c + \frac{\hbar^2 \beta V}{24 m} \int \left(\beta \frac{\partial n_2^c}{\partial \beta} + 2 n_2^c \right) \frac{\partial^2 \phi}{\partial \mathbf{r} \cdot \partial \mathbf{r}} \, d\mathbf{r} + \dots.$$

Also, from (2.12) it can be seen that the quantum correction to the first virial coefficient is

$$(2.14) \qquad \tfrac{1}{2} \delta\beta_1 = \frac{\hbar^2 \beta}{24 m} \int \exp(-\beta\phi) \frac{\partial^2 \phi}{\partial \mathbf{r} \cdot \partial \mathbf{r}} \, d\mathbf{r} + \dots.$$

This expression is easily evaluated numerically; using a potential function of the form

$$\phi(r) = 4\varepsilon \{(\sigma/r)^{12} - (\sigma/r)^6\},$$

one obtains, in fact,

$$(2.15) \qquad \tfrac{1}{2} \delta\beta_1 = \frac{2\pi\sigma\hbar^2}{3m} \sum_{s=0}^{\infty} \frac{36s-11}{192 \cdot s!} \, \Gamma(\tfrac{1}{2}s - \tfrac{1}{12}) \, (4\varepsilon\beta)^{1/2 s + 1/12}.$$

The correction just obtained must be taken into account in determining the constants ε and σ from the empirical data concerning the first virial coefficient, as described in § 4 of chapter VII. It is, however, large enough to be important in gases only for hydrogen and helium. The molecular diameter (σ) in helium, computed from data for the first virial coefficient, is 2.56×10^{-8} cm when account is taken of the quantum correction, against 2.60×10^{-8} cm in the classical theory; the constant ε in (2.15) is found to have the value 14.03×10^{-16} erg.

In the classical theory, the mean kinetic energy of a molecule is $\tfrac{3}{2}kT$; the same, however, is no longer true in the quantum theory. To obtain the deviation at moderately low temperatures, one may proceed first from the formula (2.7), and determine the velocity distribution function f_1. If one integrates over all positions and velocities except $\boldsymbol{\xi}^{(1)}$ and $\mathbf{x}^{(1)}$ one obtains

$$(N-1)! f_1 = \lambda^{-1} \exp(-\tfrac{1}{2} m \boldsymbol{\xi}^{(1)2}) \int \overset{(N-1)}{\dots} \int n_N \bigg[1 + \frac{\hbar^2 \beta^3}{24} \left\{ \left(\boldsymbol{\xi}^{(1)} \cdot \frac{\partial}{\partial \mathbf{x}^{(1)}} \right)^2 \Phi - \frac{1}{m\beta} \frac{\partial^2 \Phi}{\partial \mathbf{x}^{(1)} \cdot \partial \mathbf{x}^{(1)}} \right\} + \dots \bigg] d\mathbf{x}^{(2)} \dots d\mathbf{x}^{(N)},$$

which reduces to

16)
$$\begin{cases} f_1 = \lambda^{-1} \exp\left(-\tfrac{1}{2} m \boldsymbol{\xi}^{(1)2}\right) \left[n + \frac{\hbar^2 \beta^3}{24} \int n_2 \left\{ \left(\boldsymbol{\xi}^{(1)} \cdot \frac{\partial}{\partial \mathbf{r}}\right) \phi - \frac{1}{m\beta} \frac{\partial^2 \phi}{\partial \mathbf{r} \cdot \partial \mathbf{r}} \right\} d\mathbf{r} + \dots \right] \\ = \lambda^{-1} \exp\left(-\tfrac{1}{2} m \boldsymbol{\xi}^{(1)2}\right) \left[n + \frac{\hbar^2 \beta^3}{24 m} \left(\tfrac{1}{3} m\beta \boldsymbol{\xi}^{(1)2} - 1\right) \int n_2 \frac{\partial^2 \phi}{\partial \mathbf{r} \cdot \partial \mathbf{r}} d\mathbf{r} + \dots \right]. \end{cases}$$

The mean kinetic energy of a molecule derived from this formula is

(2. 17) $\tfrac{3}{2} kT_1 = \overline{\tfrac{1}{2} m \boldsymbol{\xi}^{(1)2}} = \tfrac{3}{2} kT \left\{ 1 + \frac{\hbar^2 \beta^2}{36 mn} \int n_2 \frac{\partial^2 \phi}{\partial \mathbf{r} \cdot \partial \mathbf{r}} d\mathbf{r} + \dots \right\}.$

It will be noticed that, at sufficiently low temperatures, the quantum corrections in the formulae (2. 11)—(2. 14) and (2. 17) all become very large. This feature is reproduced to an even greater degree in the corrections proportional to $\hbar^4$, $\hbar^6$ etc., which have not been evaluated here. It is, indeed, fairly certain that the series in powers of $\hbar^2$ become divergent below a certain temperature T_λ. For very low temperatures, the above method therefore fails, and a different approximation procedure is required.

2.1. Very low Temperatures

The quantum theory of fluids at very low temperatures has necessarily a somewhat limited application, because nearly all substances assume a crystalline state before characteristically quantum phenomena can appear. Even hydrogen, which liquefies at 20.4° K. under its own vapour pressure, crystallizes at 14° K. Helium, on the other hand, liquefies at 4.2° K., and remains fluid down to absolute zero. It is, therefore, the one fluid substance in which the macroscopic consequences of the quantum theory can be studied experimentally down to the lowest temperatures. Here, indeed, entirely novel and unexpected fluid phenomena have been observed. At 2.19° K., under the vapour pressure, nearly all macroscopic properties of liquid helium undergo a sudden change. Above this temperature, which is known as the λ-point, liquid helium exhibits properties not qualitatively different from those observed in other fluids. Below the λ-point, on the other hand, a remarkable frictionless mode of flow is exhibited under suitable conditions, on account of which the liquid is called He II, to distinguish it from the normal phase (He I) which is found above the transition temperature. The characteristics of the superfluid He II are an effectively complete lack of viscosity and an almost unlimited capacity to conduct heat. These properties are most evident in very narrow channels, such as slits and capillaries, and thin films; they are found only in a modified form in the liquid in bulk.

Other macroscopic properties of liquid helium exhibit abnormal characteristics in the neighbourhood of the λ-point. The specific heat, for example, rises from very small values near absolute zero to a maximum of about 6 cal/gm °K. at the λ-point, where it suffers an almost discontinuous change, as shown in figure 1.

The rates of change with temperature of the vapour pressure and the density also suffer a discontinuity at the λ-point, indicating a singularity in the equation of state.

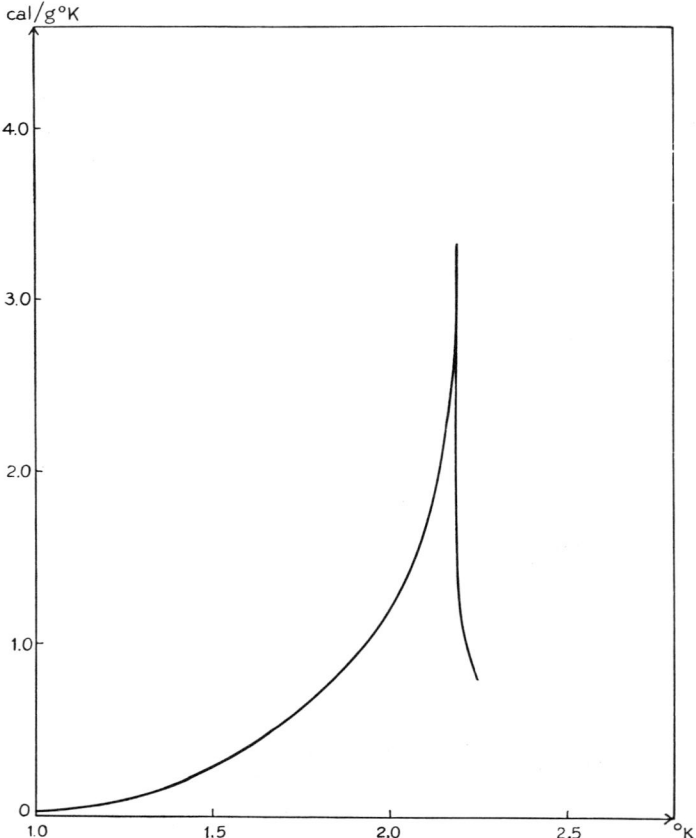

Fig. 1. Specific heat of liquid helium under the saturated vapour pressure.

These phenomena are generally attributed to a quantum 'condensation'. Such a description should not be misinterpreted as implying the occurrence of a condensation in the normal sense of the word, as there is no discontinuous change of density or internal energy in passing from He I to He II. The closest analogy with ordinary condensation can be seen by comparing the divergence of the series expansion of the pressure in powers of $\hbar^2$ at the λ-point with the divergence of the series in powers of the density which occurs in the normal condensation region.

It is evidently not practicable to use the power series in $\hbar^2$ for the theoretical study of liquid helium. An alternative procedure which suggests itself is to make an expansion in powers of the force constant [denoted by ε in (2. 15)] which is a factor in the potential energy function $\phi(r)$. The fact that this constant has an unusually small value for helium recommends such a procedure as more likely to give quickly convergent results than for most substances. There is, moreover, no difficulty in obtaining the required expansion from the fundamental equation (1. 18), with f_N independent of time. A result of the form

$$(2.18) \qquad f_N = \exp\left(F - \sum_i \tfrac{1}{2} m \boldsymbol{\xi}^{(i)2}\right) \cdot (1 + D_1 + D_2 + \ldots)$$

is to be expected, where D_1 involves Φ linearly, whilst D_2 involves Φ quadratically, and so on.

The function $X(\boldsymbol{\xi})$ defined by (1. 17) can be expressed in terms of

$$(2.19) \qquad \chi(\tfrac{1}{2}\boldsymbol{\rho}) = \left(\frac{m}{2\pi\hbar}\right)^3 \int \phi(\tfrac{1}{2}r) \cos(\tfrac{1}{2} m \mathbf{r} \cdot \boldsymbol{\rho}/\hbar) \, d\mathbf{r}$$

and the singular δ-function

$$(2.20) \qquad \delta(\boldsymbol{\xi}) = \left(\frac{m}{2\pi\hbar}\right)^3 \int \cos(m\boldsymbol{\xi} \cdot \mathbf{x}/\hbar) \, d\mathbf{x}.$$

One has in fact

$$(2.21) \qquad X(\boldsymbol{\xi}) = \sum_{j>i} \{\chi(\tfrac{1}{2}\boldsymbol{\xi}^{(i)} - \tfrac{1}{2}\boldsymbol{\xi}^{(j)}) \, \delta(\boldsymbol{\xi}^{(i)} + \boldsymbol{\xi}^{(j)}) \prod_{k \neq i \text{ or } j} \delta(\boldsymbol{\xi}^{(k)})\}.$$

Now, if $F(\boldsymbol{\xi})$ is any function which tends to zero for large values of $\boldsymbol{\xi}$, the δ-function has the property

$$\int F(\boldsymbol{\xi}) \, \delta(\boldsymbol{\xi} - \boldsymbol{\xi}_0) \, d\boldsymbol{\xi} = F(\boldsymbol{\xi}_0)$$

--- which, in conjunction with (2. 20), is actually a statement of

Fourier's theorem. Introducing (2. 21) into (1. 18), one has, therefore,

$$(2.22) \quad \sum_i \boldsymbol{\xi}^{(i)} \cdot \frac{\partial f_N}{\partial \mathbf{x}^{(i)}} + \frac{2}{\hbar} \sum_{j>i} \int \chi(\boldsymbol{\xi}_0) \sin\{2m\boldsymbol{\xi}_0 \cdot (\mathbf{x}^{(j)} - \mathbf{x}^{(i)})/\hbar\} f_N(\boldsymbol{\xi} + a_{ij}\boldsymbol{\xi}_0) d\boldsymbol{\xi}_0 = 0,$$

where $\boldsymbol{\xi} + a_{ij}\boldsymbol{\xi}_0$ stands for the variables $\boldsymbol{\xi}^{(i)} - \boldsymbol{\xi}_0$, $\boldsymbol{\xi}^{(j)} + \boldsymbol{\xi}_0$, and the $\boldsymbol{\xi}^{(k)}$ with $k \neq i$ or j. On substituting in (2. 22) from (2. 18), one obtains, for the determination of D_1,

$$\sum_i \boldsymbol{\xi}^{(i)} \cdot \frac{\partial D_1}{\partial \mathbf{x}^{(i)}} + \frac{2}{\hbar} \sum_{j>i} \int \chi(\boldsymbol{\xi}_0) \sin(2m\boldsymbol{\xi}_0 \cdot \mathbf{r}^{(ij)}/\hbar) \exp\{-\beta m(\boldsymbol{\xi}_0^2 + \boldsymbol{\xi}_0 \cdot \boldsymbol{\rho}^{(ij)})\} d\boldsymbol{\xi}_0,$$

(2. 23) $\quad\quad\quad \mathbf{r}^{(ij)} = \mathbf{x}^{(j)} - \mathbf{x}^{(i)}, \quad\quad \boldsymbol{\rho}^{(ij)} = \boldsymbol{\xi}^{(j)} - \boldsymbol{\xi}^{(i)}.$

This has the solution

$$(2.24) \quad D_1 = -\sum_{j>i} \int \chi(\boldsymbol{\xi}_0) \cos(2m\boldsymbol{\xi}_0 \cdot \mathbf{r}^{(ij)}/\hbar) \exp(-\beta m\boldsymbol{\xi}_0^2) \frac{\sinh(\beta m\boldsymbol{\xi}_0 \cdot \boldsymbol{\rho}^{(ij)})}{m\boldsymbol{\xi}_0 \cdot \boldsymbol{\rho}^{(ij)}} d\boldsymbol{\xi}_0.$$

Expressions for D_2, D_3 etc. can be obtained by a continuation of this procedure.

If one substitutes (2. 24) into (2. 18) and integrates over all velocities, one obtains for n_N,

$$(2.25) \quad n_N = \lambda^N \exp(\beta F) \{1 - \beta \sum_{j>i} \phi^+(r^{(ij)})\},$$

where

$$(2.26) \quad \begin{cases} \phi^+(r) = \left(\frac{m\beta}{4\pi}\right)^{3/2} \int \chi(\boldsymbol{\xi}) \cos(2m\boldsymbol{\xi} \cdot \mathbf{r}/\hbar) \exp(-\beta m \boldsymbol{\xi}^2) \times \\ \quad\quad\quad \times \int \frac{\sinh(\beta m \boldsymbol{\xi} \cdot \boldsymbol{\rho})}{\beta m \boldsymbol{\xi} \cdot \boldsymbol{\rho}} \exp(-\tfrac{1}{4}\beta m \boldsymbol{\rho}^2) d\boldsymbol{\rho} \, d\boldsymbol{\xi} \\ = \int \chi(\boldsymbol{\xi}) \cos(2m\boldsymbol{\xi} \cdot \mathbf{r}/\hbar) \int_0^1 \exp \beta m \boldsymbol{\xi}^2 (\sigma^2 - 1) d\sigma \, d\boldsymbol{\xi}. \end{cases}$$

Taking account of the fact that, at very low temperatures, β is very large, one may use the asymptotic formula

$$(2.27) \quad \int_0^1 \exp \beta m \boldsymbol{\xi}^2 (\sigma^2 - 1) d\sigma = (2\beta m \boldsymbol{\xi}^2)^{-1} + (2\beta m \boldsymbol{\xi}^2)^{-2} + \ldots.$$

The function $\phi^+(r)$ therefore satisfies

$$(2.28) \quad \begin{cases} \frac{\hbar^2 \beta}{2m} \frac{\partial^2 \phi^+(r)}{\partial \mathbf{r} \cdot \partial \mathbf{r}} = \int \chi(\boldsymbol{\xi})(\cos(2m\boldsymbol{\xi} \cdot \mathbf{r}/\hbar) \{1 + (2\beta m \boldsymbol{\xi}^2)^{-1} + \ldots\} d\boldsymbol{\xi} \\ = \phi(r) + O(\beta^{-1}). \end{cases}$$

It will be noticed that (2. 25) is the same expression as one would obtain classically by assuming the potential energy between two molecules at distance r was $\phi^+(r)$. The validity of the expansion

(2. 29) $$\begin{cases} \exp(-\beta\Phi^+) = 1 - \beta\Phi^+, \\ \Phi^+ = \sum_{j>i} \phi^+(r^{(ij)}) \end{cases}$$

is, of course, conditional on $\beta\phi^+(r)$ being small, but it is clear from (2. 28) that this condition is actually satisfied for all except very small distances r between the molecules, provided

$$\beta\hbar^2 > 2m\sigma^2,$$

i.e., $T < 2°$ K approximately in helium. The validity of the theory being developed is therefore confined to the temperature range below the λ-point.

It has thus been proved that the molecular distribution functions are the same as one would obtain classically on the hypothesis that the interaction between the molecules is very small, except when they are at such very close distances that the electron shells overlap.

From (2. 25) and (2. 29) one obtains

(2. 30) $$N! \exp(-\beta F) = \lambda^N \int^{(N)} \int \exp(-\beta\Phi^+)\, d\mathbf{x}^{(1)} \ldots d\mathbf{x}^{(N)}.$$

The equation of state derived from this formula should be similar to the classical equation of state for a 'fluid' consisting of rigid spherical molecules; it is

$$p = nkT - \tfrac{1}{6} \int n_2(r)\phi^{+\prime}(r)r d\mathbf{r};$$

smilarly the internal energy per molecule is

$$U_1 = \tfrac{3}{2} kT + \tfrac{1}{2} n^{-1} \int n_2(r)\phi^+(r)d\mathbf{r}.$$

The mean kinetic energy of a molecule is most easily computed by finding the average value of $\tfrac{1}{2}m\boldsymbol{\xi}^{(i)2}$, expressed where necessary in the form $\tfrac{1}{2}m(\boldsymbol{\xi}^{(ij)} - \tfrac{1}{2}\boldsymbol{\varrho}^{(ij)})^2$, in a fluid in which all the molecules have assigned positions:

(2. 31) $$\begin{cases} \int^{(N)} \int f_N \cdot \tfrac{1}{2} m\boldsymbol{\xi}^{(i)2} d\boldsymbol{\xi}^{(1)} \ldots d\boldsymbol{\xi}^{(N)} = n_N \big[\tfrac{3}{2} kT + \sum_j \iint \lambda^{-2} \exp\big\{-\tfrac{1}{2}\beta m(\boldsymbol{\xi}^{(i)2} + \boldsymbol{\xi}^{(j)2})\big\} \times \\ \qquad \times \Big\{\beta\phi^{+(ij)} - \int \chi(\boldsymbol{\xi}_0) \cos(2m\boldsymbol{\xi}_0 \cdot \mathbf{r}^{(ij)}/\hbar) \exp(-\beta m\boldsymbol{\xi}_0^2) \dfrac{\sinh \beta m\boldsymbol{\xi}_0 \cdot \boldsymbol{\varrho}^{(ij)}}{m\boldsymbol{\xi}_0 \boldsymbol{\varrho}^{(ij)}} d\boldsymbol{\xi}_0 \Big\} \times \\ \qquad \times \tfrac{1}{2} m\boldsymbol{\xi}^{(i)2} d\boldsymbol{\xi}^{(i)} d\boldsymbol{\xi}^{(j)} \big] \\ = n_N \big\{ \tfrac{3}{2} kT - \beta \sum_j \int \tfrac{1}{2} m\boldsymbol{\xi}^2 \chi(\boldsymbol{\xi}) \cos(2m\boldsymbol{\xi} \cdot \mathbf{r}^{(ij)}/\hbar) \times \\ \qquad\qquad \times \int_0^1 \sigma^2 \exp \beta m\boldsymbol{\xi}^2(\sigma^2 - 1)\, d\sigma\, d\boldsymbol{\xi} \end{cases}$$

The second term tends to the finite limit

$$-\tfrac{1}{4} n_N \sum_j \int \chi(\boldsymbol{\xi}) \cos 2\pi \boldsymbol{\xi} \cdot \mathbf{r}^{(ij)} \, d\boldsymbol{\xi} = -\tfrac{1}{4} n_N \sum_j \phi(\mathbf{r}^{(ij)})$$

as T approaches zero; hence the molecules will have a 'zero-point' energy of motion

$$-\tfrac{1}{4} n^{-1} \int n_2 \phi(r) d\mathbf{r}$$

equal to one half of the mean potential energy between the molecules, but with the positive sign. There are also other, probably smaller contributions to the zero-point energy from terms quadratic in the potential energy of interaction.

The existence of the zero-point energy has been invoked (cf. DAUNT & MENDELSSOHN [1946]) to explain the observed motion of He II in surface films and through narrow channels. It may indeed be regarded as certain that the energy of motion of the fluid in such thin layers is derived, directly or indirectly, from the zero-point energy; the task remains, however, of explaining why the latter should appear as macroscopic, rather than internal energy. Obviously the equilibrium theory alone is insufficient to throw light on this matter.

3. Quantum Statistics

The foregoing modification of the classical theory of fluids is not the only one required by the modern quantum theory of the elementary particles. Although the laws of quantum mechanics have been introduced and used in the previous section, the statistical laws employed were strictly classical. The need for *quantum statistics* arises from the fact that two particles which are in every respect identical with one another lose their identity in a way which is quite outside our normal macroscopic experience.

In classical theory, given a state of the fluid in which two identical molecules are situated at any two points, $\mathbf{x}^{(i)}$ and $\mathbf{x}^{(j)}$ respectively, a different state is obtained by interchanging their positions, though the observable properties of the fluid are unchanged. In quantum theory it cannot even be admitted that the two states differ in any way from one another, and every state function must therefore be unchanged when the coordinates $\mathbf{x}^{(i)}$ and $\mathbf{x}^{(j)}$ are interchanged, except for a possible change of sign. To formulate this principle, let $I(i, j)$ be the operator which interchanges $\mathbf{x}^{(i)}$ and $\mathbf{x}^{(j)}$; then one must have either

(3. 1) $$I(i,j)\Psi(L, \mathbf{x}) = + \Psi(L, \mathbf{x})$$

or

(3. 2) $$I(i,j)\Psi(L, \mathbf{x}) = - \Psi(L, \mathbf{x})$$

All the empirical data which have been discovered indicate that, if (3. 1) holds for one state of the system, it holds also for every state of the system; and similarly, if (3. 2) holds for one state of the system, it holds for every state. It has also been observed that if the two identical molecules have an even total number of neutrons in each of their nuclei, (3. 1) will apply; but if they each possess an odd total number of neutrons, (3. 2) will apply. For example, (3. 1) is applicable to atoms of the common isotope He^4 of helium, and the common isotope H^1 of hydrogen, but (3. 2) is applicable to the rare isotope He^3 of helium, and to atoms of the rare isotope H^2 (deuterium) of hydrogen. In general, particles for which (3. 1) is true are said to obey Bose statistics, while those for which (3. 2) is true are said to obey Fermi statistics.

All the properties of a fluid depend to some extent on whether its molecules are subject to Bose or Fermi statistics. The interaction between two molecules depends to an appreciable extent on the type of statistics which they satisfy, and will therefore be somewhat different, for example, in He^3 and He^4. Macroscopic properties arising from the average behaviour of large numbers of atoms may be expected to depend to an important extent on the statistics, particularly at low temperatures. As a matter of fact, the phenomena in He II below the λ-point have been found to be due to the He^4 atoms alone; the He^3 atoms, which occur as a very small percentage (0.00013 %) in ordinary helium, do not participate in the frictionless flow, at least down to temperatures below 1° K.

These facts lend support to an hypothesis due to LONDON, that the abnormal properties of He II arise from the circumstance that its molecules obey Bose statistics, and not classical or Fermi statistics. It can be shown that the specific heat of an assembly of non-interacting Bose particles must have a singularity, rather suggestive of that shown in figure 1, at a temperature of about 2.3° K when the molecular mass and density correspond to the values obtaining in liquid helium. On the other hand, the interaction between the molecules would generate a singularity, probably of the same kind, even if classical statistics were operative. The observed singularity is thus attributable to the resultant effect of the Bose statistics and the quantized interaction.

The singularity in the equation of state of the assembly of non-interacting Bose particles can be attributed to a Bose 'condensation' in which the particles begin to accumulate in the state of lowest energy. TISZA [1938, 1940, 1947] has made the occurrence of such a condensation the basis of a comprehensive theory of He II; the fluid is regarded as the mixture of a component with normal properties, and a second component with superfluid properties. The latter component is supposed to consist of the 'condensed' phase derived from atoms in the lowest state. LANDAU [1941, 1944], on the other hand, has proposed a theory which is similar to Tisza's in some respects, but which does not depend on the occurrence of the Bose condensation; it has features which lead to better agreement with experience than obtained from Tisza's theory below $1°$ K. It is fair to add, however, that these features could probably be incorporated into the 'condensation' theory.

The 'two-fluid' theories of Tisza and Landau will not be described in detail, because they are of a phenomenological nature, and have never been developed in a satisfactory way from the molecular theory of flow. It is, however, possible to draw certain conclusions from the experimental facts alone, and these will be enumerated.

(1) Bose statistics are somehow favourable to superfluid phenomena to a degree in which Fermi statistics are not. The question remains whether classical statistics would not also be favourable; it may be conjectured that, since classical statistics occupy a position intermediate between Bose and Fermi statistics, the application of classical statistics to the quantum theory of fluids should reproduce features similar to those derived by the application of Bose statistics, but in a lesser degree.

(2) The abnormal increase in the specific heat and internal energy of liquid helium between absolute zero and the λ-point indicates the existence of an extraordinary contribution to the entropy of the fluid. This extraordinary contribution is not carried by the superfluid in surface films and narrow slits and capillaries.

(3) Internal energy can be propagated through the fluid with a definite velocity, which increases with decreasing temperature from

zero at the λ-point, to an almost constant value between 1.5° K and 1° K, and to much higher values below 1° K.

These facts by themselves enable one to explain qualitatively and often quantitatively most of the experimental facts concerning liquid helium.

The effect of quantum statistics on the equilibrium distribution of a system of interacting molecules has never been fully investigated. The following can however be stated. If

$$\varrho_N^c(\mathbf{x}', \mathbf{x}'')$$

is the density matrix determined on the basis of classical statistics, the density matrix for particles obeying Bose statistics is

(3.3) $$\varrho_N(\mathbf{x}', \mathbf{x}'') = \sum P'(i_1 \ldots i_N)\, \varrho_N^c(\mathbf{x}', \mathbf{x}''),$$

where $P'(i_1 \ldots i_N)$ is the permutation operator which substitutes $\mathbf{x}(i_1)', \mathbf{x}(i_2)', \ldots \mathbf{x}(i_N)'$ for $\mathbf{x}^{(1)'}, \mathbf{x}^{(2)'}, \ldots \mathbf{x}^{(N)'}$ respectively in ϱ_N^c, and the summation Σ is over all the $N!$ different permutations of the numbers $1, 2, \ldots N$. Similarly the density matrix for particles obeying Fermi statistics is

(3.4) $$\varrho_N(\mathbf{x}', \mathbf{x}'') = \sum \pm P'(i_1 \ldots i_N)\, \varrho_N^c(\mathbf{x}', \mathbf{x}''),$$

where the positive sign is chosen when $i_1 \ldots i_N$ is an even permutation of the numbers $1 \ldots N$, and the negative sign otherwise.

The free energy F which appears in the formula (3.6) for ϱ_N^c is determined by the relation

(3.5) $$\begin{cases} \int \overset{(N)}{\ldots} \int [\sum P'(i_1 \ldots i_N)\, \varrho_N^c(\mathbf{x}', \mathbf{x}'')]_{\mathbf{x}'=\mathbf{x}''}\, d\mathbf{x}^{(1)} \ldots d\mathbf{x}^{(N)} = \\ \qquad = \int \overset{(N)}{\ldots} \int n_N\, d\mathbf{x}^{(1)} \ldots d\mathbf{x}^{(N)} \\ \qquad = N! \end{cases}$$

The derivation of the density matrix for a system of non-interacting particles in thermal equilibrium has been considered quite generally by Husimi [1940]. The velocity distribution function for such a system, according to classical statistics, is

$$\exp \beta(F - \sum_i \tfrac{1}{2} m \boldsymbol{\xi}^{(i)2}).$$

The corresponding density matrix, obtained with the help of the formula (1.8), is

(3.6) $$\varrho_N^c(\mathbf{x}', \mathbf{x}'') = \lambda^N \exp\{\beta F - \sum_i m(\mathbf{x}^{(i)'} - \mathbf{x}^{(i)''})^2 / (2\hbar^2\beta)\}.$$

Now consider the effect, on the right-hand side of (3. 6), of applying a typical permutation operator $P'(i_1 \ldots i_N)$. This will leave a certain number (k_1, say) of the coordinates $\mathbf{x}^{(i)\prime}$ unchanged, will interchange a certain number (k_2, say) of pairs $\mathbf{x}^{(i)\prime}$ and $\mathbf{x}^{(j)\prime}$, will induce a cyclic permutation of a certain number (k_3, say) of triads $\mathbf{x}^{(i)\prime}$, $\mathbf{x}^{(j)\prime}$ and $\mathbf{x}^{(k)\prime}$, and so on. Then the expression

$$[P'(i_1 \ldots i_N) \varrho_N^c(\mathbf{x}', \mathbf{x}'')]_{\mathbf{x}'=\mathbf{x}''}$$

obtained by applying the permutation operator and then setting all the relative coordinates $\mathbf{x}^{(i)\prime} - \mathbf{x}^{(i)\prime\prime}$ equal to zero, will consist of the factor $\lambda^N \exp \beta F$, together with k_2 factors of the type $\exp\{-2m(\mathbf{x}^{(i)} - \mathbf{x}^{(j)})^2/(2\hbar^2\beta)\}$, k_3 factors of the type

$$\exp\left[-m\{(\mathbf{x}^{(i)} - \mathbf{x}^{(j)})^2 + (\mathbf{x}^{(j)} - \mathbf{x}^{(k)})^2 + (\mathbf{x}^{(k)} - \mathbf{x}^{(i)})^2\}/(2\hbar^2\beta)\right]$$

and so on. Hence the contribution to the left-hand side of (3. 4) will be

(3. 7) $\quad \begin{cases} \lambda^N \exp(\beta F) \cdot V^{k_1} (V\mu/2^{3/2})^{k_2} (V\mu^2/3^{3/2})^{k_3} \ldots (V\mu^{l-1}/l^{3/2})^{k_l}, \\ \mu = (2\pi\hbar^2\beta/m)^{3/2}. \end{cases}$

The number of permutations of this type is, however,

$$\frac{N!}{k_2! k_3! \ldots k_l!} \cdot \frac{1}{2^{k_2} 3^{k_3} \ldots l^{k_l}},$$

where

$$k_1 + 2k_2 + \ldots + lk_l = N;$$

hence, with Bose statistics, (3. 4) reduces to

(3. 8) $\quad \exp(-\beta F) = \lambda^N \sum_{(\Sigma l k_l = N)} \frac{1}{k_1!} \left(\frac{V}{1^{3/2}}\right) \frac{1}{k_2!} \left(\frac{\mu V}{2^{3/2}}\right)^{k_2} \cdots \frac{1}{k_l!} \left(\frac{\mu^{l-1} V}{l^{3/2}}\right).$

The corresponding formula in Fermi statistics is obtained by changing the sign of μ.

Now, if (3. 8) is multiplied by $(z/\lambda\mu)^N$, and summed over all values of N, the result is

(3. 9) $\quad \begin{cases} \sum \exp\{-\beta F(N, V)\} (z/\lambda')^N = \exp V \sum b'_q z^q, \\ \lambda' = \lambda\mu, \quad b'_q = 1/(\mu q^{5/2}). \end{cases}$

This is the same formula as can be obtained from (3. 3) and (3. 4) of chapter IV by substituting λ' for λ and b'_q for b_q. One may therefore infer, by the argument leading to (3. 6) and (3. 12) of chapter IV, that

(3. 10)
$$\begin{cases} \beta p V + \tfrac{1}{2} \ln (2\pi n \zeta_T V/\beta) = V \sum b'_q z^q, \\ z = \lambda' \exp \beta \{ F_0(n) + n F'_0(n) \} \end{cases}$$

where p is the pressure, ζ_T the compressibility, and $F_0(n)$ the free energy per molecule expressed as a function of the density. By differentiating (3. 10) with respect to the density (n), keeping the volume V constant, one obtains

(3. 11) $$n + \frac{1}{V} \frac{d}{dn} \left(\frac{n \zeta_T}{2\beta} \right) = \sum q b'_q z^q,$$

which determines z as a function of n, for any given temperature.

Normally the second term on the left-hand side of (3. 11) can be neglected, since the volume V is arbitrarily large. There is then little difficulty in solving the resulting transcendental equation

(3. 12) $$\sum_{q=1}^{\infty} q^{-3/2} z^q = \mu n$$

to determine z in terms of n and β. The left-hand side of this equation, however, has a maximum value $\Sigma q^{-3/2} = 2.6124$ for $z = 1$; and, at sufficiently low temperatures, the right-hand side of (3. 12) may exceed this maximum. When this happens, no solution of (3. 8) can be found. The reason is that, in the low temperature region now considered, the term proportional to V^{-1} in (3. 11) may no longer be neglected. It has to be remembered, also, that, in the derivation of the formula (3. 8), terms proportional to V^{-1} were implicitly neglected. The best way to take this into account is to replace λ and λ' by $\lambda(1 - a/V)$ and $\lambda'(1 - a/V)$ respectively, where a is to be determined.

It may now be shown that, in the low temperature region where (3. 12) is inadequate, (3. 11) can be satisfied by expressing the activity in the form

(3. 13) $$z = 1 - a(n)/V$$

where $a(n)$ is some function of density (and temperature). For, since $a(n)/V$ is small, (3. 10) will give

$$\beta \{ F_1(n) + n F'_1(n) \} + \ln \lambda' = - a(n)/V,$$
$$(n \zeta_T)^{-1} = 2 F'_1(n) + n F''_1(n) = - a'(n)/(\beta V).$$

Thus, (3.11) leads to

$$n - \frac{1}{2}\frac{d}{dn}\left\{\frac{1}{a'(n)}\right\} = n_0 = \sum q^{-3/2}/\mu,$$
$$a'(n) = \{c^2 + (n-n_0)^2\}^{-1},$$
$$a(n) = c^{-1}\tan^{-1}(n-n_0)/c + c'.$$

where c and c' may depend on the temperature only.

This analysis is somewhat suggestive of that which arises in the theory of normal condensation. When the density is increased beyond the value n_0, the compressibility ζ_T assumes a very large value and the activity z remains almost constant. It is not, however, possible to identify anything corresponding to the condensed phase. As a result of the 'Bose–Einstein condensation', many thermodynamical quantities show a discontinuity at a certain temperature, which is near $2.3°$ K at the density of liquid helium. The discontinuity is rather suggestive of the λ-phenomenon actually observed; but it has to be borne in mind that the theory will be considerably modified by the interaction of the molecules, which has been ignored in the preceding calculation.

4. The Quantum Theory of Flow

The macroscopic behaviour of fluids in motion is determined quantitatively by the equations of continuity, motion and thermal conduction, in conjunction with the equation of state. Given a set of initial conditions, these enable one to determine the subsequent macroscopic motion and flow of energy within the fluid. This is true even in the quantum theory where, although it is impossible to make definite predictions concerning the individual molecules, the average behaviour of large numbers of molecules can be predicted with the same certainty which one might expect on the basis of the classical theory. Indeed, the fundamental equations governing the macroscopic behaviour of fluids are identical in form with those derived from the classical theory. It is only when one seeks to calculate such quantities as the pressure and the internal energy per molecule, or the coefficients of viscosity and thermal conduction, on the basis of the molecular theory, that quantitative differences will appear at low temperatures. Such calculations are beset with the difficulties already apparent in the classical theory.

It should, however, be possible to detect those qualitative features which are, for example, responsible for the abnormal behaviour of He II. Here one has to distinguish between those phenomena which are observed in narrow channels and surface films, and those which appear in the liquid in bulk. Although He II shows an almost complete absence of viscosity in sufficiently narrow capillaries, the viscosity inferred from experiments on the resistance to the motion of solid bodies in the liquid in bulk is small but normal. It can be explained in terms of the decrease in the effective intermolecular attractions predicted by the theory of § 2. The conduction of heat by the fluid, on the other hand, always has an abnormal component; as was stated in the previous section, thermal energy is propagated with a definite velocity — called the velocity of 'second sound'.

Here attention will be confined to the derivation of the macroscopic equations from the quantized molecular theory. For this purpose, only the fundamental equation (1. 18) is required, and the results are therefore independent of whether classical or quantum statistics are applied. When (1. 22) is substituted in (1. 18), the result can be written in the form

$$(4.1) \begin{cases} \dfrac{\partial f_N}{\partial t} + \sum_{i=1}^{N} \left\{ \boldsymbol{\xi}^{(i)} \cdot \dfrac{\partial f_N}{\partial \mathbf{x}^{(i)}} - \dfrac{1}{m} \dfrac{\partial \Phi}{\partial \mathbf{x}^{(i)}} \cdot \dfrac{\partial f_N}{\partial \boldsymbol{\xi}^{(i)}} \right\} = \sum_{ijk} \dfrac{\partial^3}{\partial \boldsymbol{\xi}^{(i)} \partial \boldsymbol{\xi}^{(j)} \partial \boldsymbol{\xi}^{(k)}} : \mathbf{R}^{(ijk)}, \\ \mathbf{R}^{(ijk)} = -\dfrac{\hbar^2}{4m^3} \dfrac{\partial^3 \Phi}{\partial \mathbf{x}^{(i)} \partial \mathbf{x}^{(j)} \partial \mathbf{x}^{(k)}} + \cdots . \end{cases}$$

Suppose that this equation is first multiplied by one of the quantities 1, $\boldsymbol{\xi}^{(i)}$ and $\tfrac{1}{2}m\boldsymbol{\xi}^{(i)2}$, and then integrated over all velocities and any number of the coordinates. Then the right-hand side vanishes, as one can readily ascertain by the application of Gauss' theorem to the 'space' of the velocities $\boldsymbol{\xi}^{(i)}$, $\boldsymbol{\xi}^{(j)}$ and $\boldsymbol{\xi}^{(k)}$. It follows that the equations satisfied by the molecular distribution functions n_q, the mean velocities $\mathbf{u}_q^{(i)}$, and the generalized temperatures $T_q^{(i)}$ [cf. (1. 9), (1. 10) and (1. 12)] are the same as those derived in § 8 of chapter V, within the classical theory.

In particular, the equations of motion and heat flow are

$$(4.2) \begin{cases} mn \dfrac{d}{dt} \mathbf{u} + \dfrac{\partial}{\partial \mathbf{x}} \cdot \boldsymbol{p}_1 = 0 \\ n \dfrac{dU_1}{dt} + \dfrac{\partial}{\partial \mathbf{x}} \cdot \mathbf{q} + \boldsymbol{p}_1 : \left(\dfrac{\partial}{\partial \mathbf{x}} \mathbf{u} \right) = 0, \end{cases}$$

where $\boldsymbol{p}_1$ is the mechanical pressure tensor, U_1 is the internal energy per molecule, and **q** is the thermal flux, expressed by the formulae (3. 10), (3. 13) — (3. 15) and (3. 21) respectively of chapter V.

A number of significant conclusions may be drawn from the fact that the equations (4. 2) are formally identical with the corresponding classical equations. Those parts of the classical theory which depend only on these equations require no modification in passing to the quantum theory. Thus, irreversible effects are an inevitable consequence of the existence of gradients of macroscopic velocity and temperature, even though viscosity and *normal* thermal conduction may be much smaller at low temperatures, on account of the decline of the effective interaction between the molecules. For the abnormal viscosity and thermal conduction observed in surface films and narrow channels, it is necessary that there should be no appreciable velocity gradient normal to the flow, and no appreciable temperature gradient between the two ends. Evidently liquid helium 'slips' over the surfaces with which it is in contact, below the λ-point. It may also be concluded from (4. 2) that acceleration of the fluid is possible only under the influence of a gradient of the mechanical pressure p_1, when there are no external forces. The mechanical effects induced by heating liquid helium can therefore be explained only by supposing that the mechanical pressure within the fluid suffers a sudden decline when the heat is applied. The experimental evidence indicates further that such fluctuations in pressure are propagated through the fluid, with the velocity of 'second sound'.

The velocity of propagation of any mechanical disturbance in a fluid may be inferred in the usual way from the equation of motion; it is

$$\left(\frac{1}{m}\frac{dp_1}{dn}\right)^{\frac{1}{2}},$$

where the derivative $\frac{dp_1}{dn}$ must be evaluated subject to the appropriate thermodynamical restraint. The velocity of sound is obtained by supposing that the disturbance is adiabatic. The velocity of 'second sound' in He II, on the other hand, is obtained by supposing that the thermodynamical potential $(F_1 + p/n)$ is unchanged by the disturbance, as in any other thermomechanical process. The fact that heating the fluid produces thermomechanical waves suggests further that the abnormal internal energy of He II may be of this nature.

Many of the experimental facts concerning the flow of liquid helium are thus correlated by the supposition that in He II the mechanical pressure p_1 is affected by energy exchanges within the fluid, and differs from the pressure (p) calculated on thermodynamical grounds. GREEN [1949] has obtained an expression for the difference between the two pressures p and p_1 in equilibrium, and although several attempts have been made (cf. PRICE [1950]) to show that this expression vanishes, they all depend on two assumptions:

(i) that the fluid can be assumed to be in a pure quantum state, and

(ii) that the boundary conditions are like those which prevail in a normal liquid.

Both of these assumptions are suspect, and the second in particular is known on empirical grounds to be incorrect in He II. Future progress will evidently depend on a more careful examination of conditions at the boundary of the fluid.

REFERENCES AND NAME INDEX

	Page
ACKERMANN P. G. (see MAYER)	
ANDRADE E. N. DA C. (1934) Phil. Mag. 17 (7), 497, 698	140
BAKER W. O. (see MASON)	
BETH E. (see UHLENBECK)	
BOER J. DE (1940) Amsterdam Dissertation	74, 244
BOGGS E. M. (see KIRKWOOD)	
BORN M. (1923) Atomtheorie des Festen Zustandes, Leipzig: TEUBNER	109
——— and FUCHS K. (1938) Proc. Roy. Soc. A 166, 391	91
——— and GREEN H. S. (1946) Proc. Roy. Soc. A 188, 10	126
BRAGG W. H. and BRAGG W. L. (1933) The Crystalline State, London: BELL	60
BUFF F. P. (see KIRKWOOD)	
CHAMBERLAIN O. (1950) Phys. Rev. 77, 305	57
CHANDRASEKHAR S. (1943) Rev. Mod. Phys. 15, 20	198
CHAPMAN S. (1916) Phil. Trans. Roy. Soc. A 216, 297	214
——— (1917) Phil. Trans. Roy. Soc. A 217, 115	214
——— and COWLING T. G. (1939) The Mathematical Theory of Non-Uniform Gases, Camb. Univ. Press	16, 126, 214, 225
COWLING T. G. (see CHAPMAN)	
DAUNT J. G. and MENDELSSOHN K. (1946) Phys. Rev. 69, 126	252
DAVYDOV B. (1947) J. Phys. (U.S.S.R.) 11, 33	48
DEVONSHIRE F. (see LENNARD–JONES)	
EISENSTEIN A. and GINGRICH N. S. (1942) Phys. Rev. 62, 261	62, 64
ENSKOG D. (1917) Uppsala Dissertation	214
——— (1911) Phys. Zeits. 12, 56, 533	175
EUCKEN A. (1913) Phys. Zeits. 14, 324	230
EYRING H. J. and HIRSCHFELDER O. (1937) J. Chem. Phys. 41, 250	116
FOWLER, R. H. (1937) Proc. Roy. Soc. A 159, 229	192
——— and GUGGENHEIM E. A. (1939) Statistical Thermodynamics, Camb. Univ. Press	43
FRENKEL J. (1946) Kinetic Theory of Liquids, Oxford Univ. Press	110
FUCHS K. (see BORN)	
GEDDES A. L. (see MAASS)	
GINGRICH N. S. (1943) Rev. Mod. Phys. 15, 90	61
——— (see EISENSTEIN)	
GREEN H. S. (1949) Proc. Roy. Soc. A 194	261
——— (see BORN)	
GREEN M. S. (see KIRKWOOD)	
GUGGENHEIM E. A. (see FOWLER)	
HARRISON S. F. (see MAYER)	
HEISS J. H. (see MASON)	
HENNIKER J. C. (1949) Rev. Mod. Phys. 21, 322	84, 195

	Page
HIRSCHFELDER O. (see EYRING)	
HOLBORN L. and OTTO J. (1924) Zeits. f. Phys. **30**, 320	202
———— and ———— (1926) Zeits. f. Phys. **38**, 359	202
KAHN B. (see UHLENBECK)	
KEESOM W. H. and SMEDT J. DE (1922) Proc. Amst. Acad. Sci. **25**, 118	57
———— and ———— (1923) Proc. Amst. Acad. Sci. **26**, 112	57
KELVIN, LORD (1911) Math. and Phys. Papers V, 55, Camb. Univ. Press	190
KIRKWOOD J. G. (1946) J. Chem. Phys. **14**, 180 . . . . . 126, 151,	196
———— (1950) J. Chem. Phys. **18**, 380	117
———— and BOGGS E. M. (1942) J. Chem. Phys. **10**, 394 . . . 71,	76
———— and MONROE E. (1941) J. Chem. Phys. **9**, 514	110
———— and BUFF F. P. (1949) J. Chem. Phys. **17**, 338	190
———— BUFF F. P. and GREEN M. S. (1949) J. Chem. Phys. **17**, 988, 156,	158
———— (see SLATER)	
LANDAU L. (1941) J. Phys. (U.S.S.R.) **5**, 71	254
———— (1944) J. Phys. (U.S.S.R.) **8**, 1	254
LANGEVIN P. (1908) C. R. (Paris) **146**, 530	196
LENNARD–JONES J. E. (1937) Physica **4**, 941	201
———— and DEVONSHIRE F. (1938) Proc. Roy. Soc. A **163**, 53, **165**, 1	116
LONDON F. (1930) Zeits. Phys. Chem. B **11**, 222	86
MAASS O. and GEDDES A. L. (1937) Phil. Trans. Roy. Soc. A **236**, 313	104
MASON W. P., BAKER W. O., MCSKIMIN H. J. and HEISS J. H. (1949) Phys. Rev. **75**, 936	182
MAYER J. E. (1937) J. Chem. Phys. **5**, 67	91
———— and ACKERMANN P. G. (1937) J. Chem. Phys. **5**, 74	91
———— and HARRISON S. F. (1938) J. Chem. Phys. **6**, 87, 101	91
MCLELLAN A. G. Proc. Roy. Soc. A (1952, in press) 80,	108
MCSKIMIN H. J. (see MASON)	
MENDELSSOHN K. (see DAUNT)	
MENKE H. (1932) Phys. Zeits. **33**, 593 57,	60
MONROE E. (see KIRKWOOD)	
ORNSTEIN L. S. and ZERNIKE F. (1914) Proc. Amst. Acad. Sci. **17**, 793	62
OTTO J. (see HOLBORN)	
PRICE P. J. (1950) Phil. Mag. **41**, 948	261
RODRIGUEZ A. E. (1948) Proc. Roy. Soc. A **196**, 73	106
RUSHBROOKE G. S. (1940) Proc. Roy. Soc. Edin. A **60**, 182	193
SCHRÖDINGER E. (1949) Statistical Thermodynamics Camb. Univ. Press	43
SLATER J. C. and KIRKWOOD J. G. (1931) Phys. Rev. **37**, 682 . 86,	202
SMEDT J. DE (see KEESOM)	
STEWART G. W. (1930) Rev. Mod. Phys. **2**, 116	66
TISZA L. (1938) C. R. (Paris) **207**, 1035, 1186	254
———— (1940) J. Phys. 1 (8), 164, 350	254
———— (1947) Phys. Rev. **72**, 838	254
TRILLAT J. J. (1930) Zeits. f. Phys. **64**, 191	57
UHLENBECK G. E. and KAHN B. (1938) Physica **5**, 399	91
UHLENBECK G. E. and BETH E. (1936) Physica **3**, 729	244

UHLENBECK G. E. and BETH E. (1937) Physica **4**, 915 244
URSELL H. D. (1927) Proc. Camb. Phil. Soc. **23**, 685 91
YANG L. M. (1949) Proc. Roy. Soc. A **198**, 94, 471 176
YVON J. (1937) Fluctuations en Densité; La Propagation et la Diffusion de
    la Lumière, Hermann & Cie. 62, 209
——— (1949) J. de Phys. **10**, 373 91
ZERNIKE F. and PRINS J. (1927) Zeits. f. Phys. **41**, 184 57
——— (see ORNSTEIN)

SOME DOVER SCIENCE BOOKS

SOME DOVER SCIENCE BOOKS

WHAT IS SCIENCE?,
Norman Campbell

This excellent introduction explains scientific method, role of mathematics, types of scientific laws. Contents: 2 aspects of science, science & nature, laws of science, discovery of laws, explanation of laws, measurement & numerical laws, applications of science. 192pp. 5⅜ x 8.

Paperbound $1.25

FADS AND FALLACIES IN THE NAME OF SCIENCE,
Martin Gardner

Examines various cults, quack systems, frauds, delusions which at various times have masqueraded as science. Accounts of hollow-earth fanatics like Symmes; Velikovsky and wandering planets; Hoerbiger; Bellamy and the theory of multiple moons; Charles Fort; dowsing, pseudoscientific methods for finding water, ores, oil. Sections on naturopathy, iridiagnosis, zone therapy, food fads, etc. Analytical accounts of Wilhelm Reich and orgone sex energy; L. Ron Hubbard and Dianetics; A. Korzybski and General Semantics; many others. Brought up to date to include Bridey Murphy, others. Not just a collection of anecdotes, but a fair, reasoned appraisal of eccentric theory. Formerly titled *In the Name of Science*. Preface. Index. x + 384pp. 5⅜ x 8.

Paperbound $1.85

PHYSICS, THE PIONEER SCIENCE,
L. W. Taylor

First thorough text to place all important physical phenomena in cultural-historical framework; remains best work of its kind. Exposition of physical laws, theories developed chronologically, with great historical, illustrative experiments diagrammed, described, worked out mathematically. Excellent physics text for self-study as well as class work. Vol. 1: Heat, Sound: motion, acceleration, gravitation, conservation of energy, heat engines, rotation, heat, mechanical energy, etc. 211 illus. 407pp. 5⅜ x 8. Vol. 2: Light, Electricity: images, lenses, prisms, magnetism, Ohm's law, dynamos, telegraph, quantum theory, decline of mechanical view of nature, etc. Bibliography. 13 table appendix. Index. 551 illus. 2 color plates. 508pp. 5⅜ x 8.

Vol. 1 Paperbound $2.25, Vol. 2 Paperbound $2.25,
The set $4.50

THE EVOLUTION OF SCIENTIFIC THOUGHT FROM NEWTON TO EINSTEIN,
A. d'Abro

Einstein's special and general theories of relativity, with their historical implications, are analyzed in non-technical terms. Excellent accounts of the contributions of Newton, Riemann, Weyl, Planck, Eddington, Maxwell, Lorentz and others are treated in terms of space and time, equations of electromagnetics, finiteness of the universe, methodology of science. 21 diagrams. 482pp. 5⅜ x 8.

Paperbound $2.50

CATALOGUE OF DOVER BOOKS

CHANCE, LUCK AND STATISTICS: THE SCIENCE OF CHANCE,
Horace C. Levinson
Theory of probability and science of statistics in simple, non-technical language. Part I deals with theory of probability, covering odd superstitions in regard to "luck," the meaning of betting odds, the law of mathematical expectation, gambling, and applications in poker, roulette, lotteries, dice, bridge, and other games of chance. Part II discusses the misuse of statistics, the concept of statistical probabilities, normal and skew frequency distributions, and statistics applied to various fields—birth rates, stock speculation, insurance rates, advertising, etc. "Presented in an easy humorous style which I consider the best kind of expository writing," Prof. A. C. Cohen, Industry Quality Control. Enlarged revised edition. Formerly titled *The Science of Chance*. Preface and two new appendices by the author. Index. xiv + 365pp. 5⅜ x 8. Paperbound $2.00

BASIC ELECTRONICS,
prepared by the U.S. Navy Training Publications Center
A thorough and comprehensive manual on the fundamentals of electronics. Written clearly, it is equally useful for self-study or course work for those with a knowledge of the principles of basic electricity. Partial contents: Operating Principles of the Electron Tube; Introduction to Transistors; Power Supplies for Electronic Equipment; Tuned Circuits; Electron-Tube Amplifiers; Audio Power Amplifiers; Oscillators; Transmitters; Transmission Lines; Antennas and Propagation; Introduction to Computers; and related topics. Appendix. Index. Hundreds of illustrations and diagrams. vi + 471pp. 6½ x 9¼.
Paperbound $2.75

BASIC THEORY AND APPLICATION OF TRANSISTORS,
prepared by the U.S. Department of the Army
An introductory manual prepared for an army training program. One of the finest available surveys of theory and application of transistor design and operation. Minimal knowledge of physics and theory of electron tubes required. Suitable for textbook use, course supplement, or home study. Chapters: Introduction; fundamental theory of transistors; transistor amplifier fundamentals; parameters, equivalent circuits, and characteristic curves; bias stabilization; transistor analysis and comparison using characteristic curves and charts; audio amplifiers; tuned amplifiers; wide-band amplifiers; oscillators; pulse and switching circuits; modulation, mixing, and demodulation; and additional semiconductor devices. Unabridged, corrected edition. 240 schematic drawings, photographs, wiring diagrams, etc. 2 Appendices. Glossary. Index. 263pp. 6½ x 9¼. Paperbound $1.25

GUIDE TO THE LITERATURE OF MATHEMATICS AND PHYSICS,
N. G. Parke III
Over 5000 entries included under approximately 120 major subject headings of selected most important books, monographs, periodicals, articles in English, plus important works in German, French, Italian, Spanish, Russian (many recently available works). Covers every branch of physics, math, related engineering. Includes author, title, edition, publisher, place, date, number of volumes, number of pages. A 40-page introduction on the basic problems of research and study provides useful information on the organization and use of libraries, the psychology of learning, etc. This reference work will save you hours of time. 2nd revised edition. Indices of authors, subjects, 464pp. 5⅜ x 8.
Paperbound $2.75

THE RISE OF THE NEW PHYSICS (formerly THE DECLINE OF MECHANISM), A. d'Abro
This authoritative and comprehensive 2-volume exposition is unique in scientific publishing. Written for intelligent readers not familiar with higher mathematics, it is the only thorough explanation in non-technical language of modern mathematical-physical theory. Combining both history and exposition, it ranges from classical Newtonian concepts up through the electronic theories of Dirac and Heisenberg, the statistical mechanics of Fermi, and Einstein's relativity theories. "A must for anyone doing serious study in the physical sciences," *J. of Franklin Inst.* 97 illustrations. 991pp. 2 volumes.

T3, T4 Two volume set, paperbound $5.50

THE STRANGE STORY OF THE QUANTUM, AN ACCOUNT FOR THE GENERAL READER OF THE GROWTH OF IDEAS UNDERLYING OUR PRESENT ATOMIC KNOWLEDGE, B. Hoffmann
Presents lucidly and expertly, with barest amount of mathematics, the problems and theories which led to modern quantum physics. Dr. Hoffmann begins with the closing years of the 19th century, when certain trifling discrepancies were noticed, and with illuminating analogies and examples takes you through the brilliant concepts of Planck, Einstein, Pauli, de Broglie, Bohr, Schroedinger, Heisenberg, Dirac, Sommerfeld, Feynman, etc. This edition includes a new, long postscript carrying the story through 1958. "Of the books attempting an account of the history and contents of our modern atomic physics which have come to my attention, this is the best," H. Margenau, Yale University, in *American Journal of Physics*. 32 tables and line illustrations. Index. 275pp. 5⅜ x 8.

T518 Paperbound $2.00

GREAT IDEAS AND THEORIES OF MODERN COSMOLOGY, Jagjit Singh
The theories of Jeans, Eddington, Milne, Kant, Bondi, Gold, Newton, Einstein, Gamow, Hoyle, Dirac, Kuiper, Hubble, Weizsäcker and many others on such cosmological questions as the origin of the universe, space and time, planet formation, "continuous creation," the birth, life, and death of the stars, the origin of the galaxies, etc. By the author of the popular *Great Ideas of Modern Mathematics*. A gifted popularizer of science, he makes the most difficult abstractions crystal-clear even to the most non-mathematical reader. Index. xii + 276pp. 5⅜ x 8½ T925 Paperbound $2.00

GREAT IDEAS OF MODERN MATHEMATICS: THEIR NATURE AND USE, Jagjit Singh
Reader with only high school math will understand main mathematical ideas of modern physics, astronomy, genetics, psychology, evolution, etc., better than many who use them as tools, but comprehend little of their basic structure. Author uses his wide knowledge of non-mathematical fields in brilliant exposition of differential equations, matrices, group theory, logic, statistics, problems of mathematical foundations, imaginary numbers, vectors, etc. Original publications, appendices. indexes. 65 illustr. 322pp. 5⅜ x 8. T587 Paperbound $2.00

THE MATHEMATICS OF GREAT AMATEURS, Julian L. Coolidge
Great discoveries made by poets, theologians, philosophers, artists and other non-mathematicians: Omar Khayyam, Leonardo da Vinci, Albrecht Dürer, John Napier, Pascal, Diderot, Bolzano, etc. Surprising accounts of what can result from a non-professional preoccupation with the oldest of sciences. 56 figures. viii + 211pp. 5⅜ x 8½. S1009 Paperbound $2.00

CATALOGUE OF DOVER BOOKS

COLLEGE ALGEBRA, *H. B. Fine*
Standard college text that gives a systematic and deductive structure to algebra; comprehensive, connected, with emphasis on theory. Discusses the commutative, associative, and distributive laws of number in unusual detail, and goes on with undetermined coefficients, quadratic equations, progressions, logarithms, permutations, probability, power series, and much more. Still most valuable elementary-intermediate text on the science and structure of algebra. Index. 1560 problems, all with answers. x + 631pp. 5⅜ x 8. Paperbound $2.75

HIGHER MATHEMATICS FOR STUDENTS OF CHEMISTRY AND PHYSICS, *J. W. Mellor*
Not abstract, but practical, building its problems out of familiar laboratory material, this covers differential calculus, coordinate, analytical geometry, functions, integral calculus, infinite series, numerical equations, differential equations, Fourier's theorem, probability, theory of errors, calculus of variations, determinants. "If the reader is not familiar with this book, it will repay him to examine it," *Chem. & Engineering News*. 800 problems. 189 figures. Bibliography. xxi + 641pp. 5⅜ x 8. Paperbound $2.50

TRIGONOMETRY REFRESHER FOR TECHNICAL MEN, *A. A. Klaf*
A modern question and answer text on plane and spherical trigonometry. Part I covers plane trigonometry: angles, quadrants, trigonometrical functions, graphical representation, interpolation, equations, logarithms, solution of triangles, slide rules, etc. Part II discusses applications to navigation, surveying, elasticity, architecture, and engineering. Small angles, periodic functions, vectors, polar coordinates, De Moivre's theorem, fully covered. Part III is devoted to spherical trigonometry and the solution of spherical triangles, with applications to terrestrial and astronomical problems. Special time-savers for numerical calculation. 913 questions answered for you! 1738 problems; answers to odd numbers. 494 figures. 14 pages of functions, formulae. Index. x + 629pp. 5⅜ x 8.
Paperbound $2.00

CALCULUS REFRESHER FOR TECHNICAL MEN, *A. A. Klaf*
Not an ordinary textbook but a unique refresher for engineers, technicians, and students. An examination of the most important aspects of differential and integral calculus by means of 756 key questions. Part I covers simple differential calculus: constants, variables, functions, increments, derivatives, logarithms, curvature, etc. Part II treats fundamental concepts of integration: inspection, substitution, transformation, reduction, areas and volumes, mean value, successive and partial integration, double and triple integration. Stresses practical aspects! A 50 page section gives applications to civil and nautical engineering, electricity, stress and strain, elasticity, industrial engineering, and similar fields. 756 questions answered. 556 problems; solutions to odd numbers. 36 pages of constants, formulae. Index. v + 431pp. 5⅜ x 8. Paperbound $2.00

INTRODUCTION TO THE THEORY OF GROUPS OF FINITE ORDER, *R. Carmichael*
Examines fundamental theorems and their application. Beginning with sets, systems, permutations, etc., it progresses in easy stages through important types of groups: Abelian, prime power, permutation, etc. Except 1 chapter where matrices are desirable, no higher math needed. 783 exercises, problems. Index. xvi + 447pp. 5⅜ x 8. Paperbound $3.00

FIVE VOLUME "THEORY OF FUNCTIONS" SET BY KONRAD KNOPP

This five-volume set, prepared by Konrad Knopp, provides a complete and readily followed account of theory of functions. Proofs are given concisely, yet without sacrifice of completeness or rigor. These volumes are used as texts by such universities as M.I.T., University of Chicago, N. Y. City College, and many others. "Excellent introduction . . . remarkably readable, concise, clear, rigorous," *Journal of the American Statistical Association*.

ELEMENTS OF THE THEORY OF FUNCTIONS,
Konrad Knopp

This book provides the student with background for further volumes in this set, or texts on a similar level. Partial contents: foundations, system of complex numbers and the Gaussian plane of numbers, Riemann sphere of numbers, mapping by linear functions, normal forms, the logarithm, the cyclometric functions and binomial series. "Not only for the young student, but also for the student who knows all about what is in it," *Mathematical Journal*. Bibliography. Index. 140pp. 5⅜ x 8. Paperbound $1.50

THEORY OF FUNCTIONS, PART I,
Konrad Knopp

With volume II, this book provides coverage of basic concepts and theorems. Partial contents: numbers and points, functions of a complex variable, integral of a continuous function, Cauchy's integral theorem, Cauchy's integral formulae, series with variable terms, expansion of analytic functions in power series, analytic continuation and complete definition of analytic functions, entire transcendental functions, Laurent expansion, types of singularities. Bibliography. Index. vii + 146pp. 5⅜ x 8. Paperbound $1.35

THEORY OF FUNCTIONS, PART II,
Konrad Knopp

Application and further development of general theory, special topics. Single valued functions. Entire, Weierstrass, Meromorphic functions. Riemann surfaces. Algebraic functions. Analytical configuration, Riemann surface. Bibliography. Index. x + 150pp. 5⅜ x 8. Paperbound $1.35

PROBLEM BOOK IN THE THEORY OF FUNCTIONS, VOLUME 1.
Konrad Knopp

Problems in elementary theory, for use with Knopp's *Theory of Functions*, or any other text, arranged according to increasing difficulty. Fundamental concepts, sequences of numbers and infinite series, complex variable, integral theorems, development in series, conformal mapping. 182 problems. Answers. viii + 126pp. 5⅜ x 8. Paperbound $1.35

PROBLEM BOOK IN THE THEORY OF FUNCTIONS, VOLUME 2,
Konrad Knopp

Advanced theory of functions, to be used either with Knopp's *Theory of Functions*, or any other comparable text. Singularities, entire & meromorphic functions, periodic, analytic, continuation, multiple-valued functions, Riemann surfaces, conformal mapping. Includes a section of additional elementary problems. "The difficult task of selecting from the immense material of the modern theory of functions the problems just within the reach of the beginner is here masterfully accomplished," *Am. Math. Soc.* Answers. 138pp. 5⅜ x 8.
Paperbound $1.50

NUMERICAL SOLUTIONS OF DIFFERENTIAL EQUATIONS,
H. Levy & E. A. Baggott

Comprehensive collection of methods for solving ordinary differential equations of first and higher order. All must pass 2 requirements: easy to grasp and practical, more rapid than school methods. Partial contents: graphical integration of differential equations, graphical methods for detailed solution. Numerical solution. Simultaneous equations and equations of 2nd and higher orders. "Should be in the hands of all in research in applied mathematics, teaching," *Nature*. 21 figures. viii + 238pp. 5⅜ x 8. Paperbound $1.85

ELEMENTARY STATISTICS, WITH APPLICATIONS IN MEDICINE AND THE BIOLOGICAL SCIENCES, *F. E. Croxton*

A sound introduction to statistics for anyone in the physical sciences, assuming no prior acquaintance and requiring only a modest knowledge of math. All basic formulas carefully explained and illustrated; all necessary reference tables included. From basic terms and concepts, the study proceeds to frequency distribution, linear, non-linear, and multiple correlation, skewness, kurtosis, etc. A large section deals with reliability and significance of statistical methods. Containing concrete examples from medicine and biology, this book will prove unusually helpful to workers in those fields who increasingly must evaluate, check, and interpret statistics. Formerly titled "Elementary Statistics with Applications in Medicine." 101 charts. 57 tables. 14 appendices. Index. vi + 376pp. 5⅜ x 8. Paperbound $2.00

INTRODUCTION TO SYMBOLIC LOGIC,
S. Langer

No special knowledge of math required — probably the clearest book ever written on symbolic logic, suitable for the layman, general scientist, and philosopher. You start with simple symbols and advance to a knowledge of the Boole-Schroeder and Russell-Whitehead systems. Forms, logical structure, classes, the calculus of propositions, logic of the syllogism, etc. are all covered. "One of the clearest and simplest introductions," *Mathematics Gazette*. Second enlarged, revised edition. 368pp. 5⅜ x 8. Paperbound $2.00

A SHORT ACCOUNT OF THE HISTORY OF MATHEMATICS,
W. W. R. Ball

Most readable non-technical history of mathematics treats lives, discoveries of every important figure from Egyptian, Phoenician, mathematicians to late 19th century. Discusses schools of Ionia, Pythagoras, Athens, Cyzicus, Alexandria, Byzantium, systems of numeration; primitive arithmetic; Middle Ages, Renaissance, including Arabs, Bacon, Regiomontanus, Tartaglia, Cardan, Stevinus, Galileo, Kepler; modern mathematics of Descartes, Pascal, Wallis, Huygens, Newton, Leibnitz, d'Alembert, Euler, Lambert, Laplace, Legendre, Gauss, Hermite, Weierstrass, scores more. Index. 25 figures. 546pp. 5⅜ x 8. Paperbound $2.25

INTRODUCTION TO NONLINEAR DIFFERENTIAL AND INTEGRAL EQUATIONS,
Harold T. Davis

Aspects of the problem of nonlinear equations, transformations that lead to equations solvable by classical means, results in special cases, and useful generalizations. Thorough, but easily followed by mathematically sophisticated reader who knows little about non-linear equations. 137 problems for student to solve. xv + 566pp. 5⅜ x 8½. Paperbound $2.00

CATALOGUE OF DOVER BOOKS

AN INTRODUCTION TO THE GEOMETRY OF N DIMENSIONS,
D. H. Y. Sommerville

An introduction presupposing no prior knowledge of the field, the only book in English devoted exclusively to higher dimensional geometry. Discusses fundamental ideas of incidence, parallelism, perpendicularity, angles between linear space; enumerative geometry; analytical geometry from projective and metric points of view; polytopes; elementary ideas in analysis situs; content of hyper-spacial figures. Bibliography. Index. 60 diagrams. 196pp. 5⅜ x 8.

Paperbound $1.50

ELEMENTARY CONCEPTS OF TOPOLOGY, *P. Alexandroff*

First English translation of the famous brief introduction to topology for the beginner or for the mathematician not undertaking extensive study. This unusually useful intuitive approach deals primarily with the concepts of complex, cycle, and homology, and is wholly consistent with current investigations. Ranges from basic concepts of set-theoretic topology to the concept of Betti groups. "Glowing example of harmony between intuition and thought," David Hilbert. Translated by A. E. Farley. Introduction by D. Hilbert. Index. 25 figures. 73pp. 5⅜ x 8.

Paperbound $1.00

ELEMENTS OF NON-EUCLIDEAN GEOMETRY,
D. M. Y. Sommerville

Unique in proceeding step-by-step, in the manner of traditional geometry. Enables the student with only a good knowledge of high school algebra and geometry to grasp elementary hyperbolic, elliptic, analytic non-Euclidean geometries; space curvature and its philosophical implications; theory of radical axes; homothetic centres and systems of circles; parataxy and parallelism; absolute measure; Gauss' proof of the defect area theorem; geodesic representation; much more, all with exceptional clarity. 126 problems at chapter endings provide progressive practice and familiarity. 133 figures. Index. xvi + 274pp. 5⅜ x 8.

Paperbound $2.00

INTRODUCTION TO THE THEORY OF NUMBERS, *L. E. Dickson*

Thorough, comprehensive approach with adequate coverage of classical literature, an introductory volume beginners can follow. Chapters on divisibility, congruences, quadratic residues & reciprocity. Diophantine equations, etc. Full treatment of binary quadratic forms without usual restriction to integral coefficients. Covers infinitude of primes, least residues. Fermat's theorem. Euler's phi function, Legendre's symbol, Gauss's lemma, automorphs, reduced forms, recent theorems of Thue & Siegel, many more. Much material not readily available elsewhere. 239 problems. Index. I figure. viii + 183pp. 5⅜ x 8.

Paperbound $1.75

MATHEMATICAL TABLES AND FORMULAS,
compiled by Robert D. Carmichael and Edwin R. Smith

Valuable collection for students, etc. Contains all tables necessary in college algebra and trigonometry, such as five-place common logarithms, logarithmic sines and tangents of small angles, logarithmic trigonometric functions, natural trigonometric functions, four-place antilogarithms, tables for changing from sexagesimal to circular and from circular to sexagesimal measure of angles, etc. Also many tables and formulas not ordinarily accessible, including powers, roots, and reciprocals, exponential and hyperbolic functions, ten-place logarithms of prime numbers, and formulas and theorems from analytical and elementary geometry and from calculus. Explanatory introduction. viii + 269pp. 5⅜ x 8½.

Paperbound $1.25